STEM CELLS
FROM HYPE TO HOPE

Other Related Titles from World Scientific

Revolutionary Therapies: How the California Stem Cell Program Saved Lives, Eased Suffering — and Changed the Face of Medicine Forever
by Don C. Reed
ISBN: 978-981-121-328-1

Neural Stem Cells in Health and Disease
edited by Ashok K. Shetty
ISBN: 978-981-4623-17-9

Engineering Stem Cells for Tissue Regeneration
edited by Ngan F. Huang, Nicolas L'Heureux and Song Li
ISBN: 978-981-3147-74-4

California Cures!
How the California Stem Cell Program is Fighting Your Incurable Disease!
by Don C. Reed
ISBN: 978-981-3231-36-8
ISBN: 978-981-3270-38-1 (pbk)

Stem Cells: Promise and Reality
by Lygia V. Pereira
ISBN: 978-981-3100-18-3
ISBN: 978-981-3100-19-0 (pbk)

Stem Cells, Tissue Engineering and Regenerative Medicine
edited by David Warburton
ISBN: 978-981-4612-77-7

Stem Cell Battles: Proposition 71 and Beyond
How Ordinary People Can Fight Back against the Crushing Burden of Chronic Disease — with a Posthumous Foreword by Christopher Reeve
by Don C. Reed
ISBN: 978-981-4644-01-3
ISBN: 978-981-4618-27-4 (pbk)

STEM CELLS
FROM HYPE TO HOPE

Editor

Khawaja Husnain Haider
Sulaiman AlRajhi University, Saudi Arabia

World Scientific

NEW JERSEY • LONDON • SINGAPORE • BEIJING • SHANGHAI • HONG KONG • TAIPEI • CHENNAI

Published by

World Scientific Publishing Co. Pte. Ltd.

5 Toh Tuck Link, Singapore 596224

USA office: 27 Warren Street, Suite 401-402, Hackensack, NJ 07601

UK office: 57 Shelton Street, Covent Garden, London WC2H 9HE

Library of Congress Cataloging-in-Publication Data

Names: Haider, Khawaja Husnain, editor.

Title: Stem cells : from hype to hope / editor, Khawaja Husnain Haider.

Other titles: Stem cells (Haider : 2020)

Description: Hackensack, NJ : World Scientific, [2020] |
    Includes bibliographical references and index.

Identifiers: LCCN 2019044510 | ISBN 9789811205521 (hardcover)

Subjects: MESH: Stem Cell Research | Stem Cell Transplantation

Classification: LCC QH588.S83 | NLM QU 325 | DDC 616.02/774--dc23

LC record available at https://lccn.loc.gov/2019044510

British Library Cataloguing-in-Publication Data

A catalogue record for this book is available from the British Library.

For any available supplementary material, please visit
https://www.worldscientific.com/worldscibooks/10.1142/11420#t=suppl

Typeset by Stallion Press

Email: enquiries@stallionpress.com

Dedication

I dedicate this book to my two sons: Mowahid who is the
love of my life and Anas "the angel of paradise" whose departure
from my life is a constant source of inspiration for me to do science.

Contents

About the Editor

Professor Khawaja Husnain Haider has academic, teaching and research affiliations with the University of the Punjab, Lahore, Pakistan; University of Strathclyde, Glasgow, United Kingdom; Seoul National University, South Korea; National University of Singapore, Singapore and University of Cincinnati, Ohio, USA.

Prof. Haider served as Principal and Co-Principal Investigator on multiple NIH-funded stem cell research projects and served as an invited reviewer on various national and international review panels as well as research journals. He has authored more than 250 research papers, review articles, abstracts and book chapters.

Presently serving as Professor and Chair, Department of Basic Sciences, Sulaiman AlRajhi University, Kingdom of Saudi Arabia.

Preface

Cell therapy has come of age! It is a new form of medicine preceded only by herbs, chemicals and surgery. It replenishes live cells to the degenerative human body, and in allografts, offers the normal genome for genetic complementation treatment. It regenerates tissues and organs. Exosomes secreted from cultured cells are often formulated to become molecular medicine. Since a foreign gene and its derivatives always exert its effect on the cell, cell therapy is the common pathway to health. Debilitating and fatal diseases with no known cure are often results of polygenic aberration that only cell therapy rather than individual molecular medicine can be effective.

Cell therapists harvesting the profits of developmental cell biologists should not forget their teaching. Somatic cell therapies and stem cell therapies utilize different compositions and methods of treatment. Much of the cell therapy success stems from somatic cell transplantation and not stem cell transplantation. The latter continues suffering inadequacy in cell identification, quantity, purity, viability, potency, especially uncertainty in cell differentiation because of pluripotency and carcinogenicity.

"*Stem Cells: From Hype to Hope*" is the fourth book Professor Haider has edited on various topics pertaining to stem cell and somatic cell therapies. Professor Haider has devoted more than two decades in these arenas, publishing cutting-edge research in complete honesty and dedication. In "*Stem Cells: From Hype to Hope*", he has compiled in Chapter-1, the latest development of stem cell-derived cardiomyocytes; in Chapter-2, difficulties in stem cell-based approaches for myocardial regeneration; in Chapter-3, mesenchymal stem cell differentiation; in Chapter-4: angiogenesis of limb ischemia; in Chapter-5, paracrine cardiac repair; in Chapter-6, therapeutic stem cell exosomes; in Chapter-7, mesenchymal stem cells as a biologic; in Chapter-8, renewing β-cells to treat Diabetes; an in Chapter-9, CAR T-cells for cancer treatment.

All these drops into the bucket, some representing life-long effort of the pioneers, will be collected monumentally, for this is a viable technology, a technology shared by all animals in the last 500 million years, only to be re-discovered, isolated, purified and to be re-formulated as Cell Therapy.

Peter K. Law, Ph.D.
Professor, Founder & Chairman
Cell Therapy Institute
Wuhan, CHINA

Pluripotent stem cell-derived cardiomyocytes: Current research progress and therapeutic potential

Adam T. Lynch, Stefan Hoppler*

Aberdeen Cardiovascular and Diabetes Centre,
Institute of Medical Sciences, Foresterhill Health Campus,
University of Aberdeen, Scotland, UK

ABSTRACT

Pluripotent stem cells offer tremendous potential for the treatment of various diseases, including cardiovascular disease. Myocardial infarction results in the almost irreversible loss of over 1 billion cardiomyocytes, with little endogenous replacement from the damaged heart. Subsequently, the quality of life of patients recovering from myocardial infarction is significantly reduced and is, to date, without an effective cure. Pluripotent stem cell-derived cardiomyocytes bring exciting potential for use in regenerative medicine therapies by replacing the damaged cells with healthy, lab-grown cardiomyocytes. However, safety concerns regarding the purity and maturity of pluripotent stem cell-derived heart muscle is currently still a significant barrier to the successful translation of lab-grown cells into patients. This chapter will review the current status of embryonic and induced pluripotent stem cell strategies to generate heart muscle, as well as discuss the currently remaining obstacles which must be overcome for the safe clinical application of stem cell therapy.

KEYWORDS

Cardiomyocytes; Embryonic; iPSCs; MSCs; Stem cells.

LIST OF ABBREVIATIONS

BMP4	=	Bone morphogenetic factor-4
BSA	=	Bovine serum albumin

* Corresponding author. Email: s.p.hoppler@abdn.ac.uk

CSCs   =   Cardiac stem cells
CTnT$^+$ =   Cardiac troponin-T
DMSO =   Dimethyl sulfoxide
EC   =   Embryonal carcinoma
ESCs   =   Embryonic stem cells
FACS   =   Fluorescence activated cell sorting
FBS   =   Foetal bovine serum
hESCs =   Human embryonic stem cells
iPSCs   =   Induced pluripotent stem cells
LQTS   =   Long QT syndrome
mESCs =   Mouse embryonic stem cells
SCID   =   Severe combined immunodeficient
SIRPA =   Signal regulatory protein alpha
VEGF   =   Vascular endothelial growth factor

1.1 INTRODUCTION

Cardiovascular disease is the leading cause of death in Western society, affecting 17.9 million people annually and accounting for 31% of deaths worldwide. Furthermore, congenital heart defects are the leading cause of infantile death worldwide, accounting for approximately 1% of all live births. Therefore, disorders of the heart present a large healthcare burden, and currently offer few long-term therapeutic solutions. Following myocardial infarction, over one billion cardiomyocytes are lost.[1] Given cardiomyocyte turnover in the adult heart is less than 1%,[2] natural recovery through endogenous repair mechanisms is not possible, leading patients to suffer long-term, life-altering changes with poor prognosis. Excitement within the scientific community about the existence of cardiac stem cells (CSCs) raised the possibility that endogenous, multipotent stem cells residing in the heart could be used to promote myocardial regeneration upon injury.[3] Unfortunately, more recent evidence now suggests that such CSC populations do not exist in the adult mouse heart,[4] and are therefore unlikely to hold any promise for cardiac repair of diseased or injured hearts in humans. However, it is encouraging that epicardial-derived WT1$^+$ cell progenitors can give rise to *de novo* cardiomyocytes following myocardial infarction in adult mice, suggesting that some cells in the adult heart may be able to support cardiac repair without the requirement for exogenous engraftment.[5] Whilst this offers great hope for potential drug-mediated stimulation of endogenous progenitor populations to repair the damaged heart, presently this is not possible. Therefore, renewed interest has more recently focussed on the relatively well understood pluripotent stem cells as a source of *in vitro*-grown heart muscle as a strategy to repair the damaged adult heart, particularly following myocardial infarction.

The use of pluripotent stem cell-derived cardiomyocytes has multiple applications in scientific research. Primarily, as already introduced above, generating enriched populations of cardiomyocytes for transplantation into patients suffering from myocardial infarction has tremendous potential for cardiac repair, since the quality of patient lifestyle following infarction is currently debilitating. Secondly, pluripotent stem cells can additionally be used as an *in vitro* tool to model cardiac development without the requirement for embryos.[6] Ethical considerations regarding the use of human embryos, as well as the inaccessibility of cardiac tissue prohibits their use in scientific research, therefore human pluripotent stem cells represent a powerful alternative to dissecting the genetic regulatory processes which normally guide and drive human cardiac development. Thirdly, pluripotent stem cell-derived cardiomyocytes can be used in drug-screening studies. Traditionally, pharmacological studies relied upon rodent models. However, many animal models, particularly rodent models, do not realistically represent the anatomical and physiological systems of humans, which therefore reduces their predictive application.[7] Pluripotent stem cell-derived cardiomyocytes can be used in drug discovery screens to identify compounds to improve cardiomyocyte function,[8] such as drug-induced action potential shortening in long QT syndrome (LQTS).[9] The derivation of induced pluripotent stem cells (iPSCs) from human somatic cells now also bypasses the ethical concerns previously raised from the use of human embryonic stem cells (hESCs),[10] meaning that patient-derived heart cells can be obtained from non-cardiac cell types.[11] Patient-derived iPSCs also allow patient-specific disease modelling of genetic congenital heart defects. For example, LQTS can be modelled in patients exhibiting a mutation in the *KCNQ1* gene which show reduced action potentials, a typical phenotype of LQTS.[12]

1.2 TRADITIONAL METHODS TO GENERATE CARDIOMYOCYTES

It has now been established that pluripotent stem cells are capable of differentiating into any of the many cell types that make up the adult body, including important clinically-relevant cell types such as pancreatic β-cells, dopaminergic neurons and cardiomyocytes.[13] Over the past ten years, multiple laboratories across the globe have channelled significant efforts into generating enriched populations of cardiomyocytes from various pluripotent stem cells. Early approaches using P19 embryonal carcinoma (EC) cells discovered that addition of dimethyl sulfoxide

(DMSO) enhanced cardiomyocyte differentiation[14] by up-regulating GATA4 expression in differentiating cultures.[15] After the discovery of mouse embryonic stem cells (mESCs),[16] researchers then used embryoid body (EB)-based protocols to generate differentiated cells. Aggregation of ES cells (ESCs) into three-dimensional spheroids (or embryoid bodies; EBs) by either suspension or hanging-drop cultures can induce ES cells to differentiate into a variety of cell types from across all three embryonic germ layers.[17,18] Whilst cardiomyocytes can be generated by spontaneous EB-based methods,[19] the proportion of cardiomyocytes is often very low, typically under 5%.[20] Traditional EB-based approaches use foetal bovine serum (FBS) which is undefined and therefore can substantially vary across batches, and additionally relies on using animal-derived products. More recent approaches have sought to mimic the normal developmental processes which occur *in vivo* by the temporal and sequential manipulation of genetic pathways, by regulating cardiomyocyte development using growth factor and small molecule inhibitors, without relying on undefined and highly variable FBS. For instance, addition of Activin and bone morphogenetic factor-4 (BMP4) growth factors to serum-free EB cultures induces the formation of cardiac mesoderm (Flk1^{+}/Pdgfrα^{+}), and can improve generation of significantly enriched populations of cardiomyocytes (typically 70%), as compared to FBS-based protocols (Figure-1.1).[21] Serum-free EB-based protocols can also be used to generate cardiomyocytes from human ES and iPS cells, by addition of BMP4, Activin, basic fibroblast growth factor (bFGF), vascular endothelial growth factor (VEGF) and Wnt inhibition.[22] First and second heart field derivatives can even be generated from cardiac organoids treated with Activin and BMP4 for

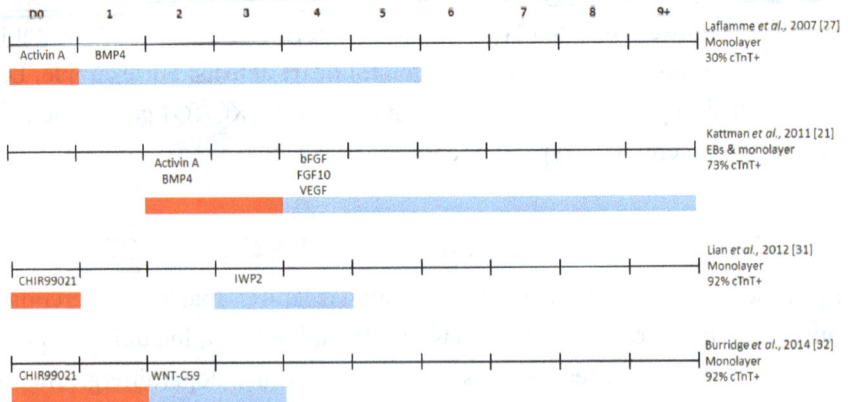

Figure-1.1.   Various differentiation protocols exist to differentiate pluripotent stem cells into cardiomyocytes. Pluripotent stem cells are differentiated either in monolayers or embryoid bodies (EBs) in basal media supplemented by the temporal addition of various growth factors and/or small molecule inhibitors.

studying cardiac development.[23] Whilst serum-free methods have significantly improved cardiomyocyte differentiation efficiencies compared to serum-based protocols, the reliance on EBs still often hinders greater cardiomyocyte yields due to the inevitable variability in the sizes of individual EBs.

1.3 CURRENT METHODS TO GENERATE CARDIOMYOCYTES FROM PLURIPOTENT STEM CELLS

The most recent methods to generate more homogenous populations of cardiomyocytes have utilized monolayer-based protocols, which dispenses with EB formation. Unlike EBs, cells cultured in 2D monolayers possess the advantage of tighter control over the differentiation process, leading ultimately to improved differentiation through exposure of all cells to the same growth factor and inhibitor concentrations, thereby avoiding reduced diffusion of molecules during EB formation. Early methods using monolayers involved co-culturing of hESCs with visceral-endoderm-like mouse END-2 cells.[24] This generates foetal-like ventricular cardiomyocytes which possess cardiac structural proteins and electrophysiological profiles of typical cardiomyocytes, albeit at low efficiency. Further progress was made when it was identified that removal of insulin during the earlier stages of cardiac differentiation can enhance cardiomyocyte differentiation.[25] Currently, two main protocols exist for generating monolayer-derived cardiomyocyte populations which carefully follow the normal developmental processes occurring during embryonic heart development (Figure-1.2A).[26] Similar to earlier efforts in serum-free EBs, addition of BMP4 and Activin to differentiating cultures of hES cells can generate more enriched cardiomyocyte populations of around 30% cTnT$^+$ (Figure-1.1), which can be enriched to over 80% following the Percoll gradient centrifugation.[27]

Such hES cell-derived cardiomyocytes have successfully muscularized experimentally-induced infarcted hearts in a non-human primate model, demonstrating that hESC-derived cardiomyocytes can function *in vivo*.[28] A similar monolayer protocol using BMP4/Activin followed by FGF stimulation has also more recently been developed in mES cells, producing cardiomyocytes of up to 78% cTnT$^+$,[29] thereby circumventing any requirement for EBs. Whilst growth factor (BMP/Activin)-based differentiation protocols are a popular method to generate cardiomyocytes, the inherent variation between batches of growth factors and their suppliers often causes significant variability during the differentiation processes. Using small molecule inhibitors in an alternative approach that circumvents the requirement for using recombinant-derived growth factors can generate

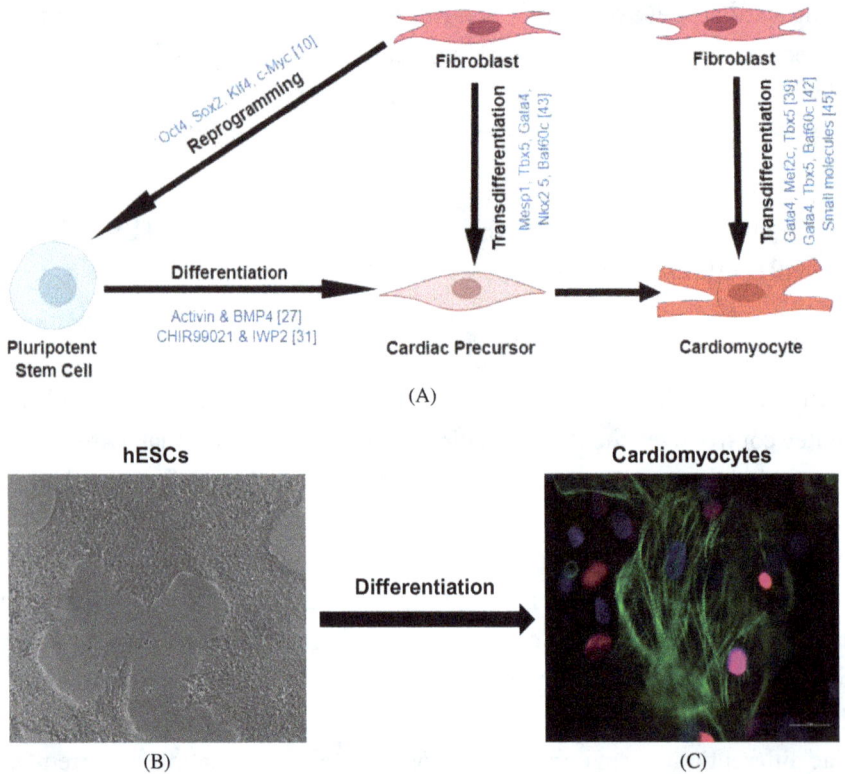

Figure-1.2. (A) Cardiomyocytes can be generated via multiple methods. Monolayer differentiation of pluripotent stem cells by manipulation of signalling pathways (Activin & BMP4 or modulation of Wnt signalling) can efficiently generate enriched populations of cardiomyocytes (see Figure-1.1). Somatic cells can also be transdifferentiated into cardiac precursors or directly to cardiomyocytes by transfection of various pro-cardiac transcription factors. (B & C) Immunocytochemistry of human embryonic stem cells (hESCs) differentiated into cardiomyocytes via temporal manipulation of Wnt signalling. hESC-derived cardiomyocytes express cardiac troponin (green) and NKX2.5 (red). The nuclei were visualized by DAPI staining (blue).

more reproducible and consistent yields of cardiomyocytes.[30] These protocols can efficiently generate cardiomyocytes of over 80% purity, even in traditionally cardiomyocyte-recalcitrant iPSC lines. Small molecule-based cardiomyocyte differentiation causes biphasic manipulation of Wnt signalling (Figure-1.2B & C), that is by first activating canonical Wnt/β-catenin signalling with the GSK3β inhibitor CHIR99021, followed by inhibition of Wnt signalling with the Porcupine inhibitors IWP2 or WNT-C59 (Figure-1.1).[31,32]

The most recent differentiation protocols have seen a shift towards fully chemically-defined media, removing the use of animal-derived products, such as bovine serum albumin (BSA), which has traditionally been used in enhancing cell survival. Such protocols preferably use human-derived albumin[33] or can be obtained in complete basal media without any exogenous factors at all (e.g., albumin-free),[34] yet still produce over 96% cardiac troponin-T (cTnT+) cardiomyocyte populations. Additionally, distinct cardiac cell types can be generated by treating mesodermal cells with retinoic acid to generate atrial cardiomyocytes,[35] or treatment with cardiac progenitors with Wnt agonists to make epicardial cells,[36] demonstrating that pluripotent stem cells can be differentiated into different cardiac subtypes.

Despite vast improvements over the last ten years in generating significantly increased cultures of pluripotent stem cell-derived cardiomyocytes, reproducibility across cell lines and robust outcome of differentiation remains a hurdle. Even intra-experimental replicates in the same tissue culture plate can sometimes generate significantly different cardiomyocyte yields, suggesting that the microenvironment and/or even experimenter personnel is a core decisive factor in the differentiation outcome.[37]

1.4 TRANSDIFFERENTIATION OF SOMATIC CELLS INTO CARDIOMYOCYTES

Pluripotent stem cell differentiation into heart muscle is the most favored method to generate cardiomyocytes at relatively high proportions. However, alternative methods bypassing the use of pluripotent cells exist, which would circumvent tumorigenicity and immune rejection concerns. Genetic conversion of one somatic cell type into another somatic cell without de-differentiation to a pluripotent intermediate (transdifferentiation) is another strategy to make heart muscle (Figure-1.2A). Cardiac fibroblasts account for over 50% of the cells comprising the heart, and therefore offer a convenient potential source of cell types for transdifferentiation.[38] Ectopic expression of the transcription factors *Tbx5*, *Mef2c*, and *Gata4* in dermal and post-natal fibroblasts can reprogram them into spontaneously contracting cardiomyocytes at approximately 25% efficiency.[39] Upon leaving these cells for another four weeks, 45% of α-MHC+ cells were present. Whilst this relatively enriched population of cardiomyocytes has the major benefit of avoiding the use of pluripotent cells, concerns about the maturity and similar gene expression profile compared to mature cardiomyocytes remain an issue. Only a fraction of reprogrammed cells display full cardiomyocyte-like characteristics such as calcium transients, therefore currently questioning the suitability of these cells for medical applications.[40] Additionally, some groups have had difficulty reprogramming fibroblasts into cardiomyocytes by *Tbx5*, *Mef2c* and *Gata4*,[41] suggesting that other external factors may be crucial for efficient transdifferentiation into cardiomyocytes. Ectopic expression of the chromatin remodelling complex *Baf60c*, together with *Gata4* and *Tbx5* has also been reported to transdifferentiate prospective non-cardiogenic mouse mesoderm into beating cardiomyocytes.[42] Additionally, mouse fibroblasts can also be reprogrammed into cardiac progenitors (Figure-1.2A).[43] Expanded cardiac progenitors can then be differentiated into cardiomyocytes at 35% efficiency (i.e., cTnT+ cells) which can improve recovery in the infarcted hearts of mice.[44]

In contrast to genetic-based reprogramming, transdifferentiation of human fibroblasts into cardiomyocytes has been reported using small molecule-based methods (Figure-1.2A). Addition of nine compounds can chemically induce human foreskin fibroblasts or human foetal lung fibroblasts into cardiomyocytes, which epigenetically, structurally and electrophysiologically resemble cardiomyocytes.[45] When the induced cardiomyocytes were incubated with pluripotent stem cell-derived cardiomyocyte-conditioned media, up to 6.9% of reprogrammed cells were cTnT+ at day 30.[39] These induced cardiomyocytes do not appear to generate teratomas when injected into immunodeficient mice and can also structurally integrate into mouse hearts following transplantation.[45]

1.5 THERAPEUTIC APPLICATIONS OF PLURIPOTENT STEM CELL-DERIVED CARDIOMYOCYTES

Pluripotent stem cell-derived cardiomyocytes hold exciting opportunities for use in regenerative medicine, as these stem cell-derived cells share many structural and genetic similarities to *bona fide* cardiomyocytes. The ultimate goal of cardiomyocyte research is the engraftment of hESCs/iPS cell-derived cardiomyocytes onto the infarcted hearts of patients to replace the substantial scarring present on the infarcted area of the heart muscle. However, several safety and practical barriers to the successful realization of this therapy currently remain a significant challenge:

The first challenge relates to the practical and economic difficulties in the generation of sufficient numbers of cells to replace the over 1 billion cardiomyocytes irreversibly lost following an infarction. Large-scale culture vessels and multiple cell culture incubators would be necessary to generate the substantial number of cardiomyocytes required for transplantation into a single patient, therefore presenting a significant financial and practical demand to generate suitable cardiomyocyte numbers. 2D culture systems employing active gas ventilation (to overcome gas diffusion issues associated with such vast quantities of plates) have generated 2.8×10^9 cardiomyocytes using stacked large-scale plates.[46] However, the differentiation conditions utilized serum at the cardiomyocyte stage, and would therefore not be suitable for patient engraftment in its current form. The application of more scalable culture, potentially via automation and/or bioreactors may prove a more long-term cost-effective solution for generating over 10^9 cardiomyocytes required for transplantation. Automation, however, has yet to become commonplace in most laboratories, and therefore remains largely a future ambition. Additionally, to generate such vast quantities of cardiomyocytes for transplantation, cryopreservation of heart muscle cells is a necessary logistical and financial requirement for generating the required quantity of cell stocks needed for transplantation. Typical formulations use serum in the cryopreservation media which increases cell viability during the stressful freezing process.[37] For patient transplantation, serum-free cryopreservation media would be necessary. Whilst serum-free formulations are commercially available, cell recovery following cryopreservation varies between 55–90%.[47] Nonetheless, cryopreserved hESC-derived cardiomyocytes did not affect engraftment during transplantation in a non-human primate model compared to non-cryopreserved cells.[28]

The second challenge for the therapeutic application of hES/iPS cell-derived cardiomyocytes is the purity of the final cardiomyocyte population after the differentiation process. Undifferentiated pluripotent stem cells residing within the terminally differentiated cardiomyocyte culture have the potential to form teratomas. As few as two hESC colonies per 10^6 cells generated teratomas when injected into severe combined immunodeficient (SCID) mice.[48] To circumvent purity issues, various strategies to enrich cardiomyocytes have been developed. Initial solutions used genetic-based purification techniques using fluorescent reporters. Live cell sorting of cardiomyocytes harbouring cardiac-specific fluorescent reporters such as $MYH6$[49] or $NKX2$-5[50] were isolated by fluorescence-activated cell sorting (FACS). However, despite significant enrichment, these genetic methods are unsuitable for human applications due to the risk of tumorigenesis and therefore are unsuitable for transplantation into human patients. Metabolic purification by switching from glucose to lactate selectively purifies cardiomyocytes from non-cardiomyocytes (and ES cells), producing up to 99% cardiomyocyte purity.[51] Cardiomyocytes can use both lactate and glucose for metabolism, whereas non-cardiomyocyte cells (e.g., cardiac fibroblasts) depend on glucose. Whilst this technique is simple, low-cost and avoids genetic manipulation, the selection process causes significant cell death, including cardiomyocytes, and therefore still requires large-scale input to generate outputs of enriched cardiomyocyte populations. An alternative to metabolic selection is FACS of differentiated cardiomyocytes based on cell-surface markers. Signal regulatory protein alpha (SIRPA) was identified using an antibody screen to uncover cardiomyocyte-specific cell-surface markers,[52] and can be used to isolate cardiomyocytes of up to 98% cTnT$^+$ cells. However, concerns about the specificity of SIRPA-expressing cells have arisen as SIRPA is also expressed on smooth muscle and endothelial cells,[50] suggesting this technology is not suitable for specifically isolating heart muscle cells. In a complementary approach, the same group used a depletion strategy to enrich for cardiomyocytes by positively identifying and sorting non-cardiomyocyte populations (i.e., fibroblasts, endothelial and smooth muscle cells) and thereby negatively selecting for cardiomyocytes. This strategy therefore obtained enriched populations of cardiomyocytes, free of any bound antibody.[52] Another method to purify cardiomyocytes utilizes the dye TMRM which binds mitochondria and can be isolated by FACS, generating up to 99% pure cultures.[53] This procedure has the advantage of isolating more mature cardiomyocytes which have high densities of mitochondria. However, this method is labour-intensive and not financially viable for large-scale applications. Furthermore, a major drawback of cell sorting is cellular stress and often results in substantial cell death during the sorting procedure. The effort to generate enriched populations of cardiomyocytes suitable for transplantation is therefore still an active area of research.

The third challenge for future transplantation studies relates to the need to consider the specific subtype of cardiomyocytes (atrial or ventricular) obtained from these differentiation protocols. Most cardiomyocyte differentiation protocols generate heterogenous mixtures of cellular subtypes, with the majority being ventricular-like cells and a small percentage of atrial- and pacemaker-like cells. Ventricular cardiomyocytes are the major cell type lost following myocardial infarction, therefore enriched populations of ventricular cardiomyocytes would be needed to be generated for transplantation into patients to ensure that engrafted cells can function effectively and correctly in the appropriate affected region. A recent study developed a protocol to differentiate pluripotent stem cells into RALDH2+ atrial cardiomyocytes (by stimulation with retinoic acid) or CD235a+ ventricular cardiomyocytes,[35] demonstrating that it is now possible to differentiate and purify populations of atrial or ventricular cardiomyocytes.

The fourth challenge is the transplantation of cardiomyocytes into patients. Concerns regarding the acceptance and integration of pluripotent stem cell-derived cardiomyocytes into the host's heart remain a major obstacle to regenerative medicine. hES cell-derived cardiomyocytes would therefore need to be allotypically matched to the patient. Additionally, patients receiving hESC-derived cardiomyocytes would require immunosuppressants throughout the remainder of their lives. Whilst iPS cell-derived cardiomyocytes would overcome the requirement for life-long immunosuppression, concerns around the genetic stability of the iPS cells during the reprogramming process still need to be addressed in order to ensure the safe transplantation of stem cell-derived cardiomyocytes into a patient.[54] Furthermore, reprogramming from patient somatic cells to iPS cells, undertaking quality-control tests, followed by differentiation into cardiomyocytes and finally, transplantation is financially intensive, and unlikely to make iPS cell-mediated therapy for individual patients generally financially viable at present.[55] Additionally, transplantation of pluripotent stem cell-derived cardiomyocytes into animal models has reported the occurrence of arrhythmias. Electrocardiograph recordings of hESC-derived cardiomyocytes transplanted into a non-human primate model observed atypical Ca^{2+} handling; a prognostic risk for arrythmias.[28] Such abnormal Ca^{2+} behaviour has been attributed to two possible causes. hESC-derived cardiomyocytes are likely not to electrically couple to the host myocardium, therefore forming localized changes in Ca^{2+} transients.[56] Alternatively, since the proportion of cTNT$^+$ cells was not determined by flow cytometry, the arrythmias could have been caused by the presence of non-cardiomyocyte cells arising during the differentiation process. However, when hiPSC-derived cardiomyocytes were transplanted into a pig myocardial infarction model, arrythmias were not detected.[57] Similarly, when ESC-derived cardiovascular progenitors were transplanted into a Rhesus macaque myocardial infarction model, arrythmias were not observed when monitored three months after transplantation.[58] Understanding the causes of these arrythmias, therefore, remains a major challenge for realizing clinical transplantation of cardiomyocytes into human patients.

A final fifth and important challenge is associated with the maturity of the differentiated cardiomyocytes following the differentiation procedure from pluripotent stem cell-derived cardiomyocytes. The foetal-like nature of pluripotent stem cell-derived cardiomyocytes remains a major challenge still needs to be overcome. Current differentiation protocols to generate cardiomyocytes from both hESCs and iPSCs typically generate cardiomyocytes which genetically and structurally resemble foetal cardiomyocytes. Similarly, transdifferentiation-derived cardiomyocytes also display maturity issues. Whilst fibroblast-derived cardiomyocytes can integrate into mouse hearts following transplantation, the cells did not exhibit organized sarcoplasmic reticulum structures.[59] Cardiac fibroblast-derived induced cardiomyocytes display closer molecular signatures to adult cardiomyocytes, as compared to iPSC-derived cardiomyocytes which display more embryonic-like molecular signatures.[60] For regenerative medicine to capitalize on pluripotent stem cell-derived cardiomyocytes, generating adult-like cardiomyocytes is a necessary challenge to overcome. This challenge is currently being tackled by multiple research groups. Treatment of stem cell-derived heart muscle cells with the growth hormone tri-iodo-L-thyronine can promote immature cardiomyocytes into more adult-like cardiomyocytes, increasing sarcomere length and sarcoplasmic reticulum ATPase expression.[61] The "Cardiopatch"; a hydrogel mould supporting growth of hESC-derived cardiomyocytes, is capable of accelerating cardiomyocyte maturation and displaying adult-like myocardial electrical and mechanical function,[62] therefore it offers a potential solution to generating more mature populations of heart muscle. This remains an intense area of research, progress of which will likely accelerate the translation of pluripotent stem cell-derived cardiomyocytes into the clinic.

1.6 FUTURE OUTLOOK

Whilst legitimate safety concerns regarding the translation of stem cells from the laboratory to the clinic still remain,[56] the past ten years have seen a remarkable progress in generating cardiomyocyte-enriched populations from pluripotent stem cells. Most clinically-relevant studies using a non-primate myocardial infarction model observe that fewer than 1% of the transplanted cells survive the transplantation procedure, and observe electrophysiologically abnormal Ca^{2+} coupling.[28] Nonetheless, trials in Japan using iPSC-derived cardiomyocytes are currently underway in patients with ischemic cardiomyopathy. Sheets of iPS cell-derived cardiomyocytes will be grafted onto patients suffering from infarctions. The cells are believed not to integrate into the heart tissue, but instead secrete factors to improve heart repair. Whilst this technology showed some increase in heart function in pigs, this technology has yet to be proven safe and effective in humans. This current technique is similar to a 2015 study whereby skeletal muscle cells were grafted onto damaged hearts (termed HeartSheet) and released growth factors to repair the heart, without integration into the heart tissue itself.[63] To date, this technology has only seen a modest improvement in heart repair, therefore concerns about the effectiveness of grafting iPS cells onto hearts are legitimate. The Scientific Community will therefore be watching the outcome of these clinical trials with intense interest and the results may spark the dawn of a new era of regenerative medicine becoming more commonplace. Additionally, CRISPR/Cas9 genome editing will enable correcting any mutations that predispose carriers to cardiac diseases and may be clinically used in the future to combat congenital diseases early during development, potentially even before any heart disease manifests itself. For example, homology-directed reparation of the MYBPC3 gene in human embryos by CRISPR/Cas9 has recently been accomplished.[64]

REFERENCES

1. Talman V, Ruskoaho H. Cardiac fibrosis in myocardial infarction — from repair and remodeling to regeneration. *Cell Tissue Res.* 2016; 365(3): 563–581.

2. Bergmann O, Zdunek S, Felker A, Salehpour M, Alkass K, Bernard S, Sjostrom SL, *et al.* Dynamics of cell generation and turnover in the human heart. *Cell.* 2015; 161(7): 1566–1575.

3. Beltrami AP, Barlucchi L, Torella D, Baker M, Limana F, Chimenti S, Kasahara H, *et al.* Adult cardiac stem cells are multipotent and support myocardial regeneration. *Cell.* 2003; 114(6): 763–776.

4. Li Y, He L, Huang X, Bhaloo SI, Zhao H, Zhang S, Pu W, *et al.* Genetic lineage tracing of nonmyocyte population by dual recombinases. *Circulation.* 2018; 138(8): 793–805.

5. Smart N, Bollini S, Dubé KN, Vieira JM, Zhou B, Davidson S, Yellon D, *et al.* De novo cardiomyocytes from within the activated adult heart after injury. *Nature.* 2011; 474(7353): 640–644.

6. Birket MJ, Mummery CL. Pluripotent stem cell derived cardiovascular progenitors — a developmental perspective. *Dev Biol.* 2015; 400(2): 169–179.

7. Breckenridge R. Heart failure and mouse models. *Dis Model Mech.* 2010; 3(3–4): 138–143.

8. Braam SR, Tertoolen L, van de Stolpe A, Meyer T, Passier R, Mummery CL. Prediction of drug-induced cardiotoxicity using human embryonic stem cell-derived cardiomyocytes. *Stem Cell Res.* 2010; 4(2): 107–116.

9. Duncan G, Denning C, Smith G, Hoang MD, Firth K, Staniforth A, George V. Drug-mediated shortening of action potentials in LQTS2 human induced pluripotent stem cell-derived cardiomyocytes. *Stem Cells Dev.* 2017; 26(23): 1695–1705.

10. Takahashi K, Tanabe K, Ohnuki M, Narita M, Ichisaka T, Tomoda K, Yamanaka S. Induction of pluripotent stem cells from adult human fibroblasts by defined factors. *Cell.* 2007; 131(5): 861–872.

11. Denning C, Borgdorff V, Crutchley J, Firth KSA, George V, Kalra S, Kondrashov A, Hoang MD, Mosqueira D, Patel A, *et al.* Cardiomyocytes from human pluripotent stem cells: from laboratory curiosity to industrial biomedical platform. *Biochim Biophys Acta — Mol. Cell Res.* 2016; 1863(7): 1728–1748.

12. Egashira T, Yuasa S, Suzuki T, Aizawa Y, Yamakawa H, Matsuhashi T, Ohno Y, *et al.* Disease characterization using LQTS-specific induced pluripotent stem cells. *Cardiovasc Res.* 2012; 95(4): 419–429.

13. Murry CE, Keller G. Differentiation of embryonic stem cells to clinically relevant populations: lessons from embryonic development. *Cell.* 2008; 132(4): 661–680.

14. McBurney MW, Jones-Villeneuve EMV, Edwards MKS, Anderson PJ. Control of muscle and neuronal differentiation in a cultured embryonal carcinoma cell line. *Nature.* 1982; 299(5879): 165–167.

15. Grépin C, Nemer G, Nemer M. Enhanced cardiogenesis in embryonic stem cells overexpressing the GATA-4 transcription factor. *Development.* 1997; 124(12): 2387–2395.

16. Evans MJ, Kaufman MH. Establishment in culture of pluripotential cells from mouse embryos. *Nature.* 1981; 292(5819): 154–156.

17. Kurosawa H. Methods for inducing embryoid body formation: in vitro differentiation system of embryonic stem cells. *J Biosci Bioeng.* 2007; 103(7): 389–398.

18. Itskovitz-Eldor J, Schuldiner M, Karsenti D, Eden A, Yanuka O, Amit M, Soreq H, Benvenisty N. Differentiation of human embryonic stem cells into embryoid bodies compromising the three embryonic germ layers. *Mol Med.* 2000; 6(2): 88–95.

19. Lynch AT, Mazzotta S, Hoppler S. Cardiomyocyte differentiation from mouse embryonic stem cells. In: Experimental Models of Cardiovascular Diseases: Methods and Protocols (ed.) K. Ishikawa. New York, NY: Springer New York, 2018, pp. 55–66.

20. Bondue A, Lapouge G, Paulissen C, Semeraro C, Lacovino M, Kyba M, Blanpain C. Mesp1 acts as a master regulator of multipotent cardiovascular progenitor specification. *Cell Stem Cell.* 2008; 3(1): 69–84.

21. Kattman SJ, Witty AD, Gagliardi M, Dubois NC, Niapour M, Hotta A, Ellis J, *et al.* Stage-specific optimization of activin/nodal and BMP signaling promotes cardiac differentiation of mouse and human pluripotent stem cell lines. *Cell Stem Cell.* 2011; 8(2): 228–240.

22. Yang L, Soonpaa MH, Adler ED, Roepke TK, Kattman SJ, Kennedy M, Henckaerts E, *et al.* Human cardiovascular progenitor cells develop from a KDR+ embryonic-stem-cell-derived population. *Nature.* 2008; 453(7194): 524–528.

23. Andersen P, Tampakakis E, Jimenez DV, Kannan S, Miyamoto M, Shin HK, Saberi A, *et al.* Precardiac organoids form two heart fields via Bmp/Wnt signaling. *Nat Commun.* 2018; 9(1): 3140.

24. Pera M, Mummery C, Spijker R, Hassink R, Doevendans P, de la Riviere AB, Opthof T, *et al.* Differentiation of human embryonic stem cells to cardiomyocytes. *Circulation.* 2003; 107(21): 2733–2740.

25. van Rooijen M, Freund C, Monshouwer-Kloots J, Mummery C, Ward-van Oostwaard D, Zweigerdt R, Passier R, *et al.* Insulin redirects differentiation from cardiogenic mesoderm and endoderm to neuroectoderm in differentiating human embryonic stem cells. *Stem Cells.* 2016; 26(3): 724–733.

26. Mazzotta S, Neves C, Bonner RJ, Bernardo AS, Docherty K, Hoppler S. Distinctive roles of canonical and noncanonical wnt signaling in human embryonic cardiomyocyte development. *Stem Cell Reports.* 2016; 7(4): 764–776.

27. Laflamme MA, Chen KY, Naumova AV, Muskheli V, Fugate JA, Dupras SK, Reinecke H, *et al.* Cardiomyocytes derived from human embryonic stem cells in pro-survival factors enhance function of infarcted rat hearts. *Nat Biotechnol.* 2007; 25(9): 1015–1024.

28. Chong JJH, Yang X, Don CW, Minami E, Liu Y-W, Weyers JJ, Mahoney MM, *et al.* Human embryonic-stem-cell-derived cardiomyocytes regenerate non-human primate hearts. *Nature.* 2014; 510(7504): 273–277.

29. Kokkinopoulos I, Ishida H, Saba R, Coppen S, Suzuki K, Yashiro K. Cardiomyocyte differentiation from mouse embryonic stem cells using a simple and defined protocol. *Dev Dyn.* 2016; 245(2): 157–165.

30. Mazzotta S, Lynch AT, Hoppler S. Cardiomyocyte Differentiation from human embryonic stem cells, In: Experimental Models of Cardiovascular Diseases: Methods and Protocols (ed.) K. Ishikawa. New York, NY: Springer New York, 2018; 67–78.

31. Lian X, Zhang J, Azarin SM, Zhu K, Hazeltine LB, Bao X, Hsiao C, Paslacek SP. Directed cardiomyocyte differentiation from human pluripotent stem cells by modulating Wnt/β-catenin signaling under fully defined conditions. *Nat Protoc.* 2012; 8(1): 162–175.

32. Burridge PW, Matsa E, Shukla P, Lin ZC, Churko JM, Ebert AD, Lan F, *et al.* Chemically defined generation of human cardiomyocytes. *Nat Methods.* 2014; 11(8): 855–860.

33. Burridge PW, Keller G, Gold JD, Wu JC. Production of de novo cardiomyocytes: human pluripotent stem cell differentiation and direct reprogramming. *Cell Stem Cell.* 2012; 10(1): 16–28.

34. LianBao X, Zilberter M, Westman M, Fisahn A, Hsiao C, Hazeltine LB, *et al.* Chemically defined, albumin-free human cardiomyocyte generation. *Nat. Methods.* 2015; 12(7): 595–596.

35. Lee JH, Protze SI, Laksman Z, Backx PH, Keller GM. Human pluripotent stem cell-derived atrial and ventricular cardiomyocytes develop from distinct mesoderm populations. *Cell Stem Cell.* 2017; 21(2): 179–194.

36. Bao X, Lian X, Qian T, Bhute VJ, Han T, Palecek SP. Directed differentiation and long-term maintenance of epicardial cells from human pluripotent stem cells under fully defined conditions. *Nature.* 2017; 12(9): 1890–1900.

37. Burridge PW, Holmström A, Wu JC. Chemically defined culture and cardiomyocyte differentiation of human pluripotent stem cells. *Curr Protoc Hum Genet.* 2015; 87(1): 1–19.

38. Baudino TA, Carver W, Giles W, Borg KT. Cardiac fibroblasts: friend or foe? *Am J Physiol Circ. Physiol.* 2006; 291(3): H1015–H1026.

39. Leda M, Fu JD, Delgado-Olguin P, Vedantham V, Hayashi Y, Bruneau BG, Srivastava D. Direct reprogramming of fibroblasts into functional cardiomyocytes by defined factors. *Cell.* 2010; 142(3): 375–386.

40. Murry CE, Pu WT. Reprogramming fibroblasts into cardiomyocytes. *N Engl J Med.* 2011; 364(2): 177–178.

41. Chen XJ, Krane M, Deutsch MA, Wang L, Rav-Acha M, Gregoire S, Engels ME, *et al.* Inefficient reprogramming of fibroblasts into cardiomyocytes using Gata4, Mef2c, and Tbx5. *Circ Res.* 2012; 111(1): 50–55.

42. Takeuchi JK, Bruneau BG. Directed transdifferentiation of mouse mesoderm to heart tissue by defined factors. *Nature.* 2009; 459(7247): 708–11.

43. Lyons GE, Kyba M, Kamp TJ, Lalit PA, Crone WC, Thomson JA, Garry DJ, *et al.* Lineage reprogramming of fibroblasts into proliferative induced cardiac progenitor cells by defined factors. *Cell Stem Cell.* 2016; 18(3): 354–367.

44. Zhang Y, Ca N, Huang Y, Spencer CI, Fu JD, Yu C, Liu K, *et al.* Expandable cardiovascular progenitor cells reprogrammed from fibroblasts. *Cell Stem Cell.* 2016; 18(3): 368–381.

45. Cao N, Huang Y, Zheng J, Spencer CI, Zhang Y, Fu JD, Nie B, *et al.* Conversion of human fibroblasts into functional cardiomyocytes by small molecules. *Science.* 2016; 352(6290): 1216–1220.

46. Tohyama S, Fujita J, Fujita C, Yamaguchi M, Kanaami S, Ohno R, Sakamoto K, *et al.* Efficient large-scale 2D culture system for human induced pluripotent stem cells and differentiated cardiomyocytes. *Stem Cell Reports.* 2017; 9(5): 1406–1414.

47. Xu C, Police S, Hassanipour M, Li Y, Chen Y, Priest C, O'Sullivan C, Laflamme MA, Zhu W-Z, Van Biber B, *et al.* Efficient generation and cryopreservation of cardiomyocytes derived from human embryonic stem cells. *Regen Med.* 2010; 6(1): 53–66.

48. Soong PL, Putti TC, Hentze H, Wang ST, Phillips BW, Dunn NR. Teratoma formation by human embryonic stem cells: evaluation of essential parameters for future safety studies. *Stem Cell Res.* 2009; 2(3): 198–210.

49. Anderson D, Self T, Mellor IR, Goh G, Hill SJ, Denning C. Transgenic enrichment of cardiomyocytes from human embryonic stem cells. *Mol Ther.* 2007; 15(11): 2027–2036.

50. Goulburn AL, Ward-van Oostwaard D, Prall OWJ, Kaye DM, Mummery CL, Khammy O, Passier R, *et al.* NKX2-5 eGFP/w hESCs for isolation of human cardiac progenitors and cardiomyocytes. *Nat Methods.* 2011; 8(12): 1037–1040.

51. Tohyama S, Hattori F, Sano M, Hishiki T, Nagahata Y, Matsuura T, Hashimoto H, *et al.* Distinct metabolic flow enables large-scale purification of mouse and human pluripotent stem cell-derived cardiomyocytes. *Cell Stem Cell.* 2013; 12(1): 127–137.

52. Dubois NC, Craft AM, Sharma P, Elliott DA, Stanley EG, Elefanty AG, Gramolini A, *et al.* SIRPA is a specific cell-surface marker for isolating cardiomyocytes derived from human pluripotent stem cells. *Nat Biotechnol.* 2011; 29(11): 1011–1018.

53. Hattori F, Chen H, Yamashita H, Tohyama S, Satoh Y, Yuasa S, Li W, *et al.* Nongenetic method for purifying stem cell–derived cardiomyocytes. *Nat Methods.* 2009; 7(1): 61–66.

54. Hayashizaki Y, Abe M, Nishida K, Yoshihara M, Araki R, Kawaji H, Sunayama M, *et al.* Hotspots of de novo point mutations in induced pluripotent stem cells. *Cell Rep.* 2017; 21(2): 308–315.

55. Silva M, Daheron L, Hurley H, Bure K, Barker R, Carr AJ, Williams D. Forum generating iPSCs: translating cell reprogramming science into scalable and robust biomanufacturing strategies. *Stem Cell.* 2015; 16(1): 13–17.

56. Joshua G, Michel P. Embryonic stem cell–derived cardiac myocytes are not ready for human trials. *Circ Res.* 2014; 115(3): 335–338.

57. Kamp TJ, Lepley M, Zhang J, Xiong Q, Kaufman QS, Ye L, Wendel JS, *et al.* Cardiac repair in a porcine model of acute myocardial infarction with human induced pluripotent stem cell-derived cardiovascular cells. *Cell Stem Cell.* 2014; 15(6): 750–761.

58. Bel A, Rücker-Martin C, Guillevic O, Stefanovic S, Desnos M, Bellamy V, Nury D, *et al.* A purified population of multipotent cardiovascular progenitors derived from primate pluripotent stem cells engrafts in postmyocardial infarcted nonhuman primates. *J Clin Invest.* 2010; 120(4): 1125–1139.

59. Liu Y, Mercola M, Schwartz RJ. The all-chemical approach a solution for converting fibroblasts into myocytes. *Circ Res.* 2016; 119(4): 505–507.

60. Zhou Y, Wang L, Liu Z, Alimohamadi S, Yin C, Liu J, Qian L. Comparative gene expression analyses reveal distinct molecular signatures between differentially reprogrammed cardiomyocytes. *Cell Rep.* 2017; 20(13): 3014–3024.

61. Rodriguez M, Sniadecki NJ, Regnier M, Pabon L, Ruohola-Baker H, Murry CE, Reinecke H, *et al.* Tri-iodo-l-thyronine promotes the maturation of human cardiomyocytes-derived from induced pluripotent stem cells. *J Mol Cell Cardiol.* 2014; 72: 296–304.

62. Shadrin IY, Bursac N, Jackman CP, Juhas ME, Allen BW, Carlson AL, Qian Y. Cardiopatch platform enables maturation and scale-up of human pluripotent stem cell-derived engineered heart tissues. *Nat Commun.* 2017; 8(1): 2017.

63. Yui Y. Author Correction: Concerns on a new therapy for severe heart failure using cell sheets with skeletal muscle or myocardial cells from iPS cells in Japan. *NPJ Regen Med.* 2018; 3(1): 2017–2018.

64. Ma H, Marti-Gutierrez N, Park SW, Wu J, Lee Y, Suzuki K, Koski A, *et al.* Correction of a pathogenic gene mutation in human embryos. *Nature.* 2017; 548: 413–419.

CHAPTER 2

Challenges in stem cell-based approaches for myocardial regeneration after myocardial infarction

Violetta A. Maltabe*,†, Theofilos M. Kolettis‡,§, Panos Kouklis*,†,¶

*Laboratory of Biology, Medical School, University of Ioannina, Greece
†Department of Biomedical Research, Institute of Molecular Biology & Biotechnology, Foundation of Research and Technology-Hellas, University of Ioannina, Greece
‡Department of Cardiology, University of Ioannina Medical School, Greece
§Cardiovascular Research Institute, Ioannina and Athens, Greece

ABSTRACT

Myocardial infarction is the leading cause of chronic heart failure, an ominous disease entity with a wide prevalence in many countries. The morbidity and mortality of chronic heart failure remains high, despite recent pharmacologic advances and cardiac resynchronization therapy. After acute coronary occlusion, the necrotic area triggers a cascade of pathophysiologic events that may lead to structural and electrophysiological left ventricular remodeling, and eventually to progressive chronic heart failure. Therapeutic strategies targeting the repair of the infarcted myocardium aim at interrupting this vicious cycle and constitute an etiological and, as such, promising approach. However, after the initial enthusiasm accompanying early reports, subsequent preclinical and clinical studies unveiled several challenges associated with cell survival and proliferation, as well as abnormal electrophysiological responses after engraftment. In this chapter, we review the main cell sources that hold promise for clinical use, either alone or combined with growth factors and biomaterials, focusing on the acute and medium-term electrophysiological effects of cardiac regeneration approaches.

KEYWORDS

Clinical; CSCs; Growth factors; Infarction; iPSCs; MSCs; Myocardial; Regeneration; Scaffold.

¶ Corresponding author. Email: pkouklis@uoi.gr

LIST OF ABBREVIATIONS

ASCs	=	Adipose tissue-derived stromal cells
BM	=	Bone marrow
BMMNCs	=	BM-derived mononuclear cells
BMP-4	=	Bone morphogenetic protein-4
CABG	=	Coronary artery bypass grafting
CEDPs	=	Cardiac/endothelial dual progenitors
CPCs	=	Cardiac progenitor cells
EPCs	=	Endothelial progenitor cells
ESCs	=	Embryonic stem cells
ESC-CM	=	Embryonic stem cells-derived cardiomyocytes
ESC-CVP	=	ESC-derived cardiovascular progenitors
FGF-2	=	Fibroblast growth factor-2
GCSF	=	Granulocyte colony-stimulating factor
hESC-CMs	=	Human embryonic stem cell-derived cardiomyocytes
IGF-1	=	Insulin-like growth factor-1
IL-6	=	Interleukin-6 (IL-6)
IPSCs	=	Induced pluripotent stem cells
LV	=	Left ventricular
LVEF	=	LV ejection fraction
MI	=	Myocardial infarction
MSCs	=	Mesenchymal stem cells
NK	=	Natural killer
TGF-β	=	Transforming growth factor-β

2.1 INTRODUCTION

Myocardial infarction (MI) remains a substantial health-related problem worldwide. Prompt reperfusion strategies have markedly decreased mortality during the acute and in-hospital phases. However, healed MI leads to progressive left ventricular (LV) enlargement and dysfunction in a significant proportion of cases, which account for the vast majority of chronic heart failure cohorts. Despite improved pharmacologic treatment and the widespread implementation of cardiac resynchronization, the morbidity and mortality of chronic heart failure remain high. Thus, the significant societal impact of the disease has fueled intensive investigations in search of newer treatments, aiming at restoring, at least in part, the function of the scarred myocardium. In this regard, cellular transplantation, often combined with growth factors and various scaffolds, has emerged as a novel therapeutic paradigm for cardiac repair. The rationale behind this therapy is based on the assumption that an increase in the number of functional cardiomyocytes within the infarcted area will improve overall contractile performance and ameliorate electrophysiological alterations.

Although early experiments have shown that transplanted cells can form viable grafts within the host myocardium, subsequent work failed to produce robust evidence that these cells can survive, engraft, differentiate, and function appropriately in the ventricular myocardium. Thus, the initial enthusiasm has been hampered by subsequent modest results, coupled with safety concerns regarding the arrhythmogenic potential of cellular transplantation.[1] Based on these considerations, cellular transplantation has entered a new, more mature era of in-depth preclinical and clinical research, addressing various aspects of cardiac repair. This chapter summarizes the current state-of-the-art, addresses present challenges, and offers some suggestions for future directions.

2.2 TIMING OF CARDIAC REPAIR POST-MI

The ischemic necrosis of a myocardial area initiates a cascade of events affecting the electrical and mechanical functions of the ischemic and non-ischemic myocardium. During the early hours following acute coronary occlusion, neutrophil accumulation occurs, triggering a long chain of wound healing responses; the subsequent release of metalloproteinases and degradation of the extracellular matrix decrease the number of myocytes across the ventricular wall. This complex structure contains structural proteins (such as collagen I and II, fibronectin, laminin) and non-structural glycoproteins, proteoglycans and glycosaminoglycans. Besides, the extracellular matrix accommodates a large number of signaling molecules, regulating cellular communication and proliferation. Thus, the disruption of extracellular matrix leads to the expansion of the infarcted area and elevates diastolic and systolic wall stress, which, in turn, leads to dilatation of the non-infarcted myocardium and distortion of the LV shape.

Studies on local infarct healing processes have offered significant advances to the current understanding of the pathophysiology of LV remodeling; within the first hours after MI, inherent repair mechanisms are activated involving various cell types, such as inflammatory cells and fibroblasts, as well as cytokines and growth factors. The healing process after MI is intertwined with remodeling and their balance largely determines the process of LV hemodynamic and topographic alterations. This complex interaction starts early post-coronary occlusion and constitutes a long and dynamic process.

It has now been established that aborting infarct expansion during the early phase of MI interferes with the entire process of LV remodeling. For example, short-term ventricular restraint with a scaffold was shown to preserve LV function and geometry and to prevent progressive LV dilatation.[2] As a result, the timing of implementation of cardiac repair strategies is important, as the *prevention* of LV remodeling appears more effective when compared to the *reversal* of major topographic alterations, as seen in established chronic heart failure.[3] For example, higher efficacy was demonstrated after early (7 days) when compared to delayed (2 months) implantation of a biomaterial scaffold in infarcted rat hearts.[4] Based on this concept, a shift towards early intervention during the acute MI-phase has been adopted by several investigators, a trend that can increase the success rates of cardiac repair strategies. On the other hand, cell-survival may be poor early post-MI, a time-period characterized by intense local inflammatory responses, forming a 'hostile' environment for the transplanted cells.

2.3 CELL SOURCES FOR CARDIAC REPAIR

Several cell-types have been evaluated for cardiac regeneration, including skeletal myoblasts, bone marrow (BM) and adipose tissue-derived mesenchymal cells, endothelial stem cells, cardiac stem cells, and pluripotent-derived cardiac cells. Each cell-type (Table-2.1) and clinical trials (Table-2.2) briefly reviewed below, presents its own advantages and disadvantages.

Table-2.1.   Stem cells for cardiac regeneration: advantages and disadvantages.

Cell types	Origin	Advantages	Disadvantages
SkMs	Skeletal muscle satellite cells	• Autologous cell source • Abundant • Easily accessible from autologous muscle biopsies • Contractile abilities • Rapidly expansion *in vitro* • Resistant to ischemia • Low ethical concerns	• Committed to skeletal muscle lineage • No trans differentiation into functional cardiomyocytes, • Ventricular arrhythmias due to the lack of electromechanical coupling

Table-2.1.   (*Continued*)

Cell types	Origin	Advantages	Disadvantages
BMMNCs	Bone Marrow	• Autologous cell source • Easily accessible from autologous bone marrow • Proof of safety in clinical trials • Promotion of neovascularization • Low ethical concerns	• Heterogeneous cell population • Low cell quantity • Inconsistent results regarding therapeutic effects, most trials show no improvement in cardiac function
EPCs (CD133+, CD34+)	a) Bone Marrow b) Peripheral blood	• Autologous cell source • Easily accessible • Angiogenic properties • Promotion of neovascularization • Paracrine effect in enhancement of cardiomyocyte survival and organization • Low ethical concerns • Proof of safety in clinical trials	• Low cell quantity • Undefined phenotype • Inconsistent results regarding therapeutic effect
MSCs	a) Bone Marrow b) Adipose tissue c) Dental pulp d) Umbilical cord	• Autologous cell source • Easily accessible from several tissues • Rapidly expansion *in vitro* • Beneficial immunomodulatory properties • Immune-privileged, allowing allogeneic use • Paracrine signalling • Anti-fibrotic effects • Promotion of neovascularization • Guided cardiopoiesis upon activation with specific factors • Proof of safety in clinical trials	• Limited differentiation potential • Differentiation to osteoblasts, adipocytes and chondrocytes • Inconsistent results regarding therapeutic effect

(*Continued*)

Table-2.1.   (*Continued*)

Cell types	Origin	Advantages	Disadvantages
CPCs	Heart	• Autologous cell source • Differentiation to cardiomyocytes *in vitro* • Suitable for autologous transplantation	• Low engraftment • Limited differentiation to cardiomyocytes *in vivo* • Access from invasive myocardial biopsies • Insufficient cell characterization • Low availability • Costly expansion • Older/autologous donors resulted in lower quality cells • Inconsistent results regarding therapeutic effect
ESCs	Embryo/Blastocyst	• Pluripotent differentiation potential • Unlimited quantities • ESC-derived cardiomyocytes integrate electromagnetically into the host myocardium	• Differentiation protocols: efficiency and purity of cells • Difficulties in generation of • mature cardiomyocytes compared to adult cardiomyocytes • Electrical instability, arrhythmias after engraftment • Ethical concerns • Risk of tumorigenicity • Genomic instability • Lack of availability • Risk of immunologic rejection and immunosuppression required
iPSCs	Fibroblasts, various adult tissue sources	• Autologous cell source • Pluripotent differentiation potential • Unlimited quantities • Differentiated to iPSCs-cardiomyocytes, functionally integrated within myocardium • No immune rejection	• Risk of tumorigenicity • Requires genetic manipulation • Differentiation protocols: efficiency and purity of cells • Difficulties in generation of • mature cardiomyocytes compared to adult cardiomyocytes • Electrical instability, arrhythmias after engraftment
Induced CMs	Trans differentiation/ Direct reprogramming of heart's fibroblasts	• Autologous cell source • No immune rejection • No malignant transformation	• Low reprogramming efficiency • Labour intensive and time-consuming

Abbreviations: BMMNCs: Bone Marrow Mononuclear Cells; CMs: Cardiomyocytes; CPCs: Cardiac Progenitor Cells; ESCs: Embryonic Stem Cells; EPCs: Endothelial Progenitor Cells; iPSCs: Induced Pluripotent Stem Cells; MSCs: Mesenchymal Stem Cells; SkMs: Skeletal Myoblasts.

Table-2.2.   Selected stem cell-based clinical trials for acute myocardial infarction and ischemic cardiomyopathy.

Trial	Disease	Timing of cell delivery after AMI	Follow-up	Cell sources	Route of delivery	Primary outcome	References
REPAIR-AMI (NCT00279175)	AMI	3–5 days	4 months	BMMNCs (autologous)	Intracoronary	- Improvement in LV systolic function - Significant reduction of MACE	[16–18]
TIME (NCT00684021)	AMI	3 vs 7 days	2 years	BMMNCs (autologous)	Intracoronary	- No difference in regional LV function - Infarct size not different from controls	[19, 21]
LateTIME (NCT00684021)	AMI	2–3 weeks	6 months	BMMNCs (autologous)	Intracoronary	- No significant increase in LVEF - No significant differences in change in global left ventricular function - infarct size not different from controls	[20]
SWISS-AMI (NCT00355186)	AMI	5–7 days vs 3–4 weeks	12 months	BMMNCs (autologous)	Intracoronary	- No improvement in LV function - Infarct size not different from controls	[22]
BAMI (NCT01569178)	AMI	2–8 days after successful reperfusion for AMI	3 years	BMMNCs (autologous)	Intracoronary	Currently ongoing	[27]
PreSERVE-AMI (NCT01495364)	AMI	4–11 days after stent placement	12 months	CD34+ cells (autologous)	Intracoronary	- No improvements in LVEF and in myocardial perfusion	[33]
CAREMI (NCT02439398)	AMI	5–7 days	12 months	CSCs (allogeneic)	Intracoronary	- Absence of immune-mediated events - No improvement in LVEDV	[49]
CADUCEUS (NCT00893360)	AMI	1.5 to 3 months	6 months	CDCs (autologous)	Intracoronary	- Moderate scar size reduction - Increase in viable cardiac muscle - Increase in regional thickening - No improvement in LVEDV, LVESV, or LVEF	[50, 51]

(Continued)

Table-2.2. (*Continued*)

Trial	Disease	Timing of cell delivery after AMI	Follow-up	Cell sources	Route of delivery	Primary outcome	References
MAGIC (NCT00102128)	ICM	–	6 months	Skeletal myoblasts (autologous)	Intra-myocardial	– No significant improvement in regional or global LV function	[11]
CARDIO133 (NCT00462774)	ICM	–	6 months	CD133+ cells (autologous)	Intra-myocardial	– Improvements in scar size and regional perfusion – No significant differences in LVEF – No effects on global function and clinical symptoms	[30]
PERFECT (NCT00950274)	ICM	–	6 months	CD133+ cells (autologous)	Intra-myocardial	– Reduction in scar size – No significant differences in LVEF – No effects on global function and clinical symptoms	[31]
PROMETHEUS (NCT00587990)	ICM	–	6 months	BM-MSCs (autologous)	Intra-myocardial	– Improvements in regional function, tissue perfusion, and fibrotic burden	[35]
POSEIDON (NCT01087996)	ICM	–	13 months	BM-MSCs (autologous vs allogeneic)	TESI	– Enhancement quality of life and functional capacity of patients, – Improvement of end-diastolic volume EDV, end-diastolic mass increased – Reduced infarct size – Improved ventricular remodeling and patient outcome. – LVEDV was significantly decreased only in the allogeneic group while the LVESV and LVEDV were reduced in both groups	[36, 37]

Trial			Duration	Cell type	Delivery	Outcomes	Ref
MSC-HF (NCT00644410)	ICM	—	6 months	BM-MSC (autologous)	Intra-myocardial	– Improvements in LVEF, stroke volume and myocardial mass – No differences in NYHA class, 6-min walking test and Kansas City cardiomyopathy questionnaire.	[38]
TAC-HFT (NCT00768066)	ICM	—	12 months	BM-MSCs vs BMMNCs (autologous)	TESI	– Improvements in both QoL and functional capacity (BM-MSCs) – No significant changes in LVEF and volumes	[39]
DREAM HF-1 (NCT02032004)	ICM	—	12 months	MPCs (Stro-3$^+$ MSCs) (allogeneic)	TESI	Currently ongoing	[40]
PRECISE (NCT00426868)	ICM	—	36 months	ASCs (autologous)	TESI	– Improved exercise capacity – preserved myocardial perfusion and left ventricular function	[43]
ATHENA II trial (NCT02052427)	ICM	—	12 months	ASCs (autologous)	Intra-myocardial	– No difference in left ventricular function or volumes – Improvement in quality of life – Significant improvement in MLHFQ	[44]
SCIENCE (NCT02673164)	ICM	—	12 months	ASCs (allogeneic)	Intra-myocardial	Currently ongoing	[45]
CHART-1 (NCT01768702)	ICM	—	39 weeks 52 weeks	Cardiopoietic cells (autologous)	Endo-myocardially	– Neutral at 39 weeks – At 52 weeks: improved reverse remodelling decrease in LVEDV and LVESV	[47, 48]

(Continued)

Table-2.2.   (*Continued*)

Trial	Disease	Timing of cell delivery after AMI	Follow-up	Cell sources	Route of delivery	Primary outcome	References
CONCERT-HF (NCT02501811)	ICM	–	12 months	MSCs and c-kit+ CSCs (autologous)	TESI	Currently ongoing	[70]
ESCORT (NCT02057900)	ICM	–	18 months	ESC-derived +/ CD15+ embedded in fibrin	Epicardial patch	– Increased systolic motion	[59]

Abbreviations: AMI: Acute Myocardial Infarction, CSCs: Cardiac Stem Cells, CDCs: Cardiosphere-Derived Stem cells, ICM: Ischemic Cardiomyopathy, MACE: major adverse cardiovascular events, BM-MSCs: Bone Marrow-derived–Mesenchymal Stem Cells, TESI: Trans-endocardial, ASCs: Adipose-derived stromal cells.

2.3.1 Skeletal Myoblasts

Skeletal myoblasts are progenitor stem cells derived from skeletal muscle satellite cells that can differentiate after appropriate stimuli; the activated myoblasts subsequently fuse to form new myofibers.[5,6] Myoblasts are abundant in the human body and are easily accessible from autologous muscle biopsies; after harvesting, these cells are remarkably resistant to ischemia and can rapidly expand in vitro.[7] Based on these properties, skeletal myoblasts were the initial cell type used in pre-clinical and clinical trials for cardiac regeneration. In *in vivo* animal models, skeletal myoblasts displayed adequate differentiation after engraftment in the ventricular myocardium, thereby improving local contractile function.[8-10] Although these findings were reproduced in small-scale non-randomized clinical trials, the larger, controlled MAGIC trial failed to demonstrate sustained clinical benefits.[11] More importantly, increased incidence of ventricular tachyarrhythmias was recorded, corroborating earlier experimental[12] and clinical observations.[11] The explanation of these findings is difficult, as many arrhythmogenic mechanisms may be operative.[1] Specifically, the paracrine effects of skeletal myoblasts may induce hypertrophy, with consequent prolongation of the action potential, leading to early afterdepolarizations and triggered activity. Second, the transplanted myoblasts differentiate into hyper-excitable myotubes, displaying different action potential duration, when compared to host cardiomyocytes, setting the stage for reentrant arrhythmias. Lastly, differentiated myotubes lack gap junctions, and, therefore, they are not coupled to surrounding ventricular cardiomyocytes or each other. To circumvent this problem, transplantation of connexin 43-overexpressing myoblasts has been proposed,[13] but the hitherto experience has yielded ambiguous results; hence, regenerative approaches focused on more promising cell types.

2.3.2 BM-Derived Mononuclear Cells

BM-derived mononuclear cells (BMMNCs) represent a highly heterogeneous population, consisting of monocytes, lymphocytes, mesenchymal stem cells, hematopoietic stem cells, and endothelial progenitor cells. Despite the initial optimism towards transdifferentiation of c-kit$^+$ cells into cardiomyocytes, it was subsequently shown that these cells invariably adopt haematopoietic fates in the ischemic myocardium.[14,15] The skepticism remained after the controversial results of four randomized clinical trials, examining the efficacy of autologous BMMNCs in post-MI patients. Specifically, in the REPAIR-AMI, improved LV systolic function was found, which was sustained during follow-up[16-18]; by contrast, no such benefit was evident in the TIME, LateTIME, and SWISS-AMI trials.[19-22] A variety of factors may account for these discrepancies, including the number of transplanted cells, the timing of intervention and the route of delivery.[23-25] The ACCRUE meta-analysis[26] examined the safety and efficacy of intracoronary BMMNCs after acute MI, including individual patient data from 12 randomized trials, of a total of n = 1252 patients. The analysis failed to show a treatment-effect on major adverse cardiac and cerebrovascular events; furthermore, no improvement was evident in ejection fraction, end-diastolic LV volume, or end-systolic LV volume. Thus, unselected BM-derived mononuclear cells may be without measurable clinical benefit or enhancement in LV function. More answers are expected from the BAMI trial, evaluating the efficacy and safety of a single intracoronary infusion of autologous BMMNCs after successful reperfusion for acute MI.[27]

2.3.3 Endothelial Progenitor Cells

Endothelial progenitor cells (EPCs) were first described in 1997, as circulating hematopoietic progenitor cells, and are characterized by the capacity to differentiate into endothelial cells *in vitro*.[28] EPCs are mobilized from the bone marrow and migrate via the circulation to various organs after injury, where they participate in local neovascularization. This is accomplished by anchoring into existing blood vessels and subsequent release of pro-angiogenic growth factors and cytokines.[29] Based on their potent angiogenic properties and the ease of access, EPCs have been studied extensively in animal models and clinical studies. These cells are obtained from the bone marrow or the blood after mobilization with granulocyte colony-stimulating factor (G-CSF); two subtypes have been used, namely CD 34[+] and the immature and less lineage-committed CD133[+].

Two double-blinded, randomized, placebo-controlled clinical trials using CD133[+], namely the CARDIO133 and PERFECT, showed improvements in scar size and regional perfusion, but no significant differences in global LV systolic function and clinical status.[30,31] In animal models, adequate engraftment of CD34[+] cells was demonstrated, leading to increased capillary density.[32] In phase II PreSERVE-AMI clinical trial, intracoronary infusion of autologous bone-marrow-derived CD34[+] cells was safe, but did not improve resting myocardial perfusion over a 6-month follow-up.[33] However, when adjusted for ischemic time, there was a significant relationship between the cell-dose and infarct size, global LV ejection fraction (LVEF) and survival. These results reiterate the importance of adequate cell-concentrations and support the concept of targeted therapy with a specific cell-type, with CD34[+] EPCs representing a promising source for cardiac repair.

2.3.4 Mesenchymal Stem Cells (MSCs)

Mesenchymal stem cells or mesenchymal stromal cells are multipotent fibroblast-like cells, playing an integral role in the hematopoietic niche. MSCs are mostly found in the BM and adipose tissue, from where they can be easily harvested, isolated and expanded *ex vivo*. Under different stimuli, these cells have the capacity to differentiate into osteoblasts, adipocytes or chondrocytes. The lack of immunogenicity, coupled with the potential to secrete proangiogenic and anti-apoptotic factors constitute additional advantages.

MSCs have been evaluated in preclinical studies, with favorable results in terms of infarct size,[34] whereas phase I clinical trials confirmed their feasibility and safety in the post-MI setting. The subsequent PROMETHEUS trial showed that the intramyocardial administration of autologous MSCs in patients undergoing coronary artery bypass grafting (CABG) improved regional function, tissue perfusion, and fibrosis.[35] The POSEIDON clinical trial compared for the first time autologous and allogeneic BM-derived MSCs, which were injected in the LV endocardium via a catheter. Both types of MSCs decreased LV dimensions, with a small advantage noted in the allogeneic group, perhaps because the cells were derived from young, healthy donors.[36,37] Examined together with the MSC-HF and the TAC-HFT trials, these results underscore the treatment-effects on functional capacity and provide evidence on reverse remodeling.[38,39] MSCs can be isolated from the BM with the use of specific markers. Particular attention has been devoted to Stro-1 and Stro-3, characterizing a subpopulation of mesenchymal precursor cells. These MSCs demonstrate a stronger mitotic and paracrine capacity and higher numbers of colony-forming unit-fibroblasts. They were used in DREAM HF-1 trial and results from phase III are awaited.[40]

2.3.5 Adipose-Derived Stromal Cells

The adipose tissue-derived stromal cells (ASCs) have similar differentiation potential, adherence and immunomodulatory properties with BM-derived MSCs, and express common cell surface molecules. ASCs can be harvested less invasively than BM-derived MSCs and yields much higher stem cells numbers.[41] Lastly, ASCs exhibit stronger inhibitory effects on CD4$^+$, CD8$^+$ and natural killer (NK) cells activation, as well as on B cells suppression.[42] Corroborating these advantages, the PRECISE clinical trial showed improved myocardial perfusion in patients with ischemic cardiomyopathy after endocardial injection of autologous ASCs.[43] Unfortunately, these results were not reproduced in the ATHENA trial; in this clinical trial, there was no demonstrable benefit in terms of functional status,

LVEF, or LV volumes, despite improved quality of life indices.[44] A phase II trial (SCIENCE) is currently evaluating the effects of direct intra-myocardial injection of large concentration of allogeneic ASCs in patients with post-MI heart failure, and is expected to provide more answers.[45]

2.3.6 Guided Cardiopoiesis

Guided cardiopoiesis is defined as the formation of cardiac stem cells by priming BM-MSCs with cardiogenic growth factors. Specifically, activation of BM-MSCs with transforming growth factor-β (TGF-β), bone morphogenetic protein-4 (BMP-4), Activin-A, together with retinoic acid, induces cytosolic expression of cardiac transcription factors; their nuclear translocation is promoted by insulin-like growth factor-1 (IGF-1) and interleukin-6 (IL-6), whereas fibroblast growth factor-2 (FGF-2) and thrombin maintain cell cycle activity.[46] However, the enthusiasm for this approach weaned after the results of the large phase III randomized, sham-controlled CHART-1 trial; in this study, no medium-term beneficial effects were evident, despite a trend towards decreased LV dimensions 12 months post-intervention.[47,48] Clearly, more data are needed before drawing any conclusions.

2.3.7 Cardiac Progenitor Cells

Several types of endogenous cardiac progenitor cells (CPCs) have been identified in adult hearts from rodents and humans, by tyrosine kinase receptor c-kit as a marker and/or formation of cardiosphere structures. The characterization of these cells gave rise to the hypothesis of enhancing the inherent, albeit limited, the regenerative capacity of the LV myocardium. Based on this rationale, CAREMI trial was initiated, evaluating the safety and efficacy of intracoronary infusion of allogeneic CPCs in patients with acute MI.[49] Similar conclusions can be drawn for CADUCEUS trial, examining the effects CD105+ cells derived from cardiospheres, spherical clusters of stem cells cultured *in vitro*.[50,51]

2.3.8 Embryonic Stem Cells-derived Cardiomyocytes (ESC-CM) and Cardiovascular Progenitors (ESC-CVP)

Embryonic stem cells (ESCs) are pluripotent cells derived from the inner side of blastocysts and can proliferate to any cell type of all three germ-layers (i.e., endoderm, mesoderm and ectoderm). ES-derived cells are isolated either as terminally differentiated cardiomyocytes or as cardiovascular progenitor populations, which can further differentiate after transplantation *in vivo*.[52] Genetic modification, specialized culture methods, and treatment with chemical and biological factors have been used to enrich, purify, and select homogeneous and functionally intact populations of ESC-CMs, generated from ESCs. The activity of developmental stage-specific promoters, such as Flk1, Isl1, and Nkx2.5, is used for the isolation of cardiac precursors.[53] These cells are characterized by the ability to self-renew and to differentiate towards cardiac lineages.

The *in vivo* electrophysiological properties of human embryonic stem cell-derived cardiomyocytes have been assessed in the rat-model after engraftment in the LV myocardium.[54] Two weeks after engraftment, positive immunostaining for connexin-43 was observed between the engrafted and host cardiomyocytes, whereas optical mapping demonstrated satisfactory electrical conduction. Furthermore, ventricular tachyarrhythmias could not be induced by programmed electrical stimulation in rats transplanted with human embryonic stem cell-derived cardiomyocytes, confirming the safety of this approach.

In a pre-clinical study using non-human primates of myocardial ischemia, intra-myocardial delivery of hESC-CMs resulted in extensive re-muscularization of infarcted tissue, although hESC-CMs maturation was incomplete over a 3-month period.[55] Another study in non-human primate models of acute myocardial infarction showed improvements in LV function in the short term and reduced native cardiac cell apoptosis, but there was no evidence of progenitors long-term engraftment and survival.[56] When the regenerative capacity of hESCs-CMs, hESC-CVPs, and hBMMCs were compared in the rat model, the former two cell types improved LV function more than the latter type; interestingly, hESC-CVPs and hESC-CMs had comparable results regarding the yield of grafting as well as the number of generated vessels.[57]

Differentiation of ESCs towards cardiac progenitors by exposure to bone morphogenetic protein-2 and a fibroblast growth factor receptor inhibitor, resulted to a novel cardiogenic population, that upon engraftment to rat and non-human primate models of MI improved LV function.[58] These promising preclinical data paved the way for the ESCORT trial, the first-in-man, which evaluated the feasibility, safety and regenerative effects of human ESC-derived cardiovascular progenitors typified by SSEA-1+(CD15+)/ISL1+ co-expression. These cells were delivered epicardially via a fibrin-gel in 6 post-MI patients undergoing surgery.[59] This trial demonstrated the short- and -medium-term safety of this treatment, that can be expanded to various ESC-derived lineage-specific progenies. Despite the positive outcome, safety concerns remain which, in conjunction with the need for immunosuppression and ethical issues, currently restrict the clinical applications of hESCs. This trial demonstrated the feasibility and safety of treatments based on different ESC-derived lineage-specific progenies. With this rationale our group has isolated a novel progenitor cell-population, exhibiting dual characteristics (CEDPs — Cardiac/Endothelial Dual Progenitors). Specifically, these cells were produced after mouse ESCs differentiation *in vitro*, and display both, cardiovascular and endothelial characteristics. For their isolation, we used a genetic approach, guided by VE-cadherin promoter activity.[60] This selection strategy was based on previous observations from our laboratory, showing that VE-cadherin promoter is transiently activated in a subset of Isl1+ cardiac progenitors. These cells could self-renew upon Wnt/b-catenin signaling pathway activation and differentiated further to cardiomyocytes and mature endothelial cells *in vitro* (Figure-2.1).

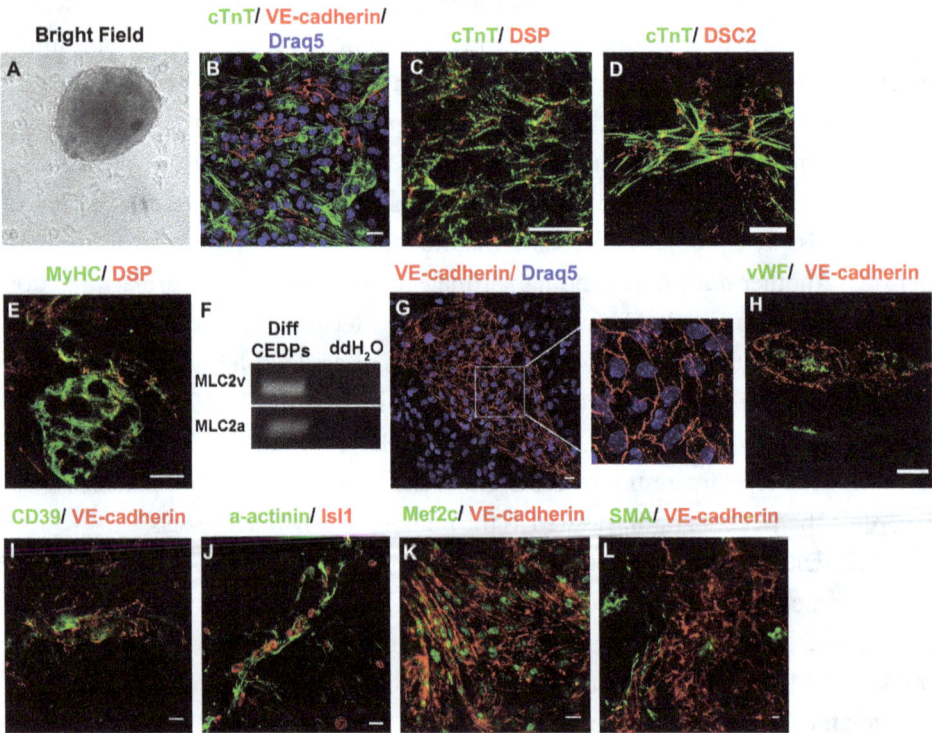

Figure-2.1.   Differentiation potential of selected novel progenitor cell-population, CEDPs. (A) Optical microscope image of a 3D-sphere formed after CEDPs differentiation. (B) CEDPs differentiation to cardiomyocytes and endothelial cells. (C–E) Formation of intercalated disk structures after CEDPs differentiation, shown by Desmoplakin (DSP), Desmocollin-2 (DSG2) and Myosin Heavy Chain (MyHC) staining. (F) Expression of MLC2v and MLC2a after 10 days of CEDPs indicated maturation of CEDPs to cardiomyocytes. (G) VE-cadherin+ endothelial cells formed extensive adherens junctions. (H-I) Expression of vWF and CD39, endothelial maturation markers after CEDPs differentiation. (J-K) Isl1 and Mef2c, cardiac transcription factors were expressed in a-actinin+ and VE-cadherin+ cells, respectively, at day 10 of CEDPs differentiation. (l) Expression of smooth muscle actin (SMA) during CEDPs differentiation. Scale bar: 20 μm. Reproduced from Maltabe *et al.*[60]

Moreover, they survived after transplantation in the myocardium of adult rats and differentiated into cardiomyocytes. Based on their unique properties CEDPs could represent a novel cell source for cardiac regeneration, currently under investigation in our institution.

2.3.9 Induced Pluripotent Stem Cells

Induced pluripotent stem cells (iPSCs) were first elicited in 2006 after somatic cells reprogramming by the introduction of a set of transcription factors (Oct3/4, Sox2, c-Myc, and Klf4, commonly referred to as Yamanaka factors). They exhibit properties similar to ESCs, including pluripotency and self-renewal and at the same time their origin circumvents ethical issues and since iPSCs are produced from autologous tissues, the risk of rejection is minimized. Experience of iPSCs as therapeutic source for cardiac regeneration was gained in many preclinical trials conducted in small and large animal models. Transplantation of an engineered tissue graft from hiPSCs-derived CM and hiPSCs-derived endothelial cells in guinea pig hearts 7 days after cryoinjury resulted in cardiomyocyte proliferation, vascularization, and electrical coupling to the host tissue in treated animals.[61] Likewise, intra-myocardial transplantation of MHC-matched allogeneic iPSC-CMs in non-human primates with MI resulted in adequate cell-survival, electrical coupling with host cardiomyocytes and improvement of cardiac contractile function, without rejection, when appropriate immunosuppression was used.[62] Implantation of sheets composed by hiPSCs- derived CM, combined with omental flap technique in a porcine model of MI improved LV function post-treatment,[63] leading to conditional approval of the first-in-human trial in Japan in May 2018, following extensive debate.

2.3.9.1 Direct reprogramming

An attractive approach for generation of new myocardium was accomplished initially *in vitro* by transdifferentiation of fibroblasts to cardiomyocytes, after delivery of GATA4, MEF2C and TBX5 transcription factors to fibroblasts.[64] Similarly, reprogramming of cardiac fibroblasts into cardiomyocytes was achieved after local delivery of these transcription factors in murine heart after coronary ligation *in vivo*.[65] Recent studies demonstrated that upregulation of Akt1, as well as inhibition of pro-fibrotic signaling enhanced significantly reprogramming.[66,67] Nonetheless, the molecular mechanisms of reprogramming are far from being understood, and further research is awaited.

2.3.10 Combination Cell Therapy

Each cell source, reviewed above, presents its distinct advantages and disadvantages, and none displays clear-cut superiority. The main hypothesis of the combination approach is to achieve a synergistic effect with the simultaneous use of two or more cell types, which is supported by the results of animal studies.[68,69] The CONCERT-HF is an underway phase II, randomized, placebo-controlled study evaluating the safety, feasibility, and efficacy of autologous MSCs and c-kit+ CSCs, alone or in combination after transendocardial administration in patients with ischemic heart failure.[70]

2.4 GROWTH FACTORS

Post-MI, growth factors play an important role in inherent cardiac repair mechanisms. Growth factors stimulate local healing of the infarcted area and exert protective actions in surviving cardiomyocytes; moreover, they can increase oxidative carbohydrate utilization, and augment sarcoplasmic reticulum calcium content. Based on these properties, the induction of physiological LV hypertrophy has been tested in animal models of chronic heart failure.[71] Extending this concept, our group has proposed the use of growth factors as a means of ameliorating LV remodeling, by intervening during the early post-MI period.[72] This approach resurfaced during the past decade, after the solid demonstration of paracrine effects of various cell types after engraftment in the myocardium. Given the widespread application of primary coronary angioplasty, the intracoronary route of administration after reperfusion during the acute phase of MI becomes attractive.

2.4.1 Growth Hormone and Insulin-Like Growth Factor-1

Growth hormone and its mediator IGF-1 have attracted considerable research interest in the post-MI setting. In a porcine model of MI, we have shown that selective intracoronary administration of growth hormone in the infarcted area after reperfusion attenuates LV remodeling; this action appeared secondary to enhanced angiogenesis and increased wall thickness in the infarcted area.[73] Moreover, growth hormone appears to exert protective actions in surviving cardiomyocytes, as shown by preserved action potential at the infarct border 24 hours post-coronary occlusion, accompanied by decreased arrhythmogenesis.[74] The beneficial effects on infarct expansion are observed after growth hormone or insulin-like growth factor-1 administration, although the former likely produces more potent effects on collagen matrix remodeling.[75] Interestingly, growth hormone not only increased cardiomyocyte volume in adult rats but was also to stimulate cardiomyocytes to re-enter the cell cycle, divide, and thereby increase their number across the LV wall.[76] In a small-scale clinical trial, patients with ST-elevation MI were randomized to intracoronary infusion of placebo or two dosages of IGF1, after successful percutaneous coronary intervention. The results were very encouraging, displaying significant improvement in LV end-diastolic volume, LV mass, stroke volume after high dose IGF-1; moreover, there was evidence of decreased fibrosis in late gadolinium enhancement at 2 months.[77]

Ghrelin, a peptide released from the stomach, has also shown cardioprotective effects after MI, by regulating intracellular calcium concentration, inhibiting pro-apoptotic cascades, and exerting antioxidant properties. Clinical reports have indicated ameliorated LV remodeling post-MI, and symptomatic improvement in patients with end-stage congestive heart failure.[78]

Growth hormone has been extensively investigated in several small-scale clinical trials including patients with a history of remote MI, but the results were equivocal.[79] This inconsistent translation from the experimental background into the clinical arena has been attributed to heterogeneous patient populations with various degrees of growth hormone-deficiency; using GH provocative tests, the percentage of patients with chronic heart failure and growth hormone-deficiency is estimated at 30%, thus constituting the target population that could benefit from such therapy.[80]

2.4.2 Fibroblast Growth Factor

The FGF family has been implicated in myocardial regeneration by promoting angiogenesis and conferring cardioprotective effects.[81] Moreover, fibroblast growth factor-2, as reported, activates resident cardiac precursors and promotes their differentiation to functional cardiomyocytes.[82]

2.4.3 Neuregulin-1 (NRG-1)

NRG-1 is a cell-adhesion molecule, essential for the normal development of the nervous system and the heart that has been implicated in cardiac repair. It is thought that NRG-1 increases the proliferation of resident cardiac precursors, which led to improved LV function in mice.[83] A clinical study in 15 patients with chronic heart failure showed favorable acute and chronic hemodynamic effects after recombinant human neuregulin-1, accompanied by serum noradrenaline after 3 months of follow-up.[84] Likewise, in 44 patients with similar characteristics, recombinant human neuregulin-1 tended to reverse LV remodeling and to improve LV function over the same time-period.[85]

2.4.4 Hepatocyte Growth Factor

HGF, secreted by mesenchymal cells, plays a major role in embryonic organ development, as well as in wound healing. Adipose-derived stem cells, transfected with lenti-hHGF were injected in rat post-MI; these cells migrated to the myocardium, resulting in increased vascular density and decreased fibrotic area by suppressing TNF-α and TGF-β1 expression.[86] These results suggest that HGF may be a useful adjunct to cellular approaches for cardiac repair.

2.5 TISSUE ENGINEERING — SCAFFOLDS

Biomaterials have been extensively evaluated in cardiac-repair strategies, based on their chemical and physical properties that permit their use as tissue engineering constructs.[87] Further to their aforementioned properties in restraining infarct expansion, biomaterials can be viewed as ideal candidates for controlled and sustained local delivery of growth factors and cells. The desired properties of a scaffold include high porosity, absence of immunogenicity, as well as optimal biodegradability and biocompatibility. Ideally, they should also promote vascularization, coupled with optimal mechanical properties, thereby providing a microenvironment similar to that of the extracellular matrix. Natural polymers (e.g., collagen, fibrin, chitosan, alginate, natural extracellular matrix, peptides) and synthetic polymers (e.g., polycaprolactone, polyglycerol sebacate, and polyurethanes) provide a choice of materials to fabricate scaffolds. They have been used alone or in combination with various cell proliferation and differentiation factors. Acellular structures present several advantages when compared to cellular structures, such as extended shelf-life, availability for immediate implantation, and limited immune reaction. Major naturally occurring acellular materials used for heart repair include collagen, hyaluronic acid, alginate, chitosan, fibrin and scaffolds derived from various decellularized organs (Table-2.3).

2.5.1 Alginate

Of the numerous biomaterials under evaluation, alginate, an anionic polysaccharide present in brown seaweed, is a particularly attractive candidate, currently approved by the Food and Drug Administration for various medical applications.[88] In cardiac repair, alginate-based scaffolds present several advantages, due to their biocompatible, non-toxic, and non-immunogenic characteristics.

Theoretically, the risk of impaired propagation of electrical depolarization may be higher after the injection of biomaterials, which lack conductive properties,

Table-2.3.   Scaffolds for cardiac regeneration: characteristics.

Acellular natural scaffolds	Characteristics
Collagen (Fibers, Hydrogel, Patch)	• Excellent biocompatibility and biodegradability • Cross-linkable to add strength • Effective cell adhesion • Low immunogenicity • Soluble at low pH and temperature • Low elastic modulus
Hyaluronic acid (Hydrogel, Patch)	• Weak mechanical properties • Easily cross-linkable to add strength
Alginate (Hydrogels, Patch, Film, Beads)	• Large pore size (50–200 μm) • Ideal for hydrogels due to its viscosity • Medium cell adhesion • Gelation capacity • Non-thrombogenic • Lack of integration with cardiac cells
Fibrin (Hydrogel, Patch)	• Porosity and stiffness depend on composition • Forms nets of fibers • Medium cell adhesion
Chitosan (Hydrogel, Patch, Fibers)	• Easy to alter degradation rate • Low cell adhesion • Porosity • High elastic modulus
Decellularized organs (ECM Hydrogel, Patch, Whole organ)	• Maintenance of the organ's ultrastructure • Poor mechanical strength of ECM • Requires optimization of the decellularization process to retain important structural, biochemical, and biomechanical properties and eliminate components which could cause an immunogenic response
Scaffolds for cellularized constructs	
Natural Scaffolds	• Native ECM components • Promotion of cell adhesion, proliferation, cell survival • High biocompatibility and degradability • Low immunogenicity • Poor mechanical properties
Synthetic Polymers	• Good workability • Improved and tuneable mechanical properties • Biocompatibility and degradability • Low risk of infection • Excellent strength and durability • Uniformity and reproducibility • Limited cell integration • Low biocompatibility • Toxicity

rather than after cellular treatments; however, the data on this field are scarce. Thus, our group has evaluated the safety of alginate injections in the peri-infarct area in the rat-model.[89] We found that such treatment was safe, evidenced by the absence of significant effects ventricular tachyarrhythmias, acute LV failure, sympathetic activation or short-term survival. We feel that this effect can be attributed to the low Ca^{2+} concentration used during biomaterial fabrication, as recently suggested by our group.[90] Based on the safety of this approach, we examined the value of sustained, local GH-release via an alginate-hydrogel on post-MI LV remodeling in the rat-model. We found that this intervention attenuated LV remodeling and improved LV function, more effectively than alginate alone. These actions, secondary to preserved scar-thickness and reduced wall-stress, can be attributed to the beneficial actions of GH on infarct healing, mediated by enhanced angiogenesis and myofibroblast-activation.[91] In the same animal-model, we assessed the treatment-effects on medium-term electrophysiological remodeling. We found decreased arrhythmogenesis in treated rats after programmed electrical stimulation, secondary to preserved electrical conduction and repolarization-dispersion at the infarct border.[92] Due to the promising results in preclinical animal models, alginate hydrogels are the only injectable biomaterial currently undergoing clinical evaluation. Recently, promising results were recently reported in patients with ischemic cardiomyopathy undergoing intramyocardial injections of alginate hydrogels; these favorable effects on LV remodeling indices were evident as early as 3 days post-injection and were sustained over 3 months.[93] Lastly, decreased collagen deposition and increased levels of VEGF were observed in a chronic rodent model of ischemic cardiomyopathy, indicating reduced scarring and neovascularization in response to RGD modified alginate.[94]

2.5.2 Collagen and Hyaluronic Acid

Collagen is the major structural protein of the extracellular matrix. As a biomaterial, collagen can form highly organized, three-dimensional scaffolds, particularly when used in combination with other biomaterials such as chitosan, elastin and gelatin. Likewise, hyaluronan is part of the extracellular matrix, forming noncovalent bonds with proteoglycans. Significant improvements in LV function were observed after epicardial injection in the infarcted region in rats *in vivo*,[95] likely attributed to enhanced angiogenesis and reduced apoptosis.[96]

2.5.3 Fibrin

Fibrin-based scaffolds, either alone or combined with skeletal myoblasts, have been tested in the subacute rat-model of MI; five weeks post-injection, both treatments preserved infarct wall thickness, reduced infarct size, and improved overall systolic function.[97] In a comparative study between fibrin, collagen type I, and matrigel as injectable biomaterials for MI repair in rats, the angiogenic potential was reported as similar for all three polymers at 5 weeks post-treatment; however, poor mechanical properties were noted as the major disadvantage of the fibrin gel, leading to the suggestion that fibrin should be combined with other high-compression components, such as chitosan or collagen.[98]

2.5.4 Decellularized Matrices

Decellularized matrices are derived from biological tissues after removal of cells, with concurrent preservation of the various protein components. Thus, decellularized matrices are promising biomaterials for cardiac repair cardiovascular tissue, as they mimic the complex array of proteins found in native tissue. Decellularized urinary bladder extracellular matrix (UB-ECM) has been evaluated as an epicardial patch in the chronic porcine MI-model at 6–8 weeks post-infarction.[99] The animals received either a UB-ECM or expanded polytetrafluoroethylene (ePTFE) patch. The decellularized matrix promoted vascularization at 3 months, whereas at the same time calcification occurred in the ePTFE patch. Moreover, UB-ECM scaffolds increase cardiac marker expression (i.e., a-smooth muscle actin (SMA$^+$) myofibroblasts, a-sarcomeric actin, myosin-HC, tropomyosin, and connexin 43 and the expression of a-SMA.

2.5.5 Polypyrrole-Chitosan Polymer

Intraventricular conduction delays result in asynchronous ventricular depolarization, which compromised LV systolic function. Cardiac resynchronization therapy is now an established treatment option in such patients when severe LV dysfunction and persistent symptoms are present; however, only a minority is suitable for cardiac resynchronization therapy, a substantial proportion of which is classified as non-responders. To address this limitation, a conductive polymer has been fabricated by grafting pyrrole to chitosan; injection of this conductive biomaterial in rats restored intraventricular conduction, leading to improved global LV function.[100]

2.5.6 Cellularized Scaffolds

Cellularized scaffolds can provide the matrix for cell homing, suitable for survival, proliferation and differentiation. Ideally, these scaffolds should be tailored for each individual cell-type. Our group has recently fabricated a biomaterial scaffold, characterized by a dynamic viscosity.[101] Specifically, this alginate-based biomaterial displays low viscosity in liquid form, enabling smoother delivery by injection; by contrast, the viscosity rises sharply during the gelation phase, occurring after intramyocardial injection. The scaffold was seeded with human adipose tissue (hAT)-derived MSCs, displaying adequate differentiation capacity *in vitro*; umbilical vein endothelial cells (HUVECs) were added, thereby enhancing its angiogenic properties. Indeed, we observed early co-localization of HUVECs and hAT-MSCs in tubular sprouts, which formed a rapidly expanding vascular network.

To address the safety concerns regarding the combined use of cellular- and or biomaterial-based therapies, we have examined the induction of ventricular tachyarrhythmias by programmed electrical stimulation two weeks post-implantation (Figure-2.2). Besides investigating the local conduction and repolarization, we recorded monophasic action potentials by means of activation mapping (Figure-2.3). We reported a comparable incidence of induced VTs in treated animals and controls; furthermore, no major differences were apparent

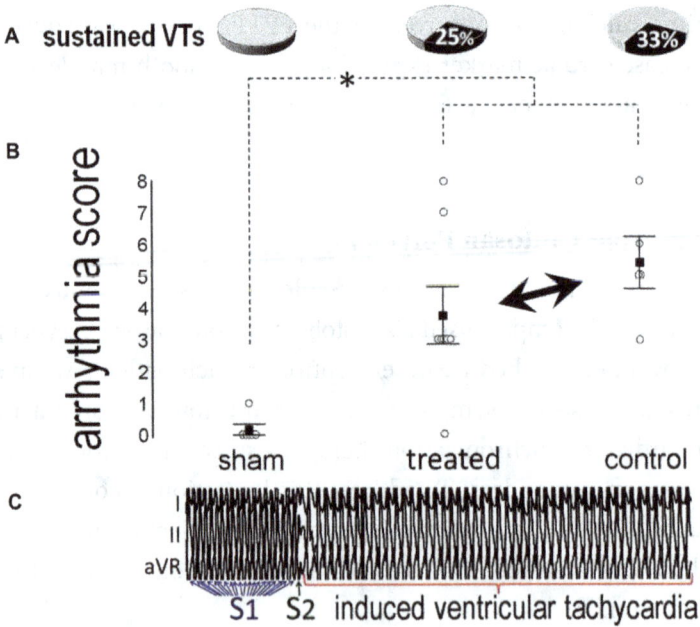

Figure-2.2. Induced ventricular tachyarrhythmias. The incidence of ventricular tachyarrhythmias after programmed electrical stimulation (A) and the arrhythmia score (B) were comparable in treated rats and controls, and higher than after sham-operation (asterisk). (C) Example of induced ventricular tachycardia with one extra stimulus (S2) after 20 paced beats (S1). Reproduced from Kolettis *et al.*[101]

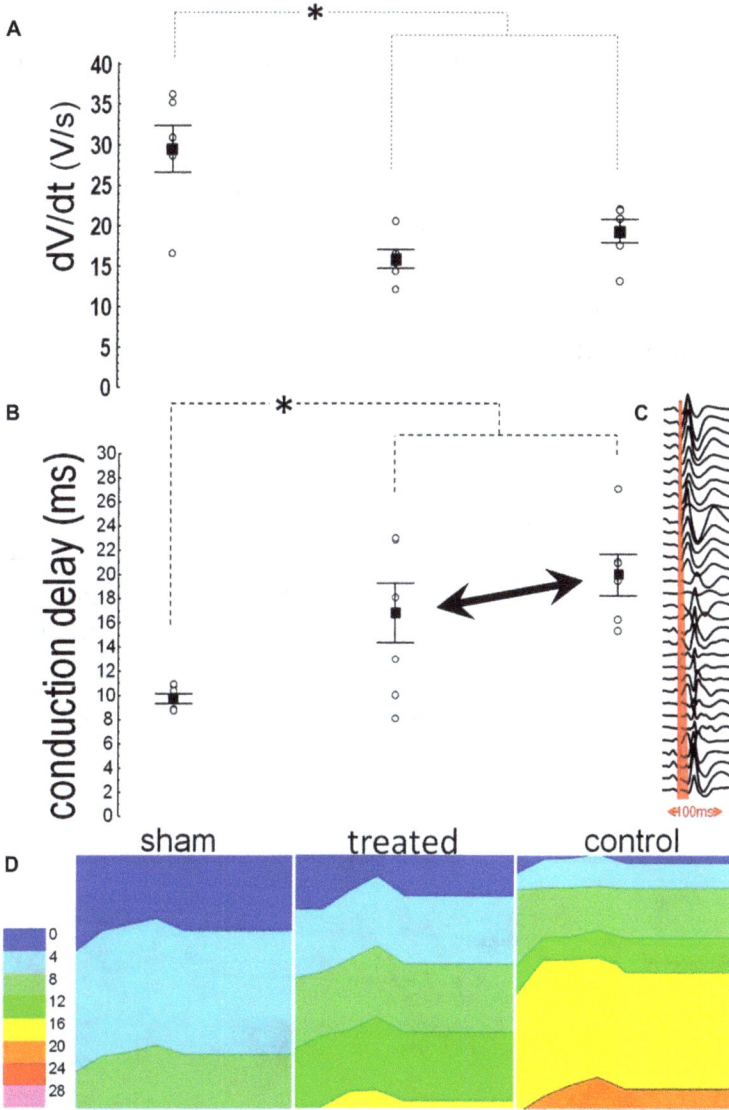

Figure-2.3.   Local conduction. (A) Voltage rise from monophasic action potentials was comparable in treated rats and controls, and lower than after sham-operation. (B) Conduction delay was comparable in treated rats and controls, and longer than after sham operation. (C) Examples of isochronal propagation maps. Asterisk denotes significant difference. Reproduced from Kolettis et al.[101]

in the electrophysiological milieu, as shown by monophasic action potentials recordings and activation mapping. However, our cellularized scaffold did not prevent repolarization dispersion, which in fact, tended to be higher in treated rats. The explanation for this diverse response is unclear, and calls for further investigation.

2.6 FURTHER CHALLENGES

On the way to successful myocardial tissue repair with stem cells several obstacles remain. One key issue towards therapeutic applications is the formation of fully functional mature cardiomyocytes, either directly after pluripotent stem cells differentiation or indirectly by the paracrine activity of multipotent adult stem cells. For using pluripotent stem cells as a source, novel external growth and differentiation factors should be identified to further optimize the differentiation conditions. Besides, the use of advanced co-culture systems with cardiac fibroblasts, cardiac endothelial cells, MSCs or combinations of them should be established (Figure-2.4).

Novel TE strategies should also be further explored, taking into consideration the heart's physiology and function. For this purpose, scaffolds with composition needed for increased biological activities, like contractile performance, paracrine activity and biophysical support will facilitate homing, survival and differentiation of stem cells. To develop the next generation tissue-engineered products, testing of three-dimensional printing, manipulation of cell differentiation by substrate stiffness, spatial surface patterning and surface topography are some attractive strategies that could be used to optimize the biological activity of tissue-engineered products necessary for clinical applications.

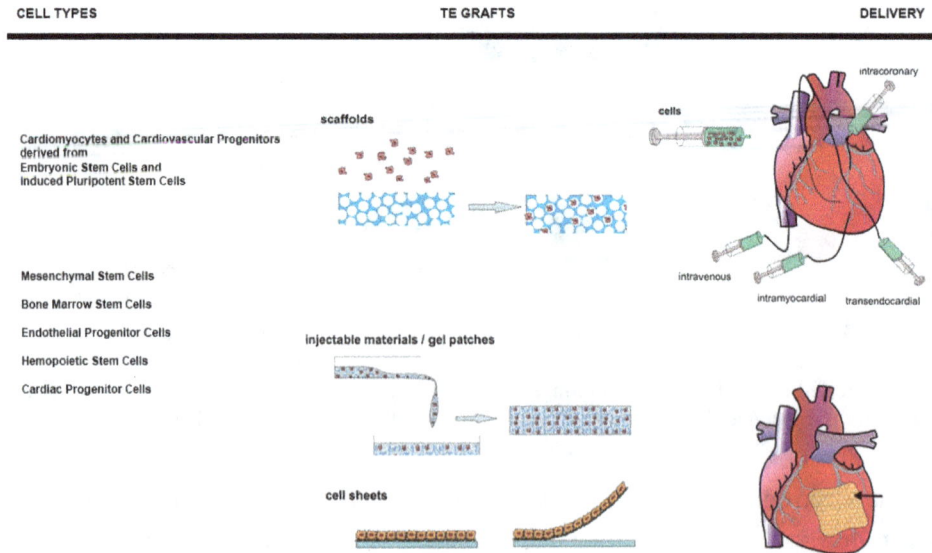

Figure-2.4.   Stem cell sources used either for direct delivery or in tissue engineering grafts in the forms of injectable materials or patches. Modified and reproduced from Arnal-Pastor *et al.*[102]

ACKNOWLEDGMENTS

This work was supported by "IKY Fellowships of Excellence for Postgraduate Studies in Greece — Siemens Program" and "Advanced Research Activities in Biomedical and Agroalimentary Technologies" (MIS 5002469), implemented under the "Action for the Strategic Development on the Research and Technological Sector", funded by the Operational Programme "Competitiveness, Entrepreneurship and Innovation" (NSRF 2014-2020) and co-financed by Greece and the European Union (European Regional Development Fund).

REFERENCES

1. Kolettis TM. Arrhythmogenesis after cell transplantation post-myocardial infarction. Four burning questions — and some answers. *Cardiovasc Res.* 2006; 69: 299–301.
2. Vilaeti AD, Dimos K, Lampri ES, Mantzouratou P, Tsitou N, Mourouzis I, Oikonomidis DL, *et al.* Short-term ventricular restraint attenuates post-infarction remodeling in rats. *Int J Cardiol.* 2013; 165: 278–284.
3. Agathopoulos S, Kolettis TM. Editorial: Novel strategies for cardiac repair post-myocardial infarction. *Curr Pharm Des.* 2014; 20: 1925–1929.
4. Landa N, Miller L, Feinberg MS, Holbova R, Shachar M, Freeman I, Cohen S, *et al.* Effect of injectable alginate implant on cardiac remodeling and function after recent and old infarcts in rat. *Circulation.* 2008; 117: 1388–1396.
5. Menasche P. Skeletal myoblasts and cardiac repair. *J Mol Cell Cardiol.* 2008; 45: 545–553.
6. Buckingham M, Montarras D. Skeletal muscle stem cells. *Curr Opin Genet Dev.* 2008; 18: 330–336.
7. Durrani S, Konoplyannikov M, Ashraf M, Haider Kh. Skeletal myoblasts for cardiac repair. *Regen Med.* 2010; 5: 919–932.
8. Khan M, Kutala VK, Vikram DS, Wisel S, Chacko SM, Kuppusamy ML, Mohan IK, *et al.* Skeletal myoblasts transplanted in the ischemic myocardium enhance in situ oxygenation and recovery of contractile function. *Am J Physiol Heart Circ Physiol.* 2007; 293: H2129–2139.
9. He KL, GH Yi, Sherman W, Zhou H, Zhang GP, Gu A, Kao R, *et al.* Autologous skeletal myoblast transplantation improved hemodynamics and left ventricular function in chronic heart failure dogs. *J Heart Lung Transplant.* 2005; 24: 1940–1949.
10. Gavira JJ, Nasarre E, Abizanda G, Perez-Ilzarbe M, de Martino-Rodriguez A, Garcia de Jalon JA, Mazo M, *et al.* Repeated implantation of skeletal myoblast in a swine model of chronic myocardial infarction. *Eur Heart J.* 2010; 31: 1013–1021.
11. Menasche P, Alfieri O, Janssens S, McKenna W, Reichenspurner H, Trinquart L, Vilquin JT, *et al.* The Myoblast Autologous Grafting in Ischemic Cardiomyopathy (MAGIC) trial: first randomized placebo-controlled study of myoblast transplantation. *Circulation.* 2008; 117: 1189–1200.

12. Fernandes S, Amirault JC, Lande G, Nguyen JM, Forest V, Bignolais O, Lamirault G, *et al*. Autologous myoblast transplantation after myocardial infarction increases the inducibility of ventricular arrhythmias. *Cardiovasc Res*. 2006; 69: 348–358.

13. Fernandes S, van Rijen HV, Forest V, Evain S, Leblond AL, Merot J, Charpentier F, *et al*. Cardiac cell therapy: overexpression of connexin43 in skeletal myoblasts and prevention of ventricular arrhythmias. *J Cell Mol Med*. 2009; 13: 3703–3712.

14. Balsam LB, Wagers AJ, Christensen JL, Kofidis T, Weissman IL, Robbins RC. Haematopoietic stem cells adopt mature haematopoietic fates in ischaemic myocardium. *Nature* 2004; 428: 668–673.

15. Murry CE, Soonpaa MH, Reinecke H, Nakajima H, Nakajima HO, Rubart M, Pasumarthi KB, Virag JI, Bartelmez SH, Poppa V, Bradford G, Dowell JD, Williams DA, Field LJ. Haematopoietic stem cells do not transdifferentiate into cardiac myocytes in myocardial infarcts. *Nature*. 2004; 428: 664–668.

16. Assmus B, Leistner DM, Schachinger V, Erbs S, Elsasser A, Haberbosch W, Hambrecht R, *et al*. Long-term clinical outcome after intracoronary application of bone marrow-derived mononuclear cells for acute myocardial infarction: migratory capacity of administered cells determines event-free survival. *Eur Heart J*. 2014; 35: 1275–1283.

17. Assmus B, Rolf A, Erbs S, Elsasser A, Haberbosch W, Hambrecht R, Tillmanns H, *et al*. Clinical outcome 2 years after intracoronary administration of bone marrow-derived progenitor cells in acute myocardial infarction. *Circ Heart Fail*. 2010; 3: 89–96.

18. Schachinger V, Erbs S, Elsasser A, Haberbosch W, Hambrecht R, Holschermann H, Yu J, *et al*. Intracoronary bone marrow-derived progenitor cells in acute myocardial infarction. *N Engl J Med*. 2006; 355: 1210–1221.

19. Traverse JH, Henry TD, Pepine CJ, Willerson JT, Chugh A, Yang PC, Zhao CXM, *et al*. TIME trial: effect of timing of stem cell delivery following ST-elevation myocardial infarction on the recovery of global and regional left ventricular function: final 2-year analysis. *Circ Res*. 2018; 122: 479–488.

20. Traverse JH, Henry TD, Vaughan DE, Ellis SG, Pepine CJ, Willerson JT, Zhao DX, *et al*. LateTIME: a phase-II, randomized, double-blinded, placebo-controlled, pilot trial evaluating the safety and effect of administration of bone marrow mononuclear cells 2 to 3 weeks after acute myocardial infarction. *Tex Heart Inst*. 2010; J37: 412–420.

21. Traverse JH, Henry TD, Pepine CJ, Willerson JT, Zhao DX, Ellis SG, Forder JR, *et al*. Effect of the use and timing of bone marrow mononuclear cell delivery on left ventricular function after acute myocardial infarction: the TIME randomized trial. *JAMA*. 2012; 308: 2380–2389.

22. Surder D, Manka D, Moccetti T, Lo Cicero V, Emmert MY, Klersy C, Soncin S, *et al*. Effect of bone marrow-derived mononuclear cell treatment, early or late after acute myocardial infarction: twelve months cmr and long-term clinical results. *Circ Res*. 2016; 119: 481–490.

23. Marban E, Malliaras K. Mixed results for bone marrow-derived cell therapy for ischemic heart disease. *JAMA*. 2012; 308: 2405–2406.

24. Simari RD, Pepine CJ, Traverse JH, Henry TD, Bolli R, Spoon DB, Yeh E, *et al.* Bone marrow mononuclear cell therapy for acute myocardial infarction: a perspective from the cardiovascular cell therapy research network. *Circ Res.* 2014; 114: 1564–1568.

25. Nowbar AN, Mielewczik M, Karavassilis M, Dehbi HM, Shun-Shin MJ, Jones S, Howard JP, *et al.* Discrepancies in autologous bone marrow stem cell trials and enhancement of ejection fraction (DAMASCENE): weighted regression and meta-analysis. *BMJ.* 2014; 348: g2688.

26. Gyongyosi M, Wojakowski W, Lemarchand P, Lunde K, Tendera M, Bartunek J, Marban E, *et al.* Meta-Analysis of Cell-based CaRdiac stUdiEs (ACCRUE) in patients with acute myocardial infarction based on individual patient data. *Circ Res.* 2015; 116: 1346–1360.

27. Mathur A, Arnold R, Assmus B, Bartunek J, Belmans A, Bonig H, Crea F, *et al.* The effect of intracoronary infusion of bone marrow-derived mononuclear cells on all-cause mortality in acute myocardial infarction: rationale and design of the BAMI trial. *Eur J Heart Fail.* 2017; 19: 1545–1550.

28. Asahara T, Murohara T, Sullivan A, Silver M, van der Zee R, Li T, Witzenbichler B, *et al.* Isolation of putative progenitor endothelial cells for angiogenesis. *Science.* 1997; 275: 964–967.

29. Bianconi V, Sahebkar A, Kovanen P, Bagaglia F, Ricciuti B, Calabro P, Patti G, Pirro M. Endothelial and cardiac progenitor cells for cardiovascular repair: a controversial paradigm in cell therapy. *Pharmacol Ther.* 2018; 181: 156–168.

30. Nasseri BA, Ebell W, Dandel M, Kukucka M, Gebker R, Doltra A, Knosalla C, *et al.* Autologous CD133+ bone marrow cells and bypass grafting for regeneration of ischaemic myocardium: the Cardio133 trial. *Eur Heart J.* 2014; 35: 1263–1274.

31. Steinhoff G, Nesteruk J, Wolfien M, Kundt G, PTI Group, Borgermann J, David R, *et al.* Cardiac function improvement and bone marrow response: outcome analysis of the randomized PERFECT phase iii clinical trial of intramyocardial CD133(+) application after myocardial infarction. *EBioMed.* 2017; 22: 208–224.

32. Kawamoto A, Iwasaki H, Kusano K, Murayama T, Oyamada A, Silver M, Hulbert C, *et al.* CD34-positive cells exhibit increased potency and safety for therapeutic neovascularization after myocardial infarction compared with total mononuclear cells. *Circulation.* 2006; 114: 2163–2169.

33. Quyyumi AA, Vasquez A, Kereiakes DJ, Klapholz M, Schaer GL, Abdel-Latif A, Frohwein S, *et al.* PreSERVE-AMI: a randomized, double-blind, placebo-controlled clinical trial of intracoronary administration of autologous CD34+ cells in patients with left ventricular dysfunction post STEMI. *Circ Res.* 2017; 120: 324–331.

34. Kanelidis AJ, Premer C, Lopez J, Balkan W, Hare JM. Route of delivery modulates the efficacy of mesenchymal stem cell therapy for myocardial infarction: a meta-analysis of preclinical studies and clinical trials. *Circ Res.* 2017; 120: 1139–1150.

35. Karantalis V, DiFede DL, Gerstenblith G, Pham S, Symes J, Zambrano JP, Fishman J, *et al.* Autologous mesenchymal stem cells produce concordant improvements in regional function, tissue perfusion, and fibrotic burden when administered to patients undergoing coronary artery bypass grafting: the prospective randomized

study of mesenchymal stem cell therapy in patients undergoing cardiac surgery (PROMETHEUS) trial. *Circ Res.* 2014; 114: 1302–1310.

36. Hare JM, Fishman JE, Gerstenblith G, DiFede Velazquez DL, Zambrano JP, Suncion VY, Tracy M, *et al.* Comparison of allogeneic vs autologous bone marrow-derived mesenchymal stem cells delivered by transendocardial injection in patients with ischemic cardiomyopathy: the POSEIDON randomized trial. *JAMA.* 2012; 308: 2369–2379.

37. Tompkins BA, Rieger AC, Florea V, Banerjee MN, Natsumeda M, Nigh ED, Landin AM, *et al.* Comparison of mesenchymal stem cell efficacy in ischemic versus nonischemic dilated cardiomyopathy. *J Am Heart Assoc.* 2018; 7. doi: 10.1161/JAHA.117.008460.

38. Mathiasen AB, Qayyum AA, Jorgensen E, Helqvist S, Fischer-Nielsen A, Kofoed KF, *et al.* Bone marrow-derived mesenchymal stromal cell treatment in patients with severe ischaemic heart failure: a randomized placebo-controlled trial (MSC-HF trial). *Eur Heart J.* 2015; 36: 1744–1753.

39. Heldman AW, DiFede DL, Fishman JE, Zambrano JP, Trachtenberg BH, Karantalis V, Mushtaq M, *et al.* Transendocardial mesenchymal stem cells and mononuclear bone marrow cells for ischemic cardiomyopathy: the TAC-HFT randomized trial. *JAMA.* 2014; 311: 62–73.

40. Westerdahl DE, Chang DH, Hamilton MA, Nakamura M, Henry TD. Allogeneic mesenchymal precursor cells (MPCs): an innovative approach to treating advanced heart failure. *Expert Opin Biol Ther.* 2016; 16: 1163–1169.

41. Camilleri ET, Gustafson MP, Dudakovic A, Riester SM, Garces CG, Paradise CR, Takai H, *et al.* Identification and validation of multiple cell surface markers of clinical-grade adipose-derived mesenchymal stromal cells as novel release criteria for good manufacturing practice-compliant production. *Stem Cell Res Ther.* 2016; 7: 107.

42. Ribeiro A, Laranjeira P, Mendes S, Velada I, Leite C, Andrade P, Santos F, *et al.* Mesenchymal stem cells from umbilical cord matrix, adipose tissue and bone marrow exhibit different capability to suppress peripheral blood B, natural killer and T cells. *Stem Cell Res Ther.* 2013; 4: 125.

43. Perin EC, Sanz-Ruiz R, Sanchez PL, Lasso J, Perez-Cano R, Alonso-Farto JC, Perez-David E, *et al.* Adipose-derived regenerative cells in patients with ischemic cardiomyopathy: the PRECISE Trial. *Am Heart J.* 2014; 168: 88–95 e2.

44. Henry TD, Pepine CJ, Lambert CR, Traverse JH, Schatz R, Costa M, Povsic TJ, *et al.* The Athena trials: autologous adipose-derived regenerative cells for refractory chronic myocardial ischemia with left ventricular dysfunction. *Catheter Cardiovasc Interv.* 2017; 89: 169–177.

45. Paitazoglou C, Bergmann MW, Vrtovec B, Chamuleau SAJ, van Klarenbosch B, Wojakowski W, Michalewska-Wludarczyk A, *et al.* Rationale and design of the European multicentre study on Stem Cell therapy in IschEmic Non-treatable Cardiac diseasE (SCIENCE). *Eur J Heart Fail.* 2019; 21: 1032–1041.

46. Terzic A, Behfar A. Stem cell therapy for heart failure: ensuring regenerative proficiency. *Trends Cardiovasc Med.* 2016; 26: 395–404.

47. Bartunek J, Terzic A, Davison BA, Filippatos GS, Radovanovic S, Beleslin B, Merkely B, *et al.* Cardiopoietic cell therapy for advanced ischaemic heart failure:

results at 39 weeks of the prospective, randomized, double blind, sham-controlled CHART-1 clinical trial. *Eur Heart J*. 2017; 38: 648–660.

48. Teerlink JR, Metra M, Filippatos GS, Davison BA, Bartunek J, Terzic A, Gersh BJ, *et al.* Benefit of cardiopoietic mesenchymal stem cell therapy on left ventricular remodelling: results from the Congestive Heart Failure Cardiopoietic Regenerative Therapy (CHART-1) study. *Eur J Heart Fail*. 2017; 19: 1520–1529.

49. Fernandez-Aviles F, Sanz-Ruiz R, Bogaert J, Plasencia AC, Gilaberte I, Belmans A, Fernandez-Santos ME, *et al.* Safety and efficacy of intracoronary infusion of allogeneic human cardiac stem cells in patients with ST-segment elevation myocardial infarction and left ventricular dysfunction. *Circ Res*. 2018; 123: 579–589.

50. Malliaras K, Makkar RR, Smith RR, Cheng K, Wu E, Bonow RO, Marban L, *et al.* Intracoronary cardiosphere-derived cells after myocardial infarction: evidence of therapeutic regeneration in the final 1-year results of the CADUCEUS trial (CArdiosphere-Derived aUtologous stem CElls to reverse ventricUlar dySfunction). *J Am Coll Cardiol*. 2014; 63: 110–122.

51. Makkar RR, Smith RR, Cheng K, Malliaras K, Thomson LE, Berman D, Czer LS, *et al.* Intracoronary cardiosphere-derived cells for heart regeneration after myocardial infarction (CADUCEUS): a prospective, randomised phase 1 trial. *Lancet*. 2012; 379: 895–904.

52. Duelen R, Sampaolesi M. Stem cell technology in cardiac regeneration: a pluripotent stem cell promise. *EBioMed*. 2017; 16: 30–40.

53. Birket MJ, Mummery CL. Pluripotent stem cell derived cardiovascular progenitors — a developmental perspective. *Dev Biol*. 2015; 400: 169–179.

54. Gepstein L, Ding C, Rahmutula D, Wilson EE, Yankelson L, Caspi O, Gepstein A, *et al.* In vivo assessment of the electrophysiological integration and arrhythmogenic risk of myocardial cell transplantation strategies. *Stem Cells*. 2010; 28: 2151–2161.

55. Chong JJ, Yang X, Don CW, Minami E, Liu YW, Weyers JJ, Mahoney WM, *et al.* Human embryonic-stem-cell-derived cardiomyocytes regenerate non-human primate hearts. *Nature*. 2014; 510: 273–277.

56. Zhu K, Wu Q, Ni C, Zhang P, Zhong Z, Wu Y, Wang Y, *et al.* Lack of remuscularization following transplantation of human embryonic stem cell-derived cardiovascular progenitor cells in infarcted nonhuman primates. *Circ Res*. 2018; 122: 958–969.

57. Fernandes S, Chong JJH, Paige SL, Iwata M, Torok-Storb B, Keller G, Reinecke H, *et al.* Comparison of human embryonic stem cell-derived cardiomyocytes, cardiovascular progenitors, and bone marrow mononuclear cells for cardiac repair. *Stem Cell Reports*. 2015; 5: 753–762.

58. Blin G, Nury D, Stefanovic S, Neri T, Guillevic O, Brinon B, Bellamy V, *et al.* A purified population of multipotent cardiovascular progenitors derived from primate pluripotent stem cells engrafts in postmyocardial infarcted nonhuman primates. *J Clin Invest*. 2010; 120: 1125–1139.

59. Menasche P, Vanneaux V, Hagege A, Bel A, Cholley B, Parouchev A, Cacciapuoti I, *et al.* Transplantation of human embryonic stem cell-derived cardiovascular progenitors for severe ischemic left ventricular dysfunction. *J Am Coll Cardiol*. 2018; 71: 429–438.

60. Maltabe VA, Barka E, Kontonika M, Florou D, Kouvara-Pritsouli M, Roumpi M, Agathopoulos S, *et al.* Isolation of an ES-derived cardiovascular multipotent cell population based on VE-cadherin promoter activity. *Stem Cells Int.* 2016; 2016: 8305624.

61. Weinberger F, Breckwoldt K, Pecha S, Kelly A, Geertz B, Starbatty J, Yorgan T, *et al.* Cardiac repair in guinea pigs with human engineered heart tissue from induced pluripotent stem cells. *Sci Transl Med.* 2016; 8: 363ra148.

62. Shiba Y, Gomibuchi T, Seto T, Wada Y, Ichimura H, Tanaka Y, Ogasawara T, *et al.* Allogeneic transplantation of iPS cell-derived cardiomyocytes regenerates primate hearts. *Nature.* 2016; 538: 388–391.

63. Kawamura M, Miyagawa S, Fukushima S, Saito A, Miki K, Funakoshi S, Yoshida Y, *et al.* Enhanced therapeutic effects of human iPS cell derived-cardiomyocyte by combined cell-sheets with omental flap technique in porcine ischemic cardiomyopathy model. *Sci Rep.* 2017; 7: 8824.

64. Ieda M, Fu JD, Delgado-Olguin P, Vedantham V, Hayashi Y, Bruneau BG, Srivastava D. Direct reprogramming of fibroblasts into functional cardiomyocytes by defined factors. *Cell.* 2010; 142: 375–386.

65. Qian L, Huang Y, Spencer CI, Foley A, Vedantham V, Liu L, Conway SJ, *et al.* In vivo reprogramming of murine cardiac fibroblasts into induced cardiomyocytes. *Nature.* 2012; 485: 593–598.

66. Zhao Y, Londono P, Cao Y, Sharpe EJ, Proenza C, O'Rourke R, Jones KL. High-efficiency reprogramming of fibroblasts into cardiomyocytes requires suppression of pro-fibrotic signalling. *Nat Commun.* 2015; 6: 8243.

67. Zhou H, Dickson ME, Kim MS, Bassel-Duby R, Olson EN. Akt1/protein kinase B enhances transcriptional reprogramming of fibroblasts to functional cardiomyocytes. *Proc Natl Acad Sci USA.* 2015; 112: 11864–11869.

68. Williams AR, Hatzistergos KE, Addicott B, McCall F, Carvalho D, Suncion V, Morales AR, *et al.* Enhanced effect of combining human cardiac stem cells and bone marrow mesenchymal stem cells to reduce infarct size and to restore cardiac function after myocardial infarction. *Circulation.* 2013; 127: 213–223.

69. Natsumeda M, Florea V, Rieger AC, Tompkins BA, Banerjee MN, Golpanian S, Fritsch J. A combination of allogeneic stem cells promotes cardiac regeneration. *J Am Coll Cardiol.* 2017; 70: 2504–2515.

70. Bolli R, Hare JM, March KL, Pepine CJ, Willerson JT, Perin EC, Yang PC, *et al.* Rationale and design of the CONCERT-HF Trial (combination of mesenchymal and c-kit(+) cardiac stem cells as regenerative therapy for heart failure). *Circ Res.* 2018; 122: 1703–1715.

71. Soppa GK, Smolenski RT, Latif N, Yuen AH, Malik A, Karbowska J, Kochan Z, *et al.* Effects of chronic administration of clenbuterol on function and metabolism of adult rat cardiac muscle. *Am J Physiol Heart Circ Physiol.* 2005; 288: H1468–1476.

72. Mitsi AC, Hatzistergos K, Baltogiannis GG, Kolettis TM. Early, selective growth hormone administration may ameliorate left ventricular remodeling after myocardial infarction. *Med Hypotheses.* 2005; 64: 582–585.

73. Mitsi AC, Hatzistergos KE, Niokou D, Pappa L, Baltogiannis GG, Tsalikakis DG, Papalois A, *et al.* Early, intracoronary growth hormone administration attenuates

ventricular remodeling in a porcine model of myocardial infarction. *Growth Horm IGF Res.* 2006; 16: 93–100.

74. Elaiopoulos DA, Tsalikakis DG, Agelaki MG, Baltogiannis GG, Mitsi AC, Fotiadis DI, Kolettis TM. Growth hormone decreases phase II ventricular tachyarrhythmias during acute myocardial infarction in rats. *Clin Sci (Lond).* 2007; 112: 385–391.

75. Hatzistergos KE, Mitsi AC, Zachariou C, Skyrlas A, Kapatou E, Agelaki MG, Fotopoulos A. Randomised comparison of growth hormone versus IGF-1 on early post-myocardial infarction ventricular remodelling in rats. *Growth Horm IGF Res.* 2008; 18: 157–165.

76. Bruel A, Christoffersen TE, Nyengaard JR. Growth hormone increases the proliferation of existing cardiac myocytes and the total number of cardiac myocytes in the rat heart. *Cardiovasc Res.* 2007; 76: 400–408.

77. Caplice NM, DeVoe MC, Choi J, Dahly D, Murphy T, Spitzer E, Van Geuns R, *et al.* Randomized placebo controlled trial evaluating the safety and efficacy of single low-dose intracoronary insulin-like growth factor following percutaneous coronary intervention in acute myocardial infarction (RESUS-AMI). *Am Heart J.* 2018; 200: 110–117.

78. Lilleness BM, Frishman WH. Ghrelin and the cardiovascular system. *Cardiol Rev.* 2016; 24: 288–297.

79. Salzano A, Marra AM, D'Assante R, Arcopinto M, Suzuki T, Bossone E, Cittadini A. Growth hormone therapy in heart failure. *Heart Fail Clin.* 2018; 14: 501–515.

80. Arcopinto M, Salzano A, Giallauria F, Bossone E, Isgaard J, Marra AM, Bobbio E, *et al.* Growth hormone deficiency is associated with worse cardiac function, physical performance, and outcome in chronic heart failure: insights from the T.O.S.CA. GHD study. *PLoS One.* 2017; 12: e0170058.

81. Buehler A, Martire A, Strohm C, Wolfram S, Fernandez B, Palmen M, Wehrens XH, *et al.* Angiogenesis-independent cardioprotection in FGF-1 transgenic mice. *Cardiovasc Res.* 2002; 55: 768–777.

82. Rosenblatt-Velin N, Lepore MG, Cartoni C, Beermann F, Pedrazzini T. FGF-2 controls the differentiation of resident cardiac precursors into functional cardiomyocytes. *J Clin Invest.* 2005; 115: 1724–1733.

83. D'Uva G, Aharonov A, Lauriola M, Kain D, Yahalom-Ronen Y, Carvalho S, Weisingeret K, *et al.* ERBB2 triggers mammalian heart regeneration by promoting cardiomyocyte dedifferentiation and proliferation. *Nat Cell Biol.* 2015; 17: 627–638.

84. Jabbour A, Hayward CS, Keogh AM, Kotlyar E, McCrohon JA, England JF, Amor R, *et al.* Parenteral administration of recombinant human neuregulin-1 to patients with stable chronic heart failure produces favourable acute and chronic haemodynamic responses. *Eur J Heart Fail.* 2011; 13: 83–92.

85. Gao R, Zhang J, Cheng L, Wu X, Dong W, Yang X, Li T, *et al.* A phase II, randomized, double-blind, multicenter, based on standard therapy, placebo-controlled study of the efficacy and safety of recombinant human neuregulin-1 in patients with chronic heart failure. *J Am Coll Cardiol.* 2010; 55: 1907–1914.

86. Zhu XY, Zhang XZ, Xu L, Zhong XY, Ding Q, Chen XY. Transplantation of adipose-derived stem cells overexpressing hHGF into cardiac tissue. *Biochem Biophys Res Commun.* 2009; 379: 1084–1090.

87. Kolettis TM, Vilaeti A, Dimos K, Tsitou N, Agathopoulos S. Tissue engineering for post-myocardial infarction ventricular remodeling. *Mini Rev Med Chem.* 2011; 11: 263–270.

88. Augst AD, Kong HJ, Mooney DJ. Alginate hydrogels as biomaterials. *Macromol Biosci.* 2006; 6: 623–633.

89. Kontonika M, Barka E, Daskalopoulos E, Vilaeti AD, Papalois A, Agathopoulos S, Kolettis TM. Effects of myocardial alginate injections on ventricular arrhythmias after experimental ischemiareperfusion. *Trends Biomater Artif Organs.* 2014; 28: 79–82.

90. Barka E, Papayannis DK, Kolettis TM, Agathopoulos S. Optimization of Ca(2+) content in alginate hydrogel injected in myocardium. *J Biomed Mater Res B Appl Biomater.* 2019; 107: 223–231.

91. Daskalopoulos EP, Vilaeti AD, Barka E, Mantzouratou P, Kouroupis D, Kontonika M, Tourmousoglou C, *et al.* Attenuation of post-infarction remodeling in rats by sustained myocardial growth hormone administration. *Growth Factors.* 2015; 33: 250–258.

92. Kontonika M, Barka E, Roumpi M, La Rocca V, Lekkas P, Daskalopoulos EP, Vilaeti AD, *et al.* Prolonged intra-myocardial growth hormone administration ameliorates post-infarction electrophysiologic remodeling in rats. *Growth Factors.* 2017; 35: 1–11.

93. Anker SD, Coats AJ, Cristian G, Dragomir D, Pusineri E, Piredda M, Bettari L, *et al.* A prospective comparison of alginate-hydrogel with standard medical therapy to determine impact on functional capacity and clinical outcomes in patients with advanced heart failure (AUGMENT-HF trial). *Eur Heart J.* 2015; 36: 2297–2309.

94. Yu J, Gu Y, Du KT, Mihardja S, Sievers RE, Lee RJ. The effect of injected RGD modified alginate on angiogenesis and left ventricular function in a chronic rat infarct model. *Biomaterials.* 2009; 30: 751–756.

95. Abdalla S, Makhoul G, Duong M, Chiu RC, Cecere R. Hyaluronic acid-based hydrogel induces neovascularization and improves cardiac function in a rat model of myocardial infarction. *Interact Cardiovasc Thorac Surg.* 2013; 17: 767–772.

96. Yoon SJ, Hong S, Fang YH, Song M, Son HS, Kim SK, Sun K, *et al.* Differential regeneration of myocardial infarction depending on the progression of disease and the composition of biomimetic hydrogel. *J Biosci Bioeng.* 2014; 118: 461–468.

97. Christman KL, Vardanian AJ, Fang Q, Sievers RE, Fok HH, Lee RJ. Injectable fibrin scaffold improves cell transplant survival, reduces infarct expansion, and induces neovasculature formation in ischemic myocardium. *J Am Coll Cardiol.* 2004; 44: 654–660.

98. Huang NF, Yu J, Sievers R, Li S, Lee RJ. Injectable biopolymers enhance angiogenesis after myocardial infarction. *Tissue Eng.* 2005; 11: 1860–1866.

99. Robinson KA, Li J, Mathison M, Redkar A, Cui J, Chronos NA, Matheny RG, Badylak SF. Extracellular matrix scaffold for cardiac repair. *Circulation.* 2005; 112: I135–143.

100. Mihic A, Cui Z, Wu J, Vlacic G, Miyagi Y, Li SH, Lu S, Sung HW, Weisel RD, Li RK. A conductive polymer hydrogel supports cell electrical signaling and improves cardiac function after implantation into myocardial infarct. *Circulation.* 2015; 132: 772–784.

101. Kolettis TM, Bagli E, Barka E, Kouroupis D, Kontonika M, Vilaeti AD, Markou M, *et al.* Medium-term electrophysiologic effects of a cellularized scaffold implanted in rats after myocardial infarction. *Cureus.* 2018; 10: e2959.

102. Arnal-Pastor M, Chachques JC, Monleón Pradas M, Vallés-Lluch A. Biomaterials for cardiac tissue engineering. 2013. DOI: 10.5772/56076.

CHAPTER 3

Non-destructive metabolomics characterization of mesenchymal stem cell differentiation

Amal Ibrahim Surrati*, Khawaja Husnain Haider, Virginie Sottile

Wolfson Centre for Stem Cell, Tissue Engineering and Modelling (STEM),
School of Medicine, The University of Nottingham,
CBS Building - University Park, Nottingham NG7 2RD, UK

ABSTRACT

Real-time monitoring of stem cells has been a growing area of interest over the past decade because of new regenerative medicine approaches. Also, the effect of culture composition on stem cell metabolic pathways and their regulation of cellular fate are of increasing importance. In this chapter, non-destructive methods for cell monitoring *in vitro* are reviewed, including recent developments enabling metabolomics monitoring using LC-MS and GC-MS. Finally, the application of metabolomics in the examination of stem cell differentiation is discussed, suggesting the potential of metabolomic profiling as a powerful tool for regenerative medicine approaches.

KEYWORDS

Mesenchymal; Mitochondria; MSCs; Metabolomics; Multipotent.

LIST OF ABBREVIATIONS

Ang-1	=	Angiopoietin-1
ASCs	=	Adipose tissue-derived stem cells
BM	=	Bone marrow
BMSC	=	Bone marrow stromal cells
BSMA	=	Broad spectrum metabolomic analysis
CRP	=	C-reactive protein
DMEM	=	Dulbecco's modified eagle medium
ECB	=	Electrochemical conductors
EGF	=	Epidermal growth factor
FAD	=	Flavin adenine dinucleotide

* Corresponding author. Email: amalsurrati@hotmail.com

FBS	=	Fetal bovine serum
FCS	=	Fetal calf serum
FGF	=	Fibroblast growth factor
FLIM	=	Fluorescence lifetime imaging
GC	=	Gas chromatography
HCECs	=	Human corneal epithelial cells
hESCs	=	Human embryonic stem cells
HSCs	=	Hematopoietic stem cells
hMSCs	=	Human MSCs
hPMSCs	=	Human placenta-derived MSCs
IHD	=	Ischemic heart disease
LC	=	Liquid chromatography
LDH	=	Lactate dehydrogenase
MS	=	Mass spectroscopy
m/z	=	Mass-to-charge ratio
NEAA	=	Non-essential amino acids
NFC	=	Nano-fibrillar cellulose
NPY	=	Neuropeptide-Y
PCR	=	Polymerase chain reaction
PDGF	=	Placenta-derived growth factor
PPARγ2	=	Peroxisome proliferator activated receptorγ2
PTHrP	=	Parathyroid hormone related protein
RGD	=	Arg-Gly-Asp
RM	=	Raman micro-spectroscopy
ROS	=	Reaction oxygen species
RPTECs	=	Renal proximal tubular epithelial cells
SDF-1α	=	Stromal cell-derived factor-1α
SPME	=	Solid-phase micro-extraction
SPR	=	Surface plasmon resonance
TCA	=	Tricarboxylic acid
TGF-1β	=	Transforming growth factor-1β
UCMSCs	=	Human cord-derived mesenchymal stem cells
UPLC-TOF-MS	=	Ultra-performance liquid chromatography-time of flight mass spectroscopy
VOC	=	Volatile organic compounds
VSEL	=	Very small embryonic-like

3.1 INTRODUCTION

The potential of stem cells for self-renewal and differentiation into specialized cell lineages has opened up promising therapeutic opportunities such as tissue repair and regeneration. Subsequent to extensive *in vitro* characterization for their biologic characteristics and pre-clinical studies in both small as well as large experimental animal models, bone marrow BM-derived stem/progenitor cells have already made significant headway in the clinical studies for multifarious applications including restoration of infarcted heart function and repair. Only for cardiovascular applications, BM cells have already reached Phase-III clinical trials for the treatment of ischemic heart disease (IHD). BM-derived mesenchymal stem cells (BMMSCs) are amongst the most well-characterized cells for their differentiation potential and reparability and hence, are currently part of many clinical studies.

3.2 MESENCHYMAL STEM CELL BIOLOGY

MSCs constitute a heterogeneous primitive cell population which forms an integral part of the stem cell niche in the BM.[1] They contribute towards the maintenance of the niche microenvironment by secreting regulatory bioactive molecules that determine the physiological activity of the hematopoietic stem cells (HSCs) in terms of their proliferative and differentiation properties in the niche.[2] More recent studies have focused on the presence of perivascular nestin+ MSCs as significant contributors to the stem cell niches that harbor HSCs.[3] Unlike their HSCs counterparts, human MSCs (hMSCs) isolated from BM aspirates of donor patients were first characterized by their ability of plastic adherence during *in vitro* culture and potential to undergo exclusive tri-lineage differentiation to adopt adipocyte, chondrocyte and osteocyte phenotype.[4] Besides, BM-derived hMSCs are characterized by the membrane expression of various clusters of differentiation which have been defined by the International Society Cell Therapy (ISCT) to express CD105, CD73, and CD90 while lacking in the expression of CD45, CD34, CD14, CD11b, CD9α or CD19 and HLA-DR.[5] The uniform criteria step up by ISCT has helped in alleviating the inconsistencies in the divergent reports from various research groups and laboratories regarding the nomenclature, as well as the biological characteristics of MSCs. It is pertinent to mention that MSCs obtained from various species and expanded under different set of culture conditions may differ in the expression of their surface markers.[7,8] On the same note, MSCs from different tissue sources may also differ in terms of concentration and the total number of CD90[+], CD44[+] and CD45[−] cells.[9,10] It is now generally established that BMMSCs also have a population of very small embryonic-like (VSEL) cells which can be identified by Oct4 and Sox2 expression.[11–15]

3.3 MESENCHYMAL STEM CELL-BASED THERAPY

BMMSCs have emerged as one of the choice cells for cell-based therapy due to their ideal characteristics.[5] Experimental animal data have shown their reparability post-engraftment with other cell types due to their multi-lineage differentiation potential including bone, cartilage, adipose tissue, muscle, heart, blood vessel, nerves, and pancreatic islet cell phenotypes.[16–20] Moreover, BMMSCs have been combined with other cell types for a combinatorial approach to enhance their reparability and regenerative potential.[21–23] They have also been reprogrammed to develop MSC-derived induced pluripotent stem cells (iPSCs) which were subsequently used for cardiomyogenic differentiation post-engraftment in experimental animal trials.[24,25] In our recent study, we have shown that combined treatment with neural stem cells (NSCs) and BMMSCs promoted the survival of NSCs post engraftment.[23] In a similar study involving combined treatment with BMMSCs and NSCs, MSCs were genetically modulated to overexpress stromal cell-derived factor-1α (SDF-1α) which further enhanced their therapeutic efficacy.[26] Molecular studies revealed a significant reduction in IL-1β and IL-6 levels in the animals receiving combined treatment with NSCs and genetically modified MSCs.

Given their robust nature, MSCs are excellent carriers of transgenes and hence, they have been genetically manipulated for single or multiple transgene/s delivery of interest to the target organs.[27–29] More recently, they have been used as carriers of microRNA delivery to the heart.[30] Besides genetic modulation, various physical manipulation strategies have been developed to enhance their paracrine activity which contributes to their therapeutic potential.[31,32] One of the important biological characteristics of BMMSCs is their partially immune-privileged or immuno-evasive status which has been attributed to their typically low-level expression of MHC class-I and absence of MHC class–II and co-stimulatory molecules.[33–35] Hence, their use in humans for cell therapy applications does not necessitate immunosuppression. Recent studies have shown that their immune-privileged properties may get compromised when they are co-cultured with allogeneic lymphocytes.[36] There are also studies that have questioned their immuno-privileged status.[37] On the other hand, BMMSCs do express non-classical MHC class-I antigen HLA-G that contributes to their immunomodulatory properties.[38] Although the exact mechanism underlying their immunomodulatory action has remained oblivious, it is now considered as multi-factorial but with a primary contribution from the soluble factors released by MSCs as part of their paracrine activity.[39,40]

Although MSCs can be obtained from almost every human body tissue, the invasiveness of the isolation protocol, logistics issues and the inadequate amount of the MSCs available from different tissue sources remain a limiting factor for their use in cell-based therapy procedures. For example, while BM remains the primary source of MSCs for use in cell-based therapy and tissue engineering, the isolation protocol is infection-prone with only a meager 0.001 to 0.01% yield of MSCs from the BM. The other adult tissue sources of hMSCs include adipose tissue, dental pulp, and peripheral blood while birth-associated tissue sources include umbilical cord, cord blood, placenta, and amnion.[41,42] Unlike BM, 1 g of adipose tissue during liposuction yields 5×10^3 stem cells and hence, adipose-derived stem cells (ASCs) are being used as an alternative source of MSCs for clinical applications.[43-45] Interestingly, MSCs from different tissue sources are identical in surface marker expression and tri-lineage differentiation capacity. MSCs derived from peripheral blood of 20 patients demonstrated similar surface markers and osteogenic, chondrogenic, and adipogenic differentiation potential as compared to the BMMSCs.[46] Given low cell yield, the direct use of unexpanded naïve MSCs will not yield optimal prognosis *post* engraftment to regenerate large tissue. Therefore, protocols are being developed for *in vitro* culture expansion of MSCs to achieve sufficient number for transplantation, making sure that the culture-expanded cells retain their stemness characteristics especially in terms of their differentiation potential and paracrine activity. Therefore, finding *ex vivo* culture conditions to improve MSCs' proliferation and differentiation proficiency is still enduring.[47-51] An un-optimized *in vitro* expansion protocol may lead to the generation of metabolic phenotypes that may significantly impact the functionality of the cells including their survival, migration, proliferation and differentiation. This necessitates the development of methods for cell metabolism analysis to explore intracellular activities and metabolic pathways that are affected by various growth factors, nutrients and cytotoxic agents throughout the cell culture process. In this regard, there are numerous primary hurdles including the very low level of the target metabolites and interference from non-target metabolites and hence, warrant detection methods having ultra-high specificity and sensitivity. In this regard, polymerase chain reaction (PCR)-based cell characterization methods for specific gene expression[52] and immunoassays for protein expression levels[53] are sensitive, precise and accurate but invasive and semi-quantitative. Therefore, alternative non-invasive, non-destructive techniques are being introduced to analyze intracellular fluorophores and also to monitor and measure the metabolites into the culture medium by viable or dead cells or. In this chapter, we provide an overview of the contemporary non-destructive methods for the evaluation of cell viability/toxicity and proliferation. Although real-time microscopy is also being used to analyze growing human embryonic stem cells (hESCs) with Cell-iQ analyzer,[54] we will only focus on the non-microscopic methods of cell monitoring.

3.4 NON-DESTRUCTIVE METHODS OF CELL MONITORING

3.4.1 Effect of Culture Composition on MSCs Differentiation *In Vitro*

The number of cells for transplantation, besides the quality of the cell preparation, remains the major determinant of optimal prognosis subsequent to a cell-based therapy. As the tissue yield of MSCs is very low, their *in vitro* expansion is important to achieve the high number of cells which require optimal *in vitro* culture conditions that simulate the BM niche microenvironment. A minor deviation from the conditions of their natural habitat will significantly affect their phenotypic and functional characteristics.[55-59] Optimal culture conditions are also important for the maintenance of their metabolic homeostasis which is controlled by the nutritional balance, growth factors and, hydrogen and oxygen concentrations of the culture environment that orchestrate the biological pathways of MSCs *in vitro*. This section presents some of the metabolic factors affecting MSCs proliferation and differentiation status.

3.4.2 MSCs Proliferation

Optimal medium composition for MSCs growth and expansion consists of (Opti-MEM), DMEM (Dulbecco's Modified Eagle Medium), 10% Fetal Calf Serum (FCS) or Fetal Bovine Serum (FBS), 1% Non-Essential Amino Acids (NEAA) and 1% antibiotic (Penicillin, Streptomycin or Gentamycin). More recently, DMEM/F12 has gained popularity as a basal medium for MSCs culture. However, a common feature of these classical media is the inclusion of ill-defined supplements such as FCS combined with the long doubling-time of the cultured cells and accentuated immunogenicity due to the presence of extraneous proteins in the culture.[60] Various strategies have been adopted to alleviate the deficiencies of traditional culture conditions in order to achieve faster proliferation by reducing their doubling time. These strategies involve the inclusion of the growth factors, i.e., fibroblast growth factor (FGF), epidermal growth factor (EGF) and placenta-derived growth factor (PDGF) to stimulate signaling pathways that promote the cell cycle. Additionally, in the next generation culture media, inclusion of human allogeneic or autogenic serum supplement and use of platelets derivatives eliminate the biosafety risk of contamination of animal serum proteins that cannot be removed by washing after expansion.[61-63] The culture medium supplements provide the cells with vitamins, amino acids, glucose, salts and other essential nutrients to regulate their growth whereas serum proteins, i.e., fibronectin, promote cell attachment and act as a source of binding proteins that facilitate the transport of nutrients, hormones and growth factors.[64-68] Although antibiotics are used in the culture medium to reduce contamination possibilities, it may conceal mycoplasma contamination.[69]

Most of the cell culture methods and culture conditions discussed so far have used a 2-dimensional (2-D) culture strategy using xenogenic media. The 2D-culture conditions controvert the *in vivo* niche microenvironment in the BM thus managing only limited *in vitro* expansion besides the loss of stemness characteristics including clonogenicity and differentiation potential. Given that MSCs are anchorage-dependent cells and hence, require adhesion to the culturing surface as well as cell-to-cell contact, the advancements in 3-D culture conditions using well-defined culture media have revolutionized the cell expansion strategy as it mimics the physiological niche microenvironment for their growth.[70] The spheroid culture of hMSCs during the hanging-drop method showed better survival in Arg-Gly-Asp (RGD) modified gel until 5 days of observation and exhibited superior paracrine behavior in terms of VEGF secretions.[71] In a recently published protocol, the upscaling of hMSCs on biocompatible 3-D nano-fibrillar cellulose (NFC) hydrogel was feasible. The culture-expanded cells were more viable and functionally better as compared to the cells cultured under conventional culture conditions.[72] Similar observations have also been reported by other researchers.[73,74]

Attempts have also been made to treat MSCs with mitogenic agents such as BIO (6-bromoindirubin-3-30xime). The compound is responsible for initiating Wnt signaling via inhibition of GSK3β.[75] In one of our published studies, we have sown that genetic modification of BM-derived MSCs for co-overexpression of Akt and angiopoietin-1 (Ang-1) was critical for cell cycle activity via transcriptional regulation of Erk1/cyclinD1 with downstream regulation of microRNA-143[76] BM-derived MSCs with co-overexpression of Akt and Ang-1 had higher phosphorylation of FoxO1, which activated Erk5, a distinct mitogen-induced MAPK that drove transcriptional activation of cyclin D1 and Cdk4. Flow cytometry showed more than 10% higher S-phase cell population that was confirmed by BrdU assay (15%) and immunohistology for Ki67 (11%) as compared to the null adenoviral vector transduced control MSCs. In another interesting study, the authors carried out a direct comparison of neonatal (2–3 weeks old rats), young (8–12 weeks old rats) and aging (24–28 months old rats) BM-derived MSCs to show that physiological aging significantly impairs the proliferative potential of MSCs.[77] We showed pro-proliferative response with the restoration of growth characteristics of the aging BM cells subsequent to neuropeptide-Y (NPY)/neuropeptide-Y Y5 receptor (NPY Y5R) ligand–receptor interaction by transgenic overexpression of NPY Y5-receptor and subsequent treatment of the cells with 5nM NPY5.

3.4.3 Differentiation of MSCs

The rate of tri-lineage differentiation potential, i.e., chondrogenic, adipogenic and osteogenic lineages, is an important characteristic of MSCs which is used in conjunction with surface marker expression to determine the purity of MSCs culture. However, transdifferentiation to the three lineages necessitates a specific set of culture conditions.[78] For example, chondrogenic differentiation is initiated when culture medium is supplemented with dexamethasone, ascorbic acid, Insulin-Transferrin-Selenium Supplement (ITS$^+$) and transforming growth factor-1β (TGF-1β). Dexamethasone activates β-catenin dependent *Runx2* expression to initiate MSCs differentiation[79] while TGF-1β promotes MSCs condensations, pre-chondrocytes proliferation, and inhibits terminal differentiation.[80–82] An improved protocol for chondrogenic differentiation has been reported which included TGF-β1, BMP2, parathyroid hormone-related protein (PTHrP), and FGF2.[83] The results showed that PTHrP and FGF2 maintained the synergistic effect of TGF-β1 and BMP2 treatment. Similarly, the standard protocol for osteogenic differentiation consists of dexamethasone, ascorbic acid and β-glycerophosphate.[79] During osteogenic differentiation, a combination of these substances orchestrate biological different differentiation signaling pathways wherein ascorbic acid couples with 2-oxoglutarate (intermediate metabolite from Tricarboxylic acid cycle; TCA) to hydrolyze proline into hydroxyproline during collagen helical structure formation and secretion into extracellular matrix[84] while alkaline phosphatase supplemented by serum utilizes β-glycerophosphate provides bone extracellular matrix with inorganic phosphate.[79]

Peroxisome proliferator activated receptorγ2 (PPARγ2) agonists and bone morphogenetic proteins are genes generally used to induce adipogenic induction.[85] Three of the main components used in this regard include dexamethasone, insulin, and IBMX. Of these components, insulin activates mitogen-activated protein kinase pathways while IBMX in combination with dexamethasone regulate PPARγ. After PPARγ2 agonist treatment, intracellular vacuoles containing lipids are induced that initiate the expression of the fat-specific transcription factor PPARγ2.[86–88] Thiazolidinedione (i.e., Indomethacin or Rosiglitazone) activate

Peroxisome proliferator-activated receptors PPARγ which regulates the expression of genes involved in lipid and glucose metabolism.[89,90] The plasticity of BMMSCs has remained an area of intense investigation with diverging opinions regarding their ability to undergo cross-lineage differentiation.

3.5 MONITORING OF CELL METABOLISM

The monitoring of the cell status and their derivative tissue structures is imperative for the successful engineering of the tissues and applications of the cultured cell preparations with special guidelines. Metabolism or the metabolic activity of a stem cell is determined by the biology of the cells and it is determinant of various processes of stem cells from survival to self-renewal, proliferation, growth, differentiation, etc. Given the lack of specific markers for each one of these processes, an assay of metabolic activity of the cells is currently being focused on gaining insight into their morphology, identification, as well as functional characteristics of the stem cells and their derivative tissue structures in tissue engineering.

3.5.1 Direct Monitoring Using Non-Destructive Assays *In Vitro*

Enzyme-based assays for detection and semi-quantification of the viable cells are commonly used *in vitro*, pre-clinical as well as clinical research. These assays primarily rely on the measurement of the enzyme activity through absorbance/emission of colorimetric/fluorometric substrates or metabolites that offer a read-out for the number of viable cells in the culture medium in which the cells have been grown. The absorbance or emission is measured at a specific wavelength using a plate reader (spectrophotometer). Cell toxicity assays such as the ones described below are also used in pre-clinical studies, particularly in pharmaceutical research. The main advantage of these techniques is that they can be applied to living cell cultures in monolayer or cell suspension cultures. Also, they are simple, non-destructive, inexpensive and environmentally safe.[91]

3.5.1.1 *Assays for glucose and lipids*

These assays identify a wide range of metabolites of live cells by analyzing their concentration in the culture medium non-destructively. For example, carbohydrate metabolism can be monitored using a variety of metabolites such as pyruvate, glucose, galactose, and amylase. Randle in 1963 first explained the mechanism of insulin resistance and diabetes mellitus by elevating fatty acid oxidation and decreasing glucose oxidation.[92] Glucose metabolism is the primary source of energy for the maintenance of cellular homeostasis through the glycolytic pathway that generates NADH, ATP and pyruvate,[93] and pentose phosphate pathway that produces NADPH, pentose and ribose 5-phosphate.[94] A recent study has confirmed that analysis of glucose can be used to monitor Ahmed's Glaucoma valve implantation *in vivo* using a rabbit model.[95] The authors collected aqueous humor samples 1 and 5 months post-operation for measurement of oxygen tension and broad-spectrum metabolomic analysis (BSMA) using ultra-performance liquid chromatography — time of flight mass spectroscopy (UPLC-TOF-MS). Besides BSMA, glucose and lactate analysis was performed using glucose and lactate assay kits. Altered glucose homeostasis has also been associated with pathologies and hence changes in glucose levels and glycated serum proteins lead to a predisposition to these pathological conditions. A recent study has shown an association between altered glucose homeostasis and structural features in patients with knee osteoarthritis.[96,97] The authors also determined changes in C-reactive protein (CRP) to determine systemic inflammation in the patients. Similarly, lipid metabolism plays an important role in the homeostasis of cell energy, especially when glucose is limited.[98–102] A variety of attributes have been implicated in lipid metabolism during starvation wherein the process begins with metabolic conversion of glycerol and fatty acids. Glycerol is oxidized via glycolysis (KEGG PATHWAY: MAP00561 and KEGG PATHWAY: MAP00010) and enters the Krebs cycle and oxidative phosphorylation; while fatty acids directly enter TCA in Acetyl CoA form and proceed to oxidative phosphorylation (KEGG PATHWAY: MAP01212). Hence, their quantifications are used to monitor the cell physiology.[103–108]

3.5.1.2 *Mitochondria and mitochondrial function*

Mitochondria are the primary site of oxidative phosphorylation to generate ATP with an integral role of oxygen which serves as an electron carrier to activate chemical reaction. Oxygen consumption rate reflects mitochondrial functionality and hence, quantification of extracellular oxygen level is inversely proportional to mitochondrial respiration. Non-destructive assay kits are commercially available now which are based on oxygen-sensitive fluorescence dyes for measurement of oxygen levels.[109]

Using Drosophila as a genetic model, Song *et al.* have reported that muscle mitochondrial dysfunction occurs due to disruption of complex-I of the electron transport chain (the entry point of an electron into the electron transport chain) that leads to concomitant mitochondrial dysfunction in the fat body with resultant increase in triglyceride accumulation.[110] These organelle-level changes led to reduced ADP-induced O_2 consumption as well as ATP production in the muscle tissue. The authors used extracellular O_2 consumption reagent and ATP determination kit during the process of analysis. O_2 consumption rate determination is also important for the culture of stem cells in 3D culture conditions. Using 3D-cultured human dermal fibroblasts and bone marrow stromal cells (BMSCs) as experimental models, a recent study has shown that the average rate constant of O_2 consumption of fibroblasts was 1.19×10^{-17} mol/cell per sec and 7.91×10^{-18} mol/cell per sec respectively when cultured on 3-D native collagen type-I scaffolds.[111] The significance of the O_2 consumption rate values dictate the parameters such as maximum cell seeding and maximum size of the construct to ensure long-term viability of the tissue models.

Pyruvate and lactate production is associated with glucose metabolism under aerobic and anaerobic conditions respectively[112,113]; however, a small amount of lactate is also produced by glutaminolysis wherein amino acid glutamine go through a series of biochemical reaction as was observed in one of the experimental studies on diploid fibroblasts.[114] When the cells are exposed to extreme conditions, glycolysis is activated, leading to excessive pyruvate synthesis which is then converted to lactate by lactate dehydrogenase (LDH) enzyme. Lactate has always been considered as a useless product in the cell and cells want to quickly rid themselves of it. Therefore, an increased level of lactate in culture medium has been used as an indication of cell damage.[115-117] In doing so, the pH of the culture remained a major determining factor. ATP and nicotinamide adenine dinucleotide (NAD⁺) concentrations have also been used to evaluate mitochondrial activity in cisplatin injured human renal proximal tubular epithelial cells (RPTECs) and their recovery by *in vitro* co-culture with human cord-derived mesenchymal stem cells (UC-MSCs).[118] While a decrease in NAD⁺ and ATP was observed in RPTECs when treated with cisplatin in their *in vitro* culture, mitochondrial activity was recovered upon co-culture with UC-MSCs which was indicated by the recovery of mitochondrial number as well as activity (indicated by an increase in NAD⁺ and ATP). Thiazolyl Blue Tetrazolium Bromide MTT and WST-1 assays analyze the absorption of blue formazan crystals of tetrazolium salt which is converted by mitochondrial dehydrogenase, which is responsible for the electron transport system, such as NADH and nicotinamide adenine dinucleotide phosphate (NADPH).[119,120] Hence, its concentration is an indicator of a cell's metabolic activity. MTT contains sulfur oxide which could explain its reported cytotoxicity by affecting cellular membrane permeability.[121] MTT assay is cytotoxic and causes loss of adherence and necrosis of the cells, thus leading to an inability for re-use after the assay.[122] Moreover, MTT reaction gets changed by medium components such as glucose and its metabolites, as the elevation of glucose concentration in the culture medium, has been reported to increase MTT reaction.[122] Consistent with these findings, alternative techniques sharing the same principle of enzymatic conversion have been introduced. WST-1 is a non-radioactive assay to monitor cell proliferation activity *in vitro*.[123] It is considered as an easy to perform and sensitive assay to ascertain cell proliferation rate *in vitro* quantitatively. The assay is based on enzymatic cleavage of tetrazolium salt to formazan by mitochondrial dehydrogenase activity in the viable cells. The amount of formazan produced is proportional to the viable cells in a sample and the data generated co-relates well with [3H]-thymidine incorporation assay. AlamarBlue is an alternative cell viability assay based on the enzymatically controlled reduction of the blue non-fluorescence dye resazurin into pink fluorescent resorufin by diaphorase such as NAD(P)H: quinone oxidoreductase and Flavin reductase.[124] Due to the solubility of both dyes in culture medium, cultivated cells are not compromised and further cellular investigation can be performed. The assay is helpful for cytokine bioassays, cell viability assay and ascertaining cytotoxicity *in vitro*. AlamarBlue has wide-ranging application in the plant, bacteria, and higher organism research. An additional potential application of AlamarBlue has been reported in bacteria after exposure to different doses of radiation to quantify resazurin concentration and to visualize colorimetric changes of resazurin during the assay.[125] It is pertinent to mention that one has to be careful in the selection of methodology to determine the proliferation of cells as the results of the assay may show cell-type dependence. For example, tendon-derived cells have a quite different metabolism as compared to the other mesenchymal tissue-derived cells.[126] Moreover, the primary culture of cells may differ from their secondary sub-culture counterparts. An alternative explanation for this phenomenon has been attributed to resazurin induced oxidative stress that impairs mitochondrial function in terms of reactive oxygen species (ROS) production.[127] These sub-cellular and molecular level changes may lead to mitochondrial dysfunction, reduced proliferation, and cell degradation. Nevertheless, due to its simplicity, ease of use and non-destructive impact on cultivated cells, AlamarBlue is a widely used assay for analyzing cellular metabolism in different cell types including primary neuronal cell, normal lymphocytes and cancer cell lines, i.e., BeWo and JEG-3 choriocarcinoma cells.[128-130]

PrestoBlue converts resazurin into resorufin and consequently measures metabolically active cells.[131,132] Lall *et al.* have reported PrestoBlue reagent to detect bacterial growth and compared it with various commercially available growth indicator reagents including iodonitrotetrazolium violet (INT) and

AlamarBlue.[131] The results showed that PrestoBlue successfully indicated 50% inhibitory concentration of positive drug controls on various cell lines. Similarly, the sensitivity of PrestoBlue was greater in assessing the viability of human corneal epithelial cells (HCEC) than MTT and AlamarBlue.[132] Also, a recent study on proliferation of the hMSCs cultured on Porous Tantalum implant biomaterial has validated the use of PrestoBlue to evaluate the cytocompatibility and osteogenic potential of the cells.[133] Furthermore, the metabolic activity of hMSCs isolated from pelvis, femur, and the tibia was assessed and compared by PrestoBlue assay.[134] Generally, PrestoBlue is a fast, simple and has less cytotoxic effect compared with MTT and AlamrBlue assays.[135]

3.5.2 Indirect Monitoring of Cell Metabolism *In Vitro*

3.5.2.1 *Optical biosensors*

As an alternative to the direct methods for monitoring of cell metabolism, optical biosensors necessitate targeting of the cellular compounds without altering their natural condition but quantitatively assessing their molecular interaction. It involves molecular interaction with biological material and the formation of light reflection changes (transitory wave) that can be measured by the close optical biosensor.[136] Therefore, unlike traditional optical reporters such as fluorescence-based techniques wherein background signals interfere and exact fluorescence quantification becomes challenging, optical biosensors can directly identify target molecules. One of these biosensors is fiber-based which has various diameters and shapes but sensitive refractive index as a result of molecular collaboration.[136]

3.5.2.2 *Fluorescence Lifetime Imaging (FLIM)*

FLIM is a non-invasive measurement of lifetime fluorescence deterioration of endogenous fluorophores which is sensitive to cellular biological environment.[137] Generally, the fluorescence emitted by the donor is stimulated by internal laser beams followed by its decay measurement on the recipient at nanosecond to picosecond. Molecules such as NAD, Flavin adenine dinucleotide (FAD), melanin, FMN and riboflavin effectively retain autofluorescence that allows the identification of fluorophores by imaging. FLIM is used to monitor not only the metabolic status of NSCs but also their lineage commitment by NADH intensity measurement as an intrinsic biomarker and as a metabolic fingerprint.[138] Since NADH is an auto-fluorescence coenzyme involved in redox reaction during cellular metabolism, its intensity measurement can be correlated with cell metabolism (Figure-3.1).

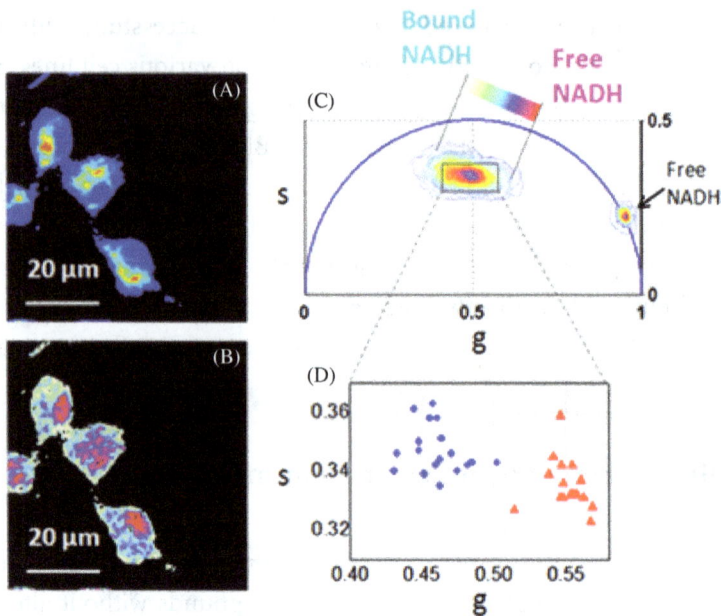

Figure-3.1. Free and bound NADH distribution within NPSCs. (A) Two-photon fluorescence intensity images. (B) Free/bound NADH FILM maps. (C) The E12 NPSCs excited at 740 nm FILM phasor plot of E12 NPSCs autofluorescence. A linear cluster represents relative concentrations of free NADH (purple) and bound NADH (cyan-white). Red-purple color indicates a high free/bound NADH ration, white violet, cyan and white indicate linearly and progressively decreasing ratios of free/bound NADH. (D) Scatter plot of the cell mitochondria (blue circle) and cell nuclei (red triangle). Nuclei contain a higher concentration of free NADH, white mitochondria contain mainly bound NADH. (Adopted from Stringari et al., 2012 [138]).

This approach offers a non-invasive label-free NADH FLIM signature to distinguish between undifferentiated and differentiated phenotypes using free-to-protein bound NADH ratio as a distinguishing feature between the two phenotypic states. Guo et al. used this approach to associate a reduction in NADH intensity with MSCs' osteogenic differentiation status.[139] These data correlated well with ATP levels which show a clear increase during differentiation. The same group of researchers also used NADH as a tissue intrinsic fluorescence chromophore for cell death detection.[140] An interesting application of FLIM has been recently reported by Evers et al. who measured pharmacologically induced metabolic changes in adipocytes.[141] Browning of adipocytes results from an extensive increase in energy utilization which was measured by the authors as a part of their novel model that quantitatively determined changes in mitochondrial and cytosolic NADH concentrations and binding sites. Although FLIM demonstrates its capability to identify intracellular biological activities non-destructively, this technique is expensive and not widely available, thus limiting its application.[142]

3.5.2.3 *Raman Microscopy (RM)*

Application of laser beams for single-cell analysis has been demonstrated by Raman Microspectroscopy (RM) which involves analysis of the chemical components of a single cell by mapping their respective vibration scattering wavelengths with a laser.[143] Cellular applications of RM have been developed not only to discriminate normal and malignant cells,[144] but also to detect their pre-cancerous stages enrooted to become cancerous.[145] Moreover, different strategies allowing the identification of cell cycle phases of stem cells,[146] cells aging assessment[147] and extracellular matrix analysis[148] have been described using Raman technology. Furthermore, the prevention of lipid accumulation related to cardiovascular disease can be obtained by this Raman imaging.[149] Ghita *et al.* have used this technique to distinguish naïve NSCs from their differentiated counterparts through the measurement of the nucleic acid concentration of each cell type.[150] Osteogenic lineage characterization of mESCs and hMSCs was also achieved *in vitro* by detecting spectrum differences at the pre-mineralized stage and after their terminal differentiation.[151–153]

3.5.2.4 *Surface plasmon resonance*

The principle of Surface Plasmon Resonance (SPR) is based on the measurement of ultraviolet (UV) light emission when surface plasmon waves interact with sample particles, the light transmission through the metal-coated nanostructure results in a strong electric field that is highly sensitive to the dielectric properties of materials immediately adjacent to the metal.[154–156] Due to its fast working, high sensitivity and specificity, it has been used to detect pathogens, toxins, drug residue, vitamins, hormones, antibodies, chemical contaminants, allergens and proteins in the given samples.[154–156]

Real-time assessment of cytokine production by epithelial cells in culture medium has been developed by SPR *in vitro*.[157] In this technology, conditioned cell medium is sampled from cell culture and delivered to the analytical device that is connected to optical fibers to deliver and collect the transmitted light which is then recorded by the computer. Although a significant difference was detected between control samples and those exposed to tumor necrosis factor alpha (TNF-α) and gold nanoparticles, an only limited number of cytokines, i.e., interleukin 8, interleukin 6 and the monocyte chemotactic protein, were identified.

3.5.2.5 *Electrochemical biosensors*

One of the major problems in the use of optical biosensors is the background noise signals originating from the endogenous fluorophores in the biological samples. The use of electrochemical conductors (ECB) curtail these signals to allow easy monitoring and analysis as ECB depends on the changes in electrical current that

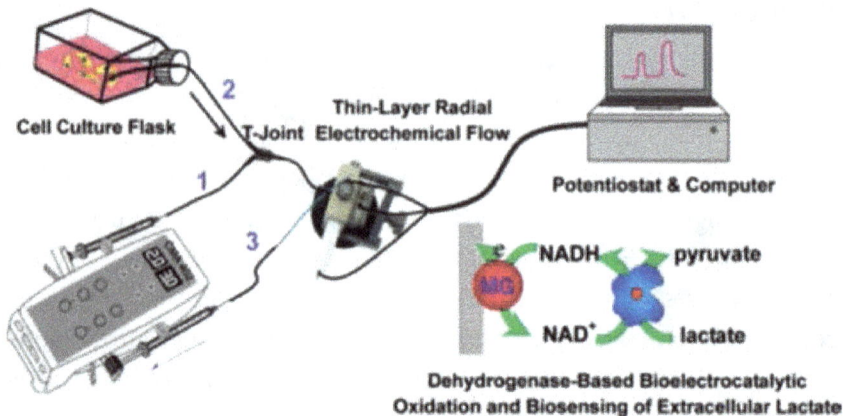

Figure-3.2.   Schematic diagram of the electrochemical detecting system for continuous monitoring of extracellular lactate production from cardiomyocyte following hypoxia. (Adopted from Li *et al.*, 2012 [162]).

are generated when a chemical interaction occurs between a biological membrane (on the working electrode) and the detected molecule.[158] When constant power is applied, current changes are compared with a wide range of reference electrodes.[159] ECB with silica proton-conducting membrane has been developed to measure metal deposits from the gas phase.[160] Real-time monitoring of the effect of trypsin on mouse 3T3 fibroblasts on organic electrochemical electrodes consists of Poly(3,4-ethylenedioxythiophene): polystyrene sulfonate (PEDOT:PSS) polymer which was achieved to prove its biocompatibility.[161] ECBs are practically the most used biosensors due to their simplicity and low cost. Moreover, unlike other biosensors; cross-contamination is excluded by the integration of a purification membrane.[158]

Electrochemical biosensors can be fabricated based on property of the tested molecules. For example, a hydrogenase-based electrochemical biosensor was used to monitor extracellular lactate production from rat cardiomyocytes post oxygen reduction (Figure-3.2).[162] Results indicated the effectiveness of such biosensors to tolerate high pH environment following hypoxia and its sensitivity to detect such physiological condition. Likewise, catecholamines which controls neurotransmitters release from adrenal chromaffin cells in a microfluidic condition, were detected by Iridium oxide fabricated electrodes.[163]

3.6 METABOLOMICS-BASED APPROACHES FOR CELL METABOLISM AND CHARACTERIZATION

Mass spectrometry (MS) integrated with chromatographic technique has been introduced to the field of cell biology due to its sensitive and rapid metabolic analysis.[164] The principle of MS is to generate multiple ions which are then separated according to their specific mass-to-charge ratio (m/z), followed by the recording of relative abundance of each ion type.[165,166] Finally, a mass spectrum of the molecule is displayed in the form of a plot of ion abundance versus the m/z ratio.

3.6.1 Liquid Chromatography-Mass Spectrometry (LC-MS)

Liquid chromatography (LC) Mass Spectrometry LC-MS is a powerful tool for the quantification of small molecules, as well as for the identification of known and unknown metabolites in a biological sample. The principal difference of ion separation in LC-MS is a liquid mobile phase delivered under high pressure (up to 400 bar (4×10^7 Pa)) to ensure a constant flow rate, and thus reproducible chromatography, while the stationary phase is packed into a column capable of resisting the high pressures which are necessary.[167]

Zhou et al. have demonstrated the efficient use of LC-MS in tracing various concentrations of mitomycin-C (MMC) in the mouse and hESC feeders which are known to induce genetic aberration.[168] In another study, saturated metabolites such as fatty acids that have considerable effects during MSCs differentiation were detected by LC-MS.[169] Quantitative time-point analysis of the key molecular changes that occur during osteogenic differentiation of MSCs has been recently demonstrated and linked to energy production and phosphate metabolism.[170] The study aimed to characterize the set of biological changes and identify the pathways regulating the formation of mature bone cells in MSCs using LC-MS based profiling of culture medium for metabolites such as D-glucarate which was upregulated during osteogenic differentiation (Figure-3.3). D-glucarate is a dicarboxylate anion of D-glucaric acid which is an intermediate metabolite in ascorbate and aldearate metabolism (KEGG PATHWAY: MAP00053). In bacteria, it is well-established that D-glucarate shares the same metabolic pathway with 2-oxoglutarate (α-ketoglutarate)[171] which is an intermediate metabolite in the TCA cycle that is known for its role in proline catabolism during bone formation. This may explain its elevation during osteogenic differentiation of MSCs (Figure-3.3A). N-Carbamoyl-L-aspartate is another compound that significantly increased with osteogenesis as compared to untreated MSCs (Figure-3.3B). This is involved in Pyridine metabolism (KEGG PATHWAY: MAP00240) and alanine, aspartate and glutamate metabolism (KEGG PATHWAY: MAP00250). The possible explanation of its elevation during osteogenic differentiation might be during orthophosphate synthesis (KEGG REACTION: R01397), the compound that was also increased during osteogenesis induction.[170]

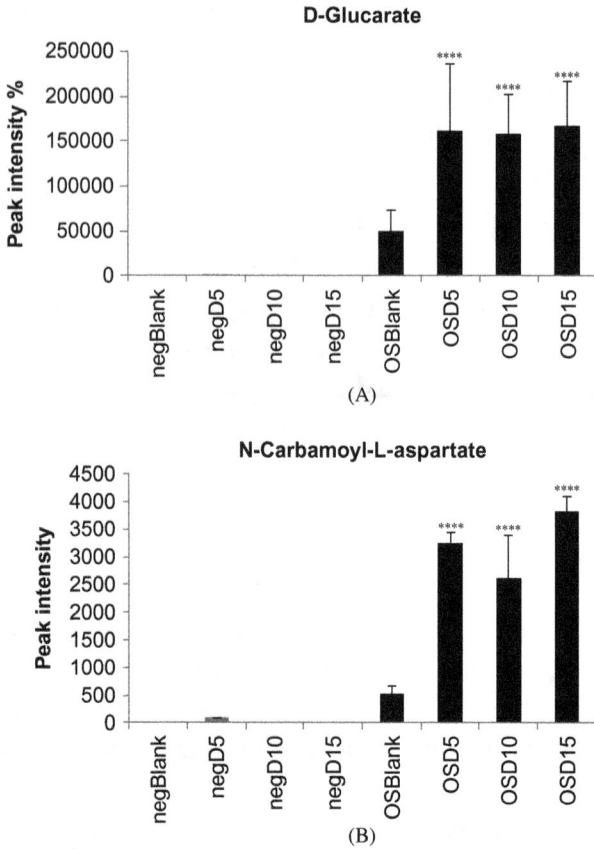

D-Glucarate

(A)

N-Carbamoyl-L-aspartate

(B)

Figure-3.3.   Metabolite changes identified in the culture medium MSC-related metabolites found to increase over time in culture medium in the presence of MSCs under OS treatment compared to control medium at day 5, 10 and 15 of treatment.

Intracellular degradation of proteins and other cytoplasmic components leads to accumulation of amino acids and fatty acids which have been identified in human placenta-derived MSCs (hPMSCs) using LC-MS isotope labeling.[172] Metabolic profiling associated with autophagy showed the degradation of intracellular components, pole of amino acids linked to arginine and proline metabolic pathway (KEGG PATHWAY: MAP00330) such as ammonia, ornithine and 4-aminobutyraldehyde ABAL were significantly elevated. On the other hands, metabolites such as L-glutamine, L-arginine, L-citrulline, and L-proline, spermine, N-acetyl putrescine and gamma-aminobutyric acid GABA were dramatically reduced. A similar reduction of L-arginine, L-citrulline, and L-proline have been reported in hPMSCs under hypoxic conditions using LC-MS.[173] Therefore, it is reasonable to speculate that differentiation of MSCs may be characterized by its distinct metabolic profile, including metabolite secretion and/or excretion and nutrient consumption due to their unique and specific metabolic pathways activated during the process. The pathways identified by LC-MS may be used to monitor stem cell differentiation through the analysis of metabolite changes not only between various cell fates, but also at different stages within specific lineage maturation.

3.6.2 Gas Chromatography-Mass Spectrometry (GC-MS)

Gas chromatography (GC)-MS segregates chemicals into discrete components in a mobile phase through solid-phase microextraction (SPME),[174] then transferred into the stationary phase by helium gas (GC-MS). In GC-MS, the stationary phase consists of a capillary column made from various materials (glass, stainless steel or wax), and of different lengths and widths which is held at a controlled temperature. Consequently, separation of molecules occurs based on the interaction between the mobile and stationary phase. Components are eluted according to their interface speed as well as the boiling rate. During the segregation process, the abundance of each detected compound is measured against elution time; followed by identification of each molecule's ion according to the mass spectral libraries and databases.[175]

Volatile organic compounds (VOCs) have been widely used to explore various physiological and pathological conditions in humans. For instance, food malabsorption, as well as sleep monitoring has been investigated by measuring the concentration of VOCs in a patient's breath.[176] Also, the detection of volatile emission from the skin has been reported to provide a fingerprinting profile of skin cells for applications in dermatology and dermatopathology.[177] Compounds analyzed by this method can indicate physiological changes in the cells as a result of their response to the microenvironment and changes in their metabolism. Interestingly, some cancers can be potentially diagnosed according to VOCs profiles, for instance, a high level of 1,1,4,4-tetramethyl-2,5-dimethylene-cyclohexane was found to be associated with colorectal malignancy by analyzing patient's blood and elevation of acetone methyl ethyl ketone, n-propanol in lung cancer patient's breath.[178, 179] Thus, VOCs analysis could potentially be useful in cancer screening and prevention programs to facilitate the detection of pre-cancerous stages, using these VOCs as biomarkers.

GC-MS has been used in cell culture to monitor VOCs affected by the presence of growth factors, nutrients, and cytotoxic agents during cell culture leading to a range of metabolic entities such as alcohols, aldehydes, furans, ketones, pyrroles, and terpenes.[175] Due to its applicability to the quantification of cellular phenotype,

GC-MS offers potential for stem cells biology, as illustrated by recent analyses of glucose metabolism in cell extracts from the human MSCs undergoing osteogenic and adipogenic differentiation.[180-182] The 3T3-L1 murine cell line was subjected to the hormonal trigger for adipogenic differentiation and the metabolic phenotype associated with the differentiation to adipocyte was assessed by GC-MS, nuclear magnetic resonance and LC-MS. The metabolite concentrations over time course of differentiation were profiled. GC-MS approach relied on the use of cell extracts thus leaving a scope to explore improvement by the implementation of a non-destructive process to enable a live analysis of stem cells during their proliferation and differentiation through the sampling of the cellular environment. Variations of cellular metabolism during osteogenic differentiation of human umbilical cord blood (UCB) MSCs were elucidated using live GC-MS analysis.[183] This study demonstrated that dexamethasone increased cell proliferation through upregulating mitochondrial activity and induced differentiation by activation of threonine metabolism, the pathway that is known for its role during stem cell differentiation.[184]

3.6.3 Nuclear Magnetic Resonance Spectroscopy (NMR)

The underlying principle of Nuclear Magnetic Resonance (NMR) technology is identification of active metabolic pathways, measuring enzyme activity and metabolic fluxes through flow tracing of nuclei, i.e., ^2H, ^{13}C, and ^{15}N, from labeled substrates as they get integrated into the downstream metabolites. ^1H (proton), ^{13}C, ^{15}N, and ^{31}P are considered to be the most critical nuclei in bimolecular NMR studies.[185] In stem cell culture, NMR was used to analyze metabolic biomarkers of hMSCs chondrogenesis in vitro.[186] The authors identified an increase in fatty acid synthesis that may indicate high ATP production. Moreover, sperms viability monitoring through evaluation and analysis of pyruvate and lactate metabolism has been reported.[187]

3.7 METABOLOMICS ANALYSIS METHODOLOGY

The texture of a biological sample justifies its metabolite extraction and measurement methodologies. Also, extraction parameters such as temperature, pH, nature of the solvent and time of extraction should be adjusted for higher range of metabolite recoveries while preventing their chemical and physical properties from degradation.[188-190]

3.7.1 Data Processing and Pathway Analysis Strategies

Appropriate experimental design (including cell culture, metabolite extraction and measurement) and application of appropriate data analysis methodologies provide

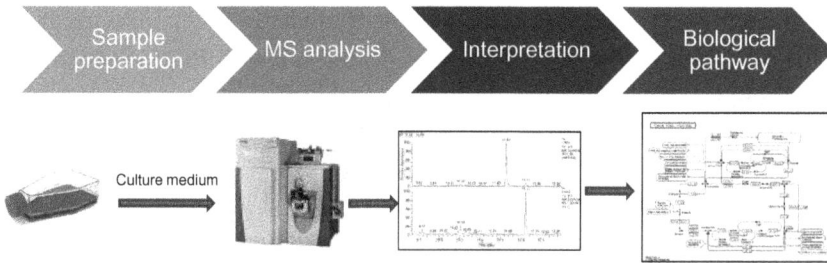

Figure-3.4. Metabolomics analysis scheme summarising the process starts from stem cell culture then MS analysis of cellular metabolites and their data interpretation followed by pathway analysis.

good coverage of metabolomic profile in a biological sample (Figure-3.4).[188] Also, biological pathway analysis provides a logical understanding of complex biological interaction orchestrating cell physiology. Key tools on metabolites identification process and multivariate analysis have been detailed in these studies.[170,192,193]

3.8 CONCLUSION AND PERSPECTIVE

This chapter has discussed and summarized metabolomics approaches of MSCs and their differentiated populations through non-destructive analysis of cellular metabolites. It started with an understanding of how culture compositions effect MSCs proliferation and differentiation potential followed by direct and indirect analysis of metabolite shift during their dual differentiation, highlighting potential new biomarkers. Metabolic pathways orchestrate the essential regulatory mechanisms that regulate stem cell fate through manipulation of the changes in the stem cell niche, physiological status, and nutrient availability based on cellular needs. Metabolomics approach has recently moved towards the large-scale pathway analysis down to molecular level to reflect cell phenotypes. However, future studies are essential to provide a comprehensive list of metabolites that link intracellular and extracellular metabolomics, so as to provide essential information for cell analysis. Put together, this chapter not only describes the new methods of non-destructive analysis of stem cells, but also identifies new therapeutic agents that may have a significant impact on stem cell biology and their use for clinical applications.

ACKNOWLEDGMENTS

I dedicate this chapter to my great parents who have never stopped giving me the best and a special word of mention to my mother whom I have always considered as my role model and a source of inspiration in my life. My special thanks to

Dr Virginie Sottile, Prof Ian Fisk and Dr Dong-Hyun Kim. And finally, my gratitude for Walaa and Yousef who have always supported me with their innocent smiles; believing in me and placing me as a role model capable of achieving the impossible.

REFERENCES

1. Kfoury Y, Scadden DT. Mesenchymal cell contributions to the stem cell niche. *Cell Stem Cell*. 2015; 16(3): 239–253.
2. Frenette PS, Pinho S, Lucas D, Scheiermann C. Mesenchymal stem cell: keystone of the hematopoietic stem cell niche and a stepping-stone for regenerative medicine. *Ann Rev Immunol*. 2013; 31: 285–316.
3. Ehninger A, Trumpp A. The bone marrow stem cell niche grows up: mesenchymal stem cells and macrophages move in. 2011; 208(3): 421–428.
4. Pittenger MF, Mackay AM, Beck SC, Jaiswal RK, Douglas R, Mosca JD, Moorman MA, *et al*. Multilineage potential of adult human mesenchymal stem cells. *Science*. 1999; 284(5411): 143–147.
5. Dominici M, Le Blanc K, Mueller I, Slaper-Cortenbach I, Marini F, Krause D, Deans R, Keating A, Prockop Dj, Horwitz E. Minimal criteria for defining multipotent mesenchymal stromal cells. The International Society for Cellular Therapy position statement. *Cytotherapy*. 2006; 8(4): 315–317.
6. Horwitz EM, Le Blanc K, Dominici M, Mueller I, Slaper-Cortenbach I, Marini FC, Deans RJ, Krause DS, Keating A; International Society for Cellular Therapy. Clarification of the nomenclature for MSC: The International Society for Cellular Therapy position statement. *Cytotherapy*. 2005; 7(5): 393–395.
7. Boxall SA, Jones E. Markers for characterization of bone marrow multipotential stromal cells. *Stem Cells Int*. 2012; 2012: Article ID 975871.
8. Jones E, Schäfer R. Where is the common ground between bone marrow mesenchymal stem/stromal cells from different donors and species? *Stem Cell Res & Ther*. 2015; 6: 143.
9. Sullivan MO, Gordon-Evans WJ, Fredericks LP, Kiefer K, Conzemius MG, Griffon DJ. Comparison of mesenchymal stem cell surface markers from bone marrow aspirates and adipose stromal vascular fraction sites. *Front Vet Sci*. 2015; 2: 82.
10. Yoshimura H, Muneta T, Nimura A, Yokoyama A, Koga H, Sekiya I. Comparison of rat mesenchymal stem cells derived from bone marrow, synovium, periosteum, adipose tissue, and muscle. *Cell Tissue Res*. 2007; 327(3): 449–462.
11. Ratajczak MZ, Zuba-Surma EK, Ratajczak J, Wysoczynski M, Kucia M. Very small embryonic like (VSEL) stem cells — characterization, developmental origin and biological significance. *Exp Hematol*. 2008; 36(6): 742–751.
12. Kuroda Y, Kitada M, Wakao S, *et al*. Unique multipotent cells in adult human mesenchymal cell populations. *Proc Natl Acad Sci USA*. 2010; 107: 8639–8643.
13. Motoi O, Haider HKh, Koichi I, Ashraf M. Existence of small juvenile cells in the aging bone marrow stromal cells and their therapeutic potential for ischemic heart disease. *Circulation*. 2011; 124(21-Supplement S).

14. Shin D-M, Suszynska M, Mierzejewska K, Ratajczak J, Ratajczak MZ. Very small embryonic-like stem-cell optimization of isolation protocols: an update of molecular signatures and a review of current in vivo applications. *Exp Mol Med.* 2013; 45(11): e56.

15. Haider HKh. Bone marrow cells for cardiac regeneration and repair: current status and issues. *Expert Rev Cardiovasc Ther.* 2006; 4(4): 557–568.

16. Caplan AI. Adult mesenchymal stem cells for tissue engineering versus regenerative medicine. *J Cell Physiol.* 2007; 213(2): 341–347.

17. Makino S, *et al.* Cardiomyocytes can be generated from marrow stromal cells in vitro. *J Clin Invest.* 1999; 103(5): 697–705.

18. Kopen GC, Prockop DJ, Phinney DG. Marrow stromal cells migrate throughout forebrain and cerebellum, and they differentiate into astrocytes after injection into neonatal mouse brains. *Proc Natl Acad Sci USA.* 1999; 96(19): 10711–10716.

19. Caplan AI, Correa D. PDGF in bone formation and regeneration: new insights into a novel mechanism involving MSCs. *J Orthop Res.* 2011; 29(12): 1795–1803.

20. Monfrini M, *et al.* Therapeutic potential of mesenchymal stem cells for the treatment of diabetic peripheral neuropathy. *Exp Neurol.* 2017; 288: 75–84.

21. Yao Y, Huang C, Gu P, Wen T. Combined MSC-secreted factors and neural stem cell transplantation promote functional recovery of PD rats. *Cell Transplant.* 2016; 25: 1101–1113.

22. Witt R, Weigand A, Boos AM, Cai A, Dippold D, Boccaccini AR, Schubert DW, Hardt M, Lange C, Arkudas A, Horch RE, Beier JP. Mesenchymal stem cells and myoblast differentiation under HGF and IGF-1 stimulation for 3-D skeletal muscle tissue engineering. *BMC Cell Biol.* 2017; 18: 15.

23. Hosseini SM, Sani M, Haider HKh, Dorvash M, Ziaee SM, Karimi A, Namavar MR. Concomitant use of mesenchymal stem cells and neural stem cells for treatment of spinal cord injury: a combo cell therapy approach. *Neurosci Lett.* 2018; 668: 138–146.

24. Buccini S, Haider HKh, Ahmed RPH, Jiang S, Ashraf M. Cardiac progenitors derived from reprogrammed mesenchymal stem cells contribute to angiomyogenic repair of the infarcted heart. *Basic Res Cardiol.* 2012; 107(6): 301–311.

25. Buccini S, Ahmed RPH, Haq A, Jiang S, Haider HKh. iPS cells derived from MSCs show superior angiogenesis following transplantation into the infarcted heart due to differential miRNA expression. *Circulation.* 128(Suppl 22): A12561.

26. Stewart AN, Kendziorski G, Deak ZM, Brown DJ, Fini MN, Copely KL, Rossignol J, Dunbar GL. Co-transplantation of mesenchymal and neural stem cells and overexpressing stromal-derived factor-1 for treating spinal cord injury. *Brain Res.* 2017; 1672: 91–105.

27. Haider HKh, Jiang S, Idris NM, Ashraf M. IGF-1–overexpressing mesenchymal stem cells accelerate bone marrow stem cell mobilization via paracrine activation of SDF-1α/CXCR4 signaling to promote myocardial repair. *Circ Res.* 2008; 103(11): 1300–1308.

28. Jiang S, Haider HKh, Idris NM, Salim A, Ashraf M. Supportive interaction between cell survival signaling and angiocompetent factors enhances donor cell survival and promotes angiomyogenesis for cardiac repair. *Circ Res.* 2006; 99(7): 776–784.

29. Kumar S, Chanda D, Ponnazhagan S. Therapeutic potential of genetically modified mesenchymal stem cells. *Gene Ther.* 2008; 15(10): 711–715.

30. Kim HW, Jiang S, Ashraf M, Haider HKh. Stem cell-based delivery of Hypoxamir-210 to the infarcted heart: implications on stem cell survival and preservation of infarcted heart function. *J Mol Med.* 2012; 90(9): 997–1010.

31. Feng Y, Haider HKh, Jiang S, *et al.* Pre-induction of Hsp70 is associated with stem cell resistance to ischemic stress via Hsf1-mir34a-hsp70 interaction. *Circulation.* 2011; 124(suppl_21): A10371.

32. Durrani S, Puthagram RA, Khach VL, Jiang S, Haider HKh. HIF-1α/HSP70 interaction is a critical determinant of stem cell survival during ischemic preconditioning. *Circulation.* 2011; 126(suppl_21): A14854.

33. Di Nicola M, Carlo-Stella C, Magni M, Milanesi M, Longoni PD, *et al.* Human bone marrow stromal cells suppress T-lymphocyte proliferation induced by cellular or nonspecific mitogenic stimuli. *Blood.* 2002; 99: 3838–3843.

34. Tse WT, Pendleton JD, Beyer WM, Egalka MC, Guinan EC. Suppression of allogeneic T-cell proliferation by human marrow stromal cells: implications in transplantation. *Transplantation.* 2003; 75: 389–397.

35. Ankrum JA, Ong JF, Karp JM. Mesenchymal stem cells: immune evasive, not immune privileged. *Nat Biotechnol.* 2014; 32(3): 252–260.

36. Kapranov NM, Davydova YO, Petinati MV, Bakshinskayte MV, Galtseva IV, Drize NI, Kuzmina LA, *et al.* Immune privileged features of multipotent mesenchymal stromal cells are lost after co-cultivation with allogeneic lymphocytes in vitro. *Blood.* 2016; 128: 5722.

37. Berglund AK, Fortier LA, Antczak DF, Schnabel LV. Immunoprivileged no more: measuring the immunogenicity of allogeneic adult mesenchymal stem cells. *Stem Cell Res Ther.* 2017; 8: 288.

38. Chapel A, Nasef A, Mathieu N, Frick J, Bouchet S, Gorin NC, Thierry D, Fouillard L. Human mesenchymal stem cells express HLA-G: role in MSC mediated immunosuppressive effect. *Blood.* 2006; 108: 4257.

39. Nautaa AJ, Fibbea WE. Immunomodulatory properties of mesenchymal stromal cells. *Blood.* 2007; 110(10): 3499–3506.

40. Gao F, Chiu SM, Motan DAL, Zhang Z, Chen L, Ji H-L, Tse H-F, *et al.* Mesenchymal stem cells and immunomodulation: current status and future prospects. *Cell Death Dis.* 2016; 7(1): e2062.

41. Hass R, Kasper C, Böhm S, Jacobs R. Different populations and sources of human mesenchymal stem cells (MSC): a comparison of adult and neonatal tissue-derived MSC. *Cell Commun Signal.* 2011; 9: 12.

42. Berebichez-Fridman R, Montero-Olverq PR. Sources and clinical applications of mesenchymal stem cells. State-of-the-art review. *Sultan Qaboos Univ Med J.* 2018; 18(3): e264–e277.

43. Kuhbier JW, *et al.* Isolation, characterization, differentiation, and application of adipose-derived stem cells. *Adv Biochem Eng Biotechnol.* 2010; 123: 55–105.

44. Yamamoto T, *et al.* Periurethral injection of autologous adipose-derived stem cells for the treatment of stress urinary incontinence in patients undergoing radical prostatectomy: report of two initial cases. *Int J Urol.* 2010; 17(1): 75–82.

45. Schaffler A, Buchler C. Concise review: adipose tissue-derived stromal cells — basic and clinical implications for novel cell-based therapies. *Stem Cells.* 2007; 25(4): 818–827.

46. Chong PP, *et al.* Human peripheral blood derived mesenchymal stem cells demonstrate similar characteristics and chondrogenic differentiation potential to bone marrow derived mesenchymal stem cells. *J Orthop Res.* 2012; 30(4): 634–642.

47. Tsutsumi S, Shimazu A, Miyazaki K, Pan H, Koike C, Yoshida E, Takagishi K, *et al.* Retention of multilineage differentiation potential of mesenchymal cells during proliferation in response to FGF. *Biochem Biophys Res Commun.* 2001; 288(2): 413–419.

48. Bianchi G, Banfi A, Mastrogiacomo M, Notaro R, Luzzatto L, Cancedda R, Quarto R. Ex vivo enrichment of mesenchymal cell progenitors by fibroblast growth factor 2. *Exp Cell Res.* 2003; 287(1): 98–105.

49. Yuan Y, Kallos MS, Hunter C, Sen A. Improved expansion of human bone marrow-derived mesenchymal stem cells in microcarrier-based suspension culture. *J Tissue Eng Regen Med.* 2014; 8(3): 210–225.

50. Miller RP, Hanley PJ. Isolation and manufacture of clinical-grade bone marrow-derived human mesenchymal stromal cells. *Methods Mol Biol.* 2016; 1416: 301–312.

51. Becherucci V, Piccini L, Casamassima S, Bisin S, Gori V, Gentile F, Ceccantini R, *et al.* Human platelet lysate in mesenchymal stromal cell expansion according to a GMP grade protocol: a cell factory experience. *Stem Cell Res Ther.* 2018; 9(1): 124.

52. Lipp M, Shillito R, Giroux R, Spiegelhalter F, Charlton S, Pinero D, Song P. Polymerase chain reaction technology as analytical tool in agricultural biotechnology. *J AOAC Int.* 2005; 88(1): 136–155.

53. Grothaus GD, Bandla M, Currier T, Giroux R, Jenkins GR, Lipp M, Shan G, *et al.* Immunoassay as an analytical tool in agricultural biotechnology. *J AOAC Int.* 2006; 89(4): 913–928.

54. Smith D, Glen K, Thomas R. Automated image analysis with the potential for process quality control applications in stem cell maintenance and differentiation. *Biotechnol Prog.* 2016; 32(1): 215–223.

55. Azouna NB, Jenhani F, Regaya Z, Berraeis L, Ben Othman T, Ducrocq E, Domenech J. Phenotypical and functional characteristics of mesenchymal stem cells from bone marrow: comparison of culture using different media supplemented with human platelet lysate or foetal bovine serum. *Stem Cell Res Ther.* 2012; 3(1): 6.

56. Bernardi M, Albiero E, Alghisi A, Chieregato K, Lievore C, Madeo D, Rodeghiero F, Astori G. Production of human platelet lysate by use of ultrasound for ex vivo expansion of human bone marrow-derived mesenchymal stromal cells. *Cytotherapy.* 2013; 15(8): 920–929.

57. Hagmann S, Babak Moradi B, Frank S, Dreher T, Kämmerer PW, Richter W, Gotterbarm T. Different culture media affect growth characteristics, surface marker distribution and chondrogenic differentiation of human bone marrow-derived mesenchymal stromal cells. *BMC Musculoskelet Disord.* 2013; 14: 223.

58. Liu Y, Li YQ, Wang HY, Li YJ, Liu GY, Xu X, Wu XB, *et al.* Effect of serum choice on replicative senescence in mesenchymal stromal cells. *Cytotherapy.* 2015; 17(7): 874–884.

59. Laitinen A, Lampinen M, Liedtke S, Kilpinen L, Kerkelä E, Sarkanen JR, Heinonen T, et al. The effects of culture conditions on the functionality of efficiently obtained mesenchymal stromal cells from human cord blood. *Cytotherapy*. 2016; 18(3): 423–437.

60. Jung S, Panchalingam KM, Rosenberg L, Behie LA. Ex vivo expansion of human mesenchymal stem cells in defined serum-free media. *Stem Cells Int*. 2012; 2012: 123030.

61. Bieback K, et al. Human alternatives to fetal bovine serum for the expansion of mesenchymal stromal cells from bone marrow. *Stem Cells*, 2009; 27(9): 2331–2341.

62. Gottipamula S, Sharma A, Krishnamurthy S, Majumdar AS, Seetharam RN. Human platelet lysate is an alternative to fetal bovine serum for large-scale expansion of bone marrow-derived mesenchymal stromal cells. *Biotechnol Lett*. 2012; 34: 1367–1374.

63. Naskou MC, Sumner SM, Chocallo A, Kemelmakher H, Thoresen M, Copland I, Galipeau J, et al. Platelet lysate as a novel serum-free media supplement for the culture of equine bone marrow-derived mesenchymal stem cells. *Stem Cell Res & Ther*. 2018; 9: 75.

64. Ham RG, McKeehan WL. Media and growth requirements. *Methods Enzymol*. 1979; 58: 44–93.

65. Eagle H. Nutrition needs of mammalian cells in tissue culture. *Science*. 1955; 122(3168): 501–514.

66. Freshney RI. *Culture of Animal Cells: A Manual of Basic Technique*. Hoboken, NJ: Wiley-Liss, 2005.

67. Lane BP, Miller SL. Preparation of large numbers of uniform tracheal organ cultures for long term studies: effects of serum on establishment in culture. *In Vitro*. 1976; 12(2): 147–154.

68. Kragh-Hansen U. Molecular aspects of ligand binding to serum albumin. *Pharmacol Rev*. 1981; 33(1): 17–53.

69. McGarrity GJ. Spread and control of mycoplasmal infection of cell cultures. *In Vitro*. 1976; 12(9): 643–648.

70. McKee C, Chaudhry GR. Advances and challenges in stem cell culture. *Colloids and Surfaces B: Biointerfaces*. 2017; 159: 62–77.

71. Ho SS, Murphy KC, Binder BYK, Vissers CB, Leach JK. Increased survival and function of mesenchymal stem cell spheroids entrapped in instructive alginate hydrogels. *Stem Cells Trans Med*. 2016; 5(6): 773–781.

72. Azoidis I, Metcalfe J, Reynolds J, Keeton S. Three-dimensional cell culture of human mesenchymal stem cells in nanofibrillar cellulose hydrogels. *MRS Communications*. 2017; 7(3): 458–465.

73. Favi PM, Benson RS, Neilsen NR, Hammonds RL, Bates CC, Stephens CP, Dhar MS. Cell proliferation, viability, and in vitro differentiation of equine mesenchymal stem cells seeded on bacterial cellulose hydrogel scaffolds. *Mater Sci Eng C Mater Biol Appl*. 2013; 33: 1935.

74. Cochis A, Grad S, Stoddart MJ, Fare S, Altomare L, Azzimonti B, Alini M, Rimondini L. Bioreactor mechanically guided 3D mesenchymal stem cell chondrogenesis using

a biocompatible novel thermo-reversible methylcellulose-based hydrogel. *Sci Rep*. 2017; 7: 45018.

75. Eslaminejad MB, Fallah N. Effects of BIO on proliferation and chondrogenic differentiation of mouse marrow-derived mesenchymal stem cells. *Vet Res Forum*. 2013; 4(2): 69–76.

76. Lai VK, Ashraf M, Jiang S, Haider HKh. MicroRNA-143 is a critical regulator of cell cycle activity in stem cells with co-overexpression of Akt and angiopoietin-1 via transcriptional regulation of Erk5/cyclin D1 signaling. *Cell Cycle*. 2012; 11(4): 767–777.

77. Igura K, Haider HKh, Ahmed RPH, Sheriff S, Ashraf M. Neuropeptide y and neuropeptide y y5 receptor interaction restores impaired growth potential of aging bone marrow stromal cells. *Rejuv Res*. 2011; 14(4): 393–403.

78. Ciuffreda MC, Malpasso G, Musarò P, Turco V, Gnecchi M. Protocols for in vitro differentiation of human mesenchymal stem cells into osteogenic, chondrogenic and adipogenic lineages. *Methods Mol Biol*. 2016; 1416: 149–158.

79. Langenbach F, Handschel J. Effects of dexamethasone, ascorbic acid and beta-glycerophosphate on the osteogenic differentiation of stem cells in vitro. *Stem Cell Res Ther*. 2013; 4(5): 117.

80. Canalis E, McCarthy TL, Centrella M. Effects of platelet-derived growth factor on bone formation in vitro. *J Cell Physiol*. 1989; 140(3): 530–537.

81. Bonewald LF, Dallas SL. Role of active and latent transforming growth factor beta in bone formation. *J Cell Biochem*, 1994; 55(3): 350–357.

82. Weiss S, *et al*. Impact of growth factors and PTHrP on early and late chondrogenic differentiation of human mesenchymal stem cells. *J Cell Physiol*. 2010; 223(1): 84–93.

83. Nasrabadi D, Rezaeiani S, Eslaminejad MB, Shabani A. Improved protocol for chondrogenic differentiation of bone marrow derived mesenchymal stem cells — effect of PTHrP and FGF-2 on TGFβ1/BMP2-induced chondrocytes hypertrophy. *Stem Cell Rev*. 2018; 14(5): 755–766.

84. Franceschi RT, Iyer BS. Relationship between collagen synthesis and expression of the osteoblast phenotype in MC3T3-E1 cells. *J Bone Miner Res*. 1992; 7(2): 235–246.

85. Scott MA, Nguyen VT, Levi B, James AW. Current methods of adipogenic differentiation of mesenchymal stem cells. *Stem Cells Dev*. 2011; 20(10): 1793–1804.

86. Picard F, Auwerx J. PPAR (gamma) and glucose homeostasis. *Annu Rev Nutr*. 2002; 22: 167–197.

87. Tontonoz P, Hu E, Spiegelman BM. Stimulation of adipogenesis in fibroblasts by PPAR gamma 2, a lipid-activated transcription factor. *Cell*. 1994; 79(7): 1147–1156.

88. Sears IB, *et al*. Differentiation-dependent expression of the brown adipocyte uncoupling protein gene: regulation by peroxisome proliferator-activated receptor gamma. *Mol Cell Biol*. 1996; 16(7): 3410–3419.

89. Petersen RK, *et al*. Cyclic AMP (cAMP)-mediated stimulation of adipocyte differentiation requires the synergistic action of Epac- and cAMP-dependent protein kinase-dependent processes. *Mol Cell Biol*. 2008; 28(11): 3804–3816.

90. Schoonjans K, Staels B, Auwerx A. The peroxisome proliferator activated receptors (PPARS) and their effects on lipid metabolism and adipocyte differentiation. *Biochim Biophys Acta.* 1996; 1302(2): 93–109.

91. Silbereisen A, Tritten L, Keiser J. Exploration of novel in vitro assays to study drugs against trichuris spp. *J Microbiol Methods.* 2011; 87(2): 169–175.

92. Randle PJ, Garland PB, Hales CN, Newsholme EA. The glucose fatty-acid cycle. Its role in insulin sensitivity and the metabolic disturbances of diabetes mellitus. *Lancet.* 1963; 1(7285): 785–789.

93. Romano AH, Conway T. Evolution of carbohydrate metabolic pathways. *Res Microbiol.* 1996; 147(6–7): 448–455.

94. Kruger NJ, von Schaewen A. The oxidative pentose phosphate pathway: structure and organisation. *Curr Opin Plant Biol.* 2003; 6(3): 236–246.

95. Williamson BK, *et al.* The effects of glaucoma drainage devices on oxygen tension, glycolytic metabolites, and metabolomics profile of aqueous humor in the rabbit. *Transl Vis Sci Technol.* 2018; 7(1): 14.

96. Stout AC, Barbe MF, Eaton CB, Amin M, Al-Eid F, Price LL, Lu B, *et al.* Inflammation and glucose homeostasis are associated with specific structural features among adults without knee osteoarthritis: a cross-sectional study from the osteoarthritis initiative. *BMC Musculoskelet Disord.* 2018; 19(1): 1.

97. Driban JB, Eaton CB, Amin M, Stout AC, Price LL, Lu B, Lo GH, *et al.* Glucose homeostasis influences the risk of incident knee osteoarthritis: data from the osteoarthritis initiative. *J Orthop Res.* 2017; 35(10): 2282–2287.

98. Trayhurn P. Endocrine and signalling role of adipose tissue: new perspectives on fat. *Acta Physiol Scand.* 2005; 184(4): 285–293.

99. Rosen ED, Spiegelman BM. Adipocytes as regulators of energy balance and glucose homeostasis. *Nature.* 2006; 444(7121): 847–853.

100. Fruhbeck G, Salvador J. Relation between leptin and the regulation of glucose metabolism. *Diabetologia.* 2000; 43(1): 3–12.

101. Saltiel AR, Kahn CR. Insulin signalling and the regulation of glucose and lipid metabolism. *Nature.* 2001; 414(6865): 799–806.

102. Nonogaki K. New insights into sympathetic regulation of glucose and fat metabolism. *Diabetologia.* 2000; 43(5): 533–549.

103. Lee SW, Rho JH, Lee SY, Chung WT, Oh YJ, Kim JH, *et al.* Dietary fat-associated osteoarthritic chondrocytes gain resistance to lipotoxicity through PKCK2/STAMP2/FSP27. *Bone Res.* 2018; 6: 20.

104. Mobley CB, Fox CD, Ferguson BS, Pascoe CA, Healy JC, McAdam JS, Lockwood CM, Roberts MD. Effects of protein type and composition on postprandial markers of skeletal muscle anabolism, adipose tissue lipolysis, and hypothalamic gene expression. *J Int Soc Sports Nutr.* 2015; 12: 14.

105. Domingo-Espín J, Lindal M, Nilsson-Wolanin O, Cushman SW, Stenkula KG, Lagersted JO. Dual actions of apolipoprotein A-I on glucose-stimulated insulin secretion and insulin-independent peripheral tissue glucose uptake lead to increased heart and skeletal muscle glucose disposal. *Diabetes.* 2016; 65(7): 1838–1848.

106. Choudhury GR, Winters A, Rich RM, Ryou M-G, Gryczynski Z, Yuan F, Yang S-H, et al. Methylene blue protects astrocytes against glucose oxygen deprivation by improving cellular respiration. *PLoS One.* 2015; 10(4): e0123096.

107. Mukherjee S, Zhang T, Lackoa LA, Jenny LT, Xiang Z, Butlerade JM, Chen S. Derivation and characterization of a UCP1 reporter human ES cell line. *Stem Cell Res.* 2018; 30: 12–21.

108. Maulucci G, Di Giacinto F, De Angelis C, Cohen O, Daniel B, Ferreri C, De Spirito M, et al. Real time quantitative analysis of lipid storage and lipolysis pathways by confocal spectral imaging of intracellular micropolarity. *Biochim Biophys Acta Mol Cell Biol Lipids.* 2018; 1863(7): 783–793.

109. Plitzko B, Loesgen S. Measurement of oxygen consumption rate (OCR) and extracellular acidification rate (ECAR) in culture cells for assessment of energy metabolism. *Bioprotocols.* 2018; 8(10): e2850.

110. Song W, Owusu-Ansah E, Hu Y, Cheng D, Ni X, Zirin J, Perrimon N. Activin signaling mediates muscle-to-adipose communication in a mitochondria dysfunction-associated obesity model. *Proc Natl Acad Sci USA.* 2017; 114(32): 8596–6601.

111. Streeterab I, Cheema U. Oxygen consumption rate of cells in 3D culture: the use of experiment and simulation to measure kinetic parameters and optimise culture conditions. *Analyst.* 2011; 136: 4013–4019.

112. Wlaschin KF, Hu WS. Engineering cell metabolism for high-density cell culture via manipulation of sugar transport. *J Biotechnol.* 2007; 131(2): 168–176.

113. Martinez-Monge I, Albiol J, Lecina M, Calleja LL, Miret J, Sola C, Cairo JJ. Metabolic flux balance analysis during lactate and glucose concomitant consumption in HEK293 cell cultures. *Biotechnol Bioeng.* 2019; 116(2): 388–404.

114. Zielke HR, Sumbilla CM, Sevdlian DA, Hawkins RL, Ozand PT. Lactate: a major product of glutamine metabolism by human diploid fibroblasts. *J Cell Physiol.* 1980; 104(3): 433–441.

115. Hirusaki K, Yokohama K, Cho K, Ohta Y. Temporal depolarization of mitochondria during M phase. *Sci Rep.* 2017; 7(1): 16044.

116. Girardot T, Rimmele T, Monneret G, Textoris J, Venet F. Intra-cellular lactate concentration in T lymphocytes from septic shock patients—a pilot study. *Intensive Care Med Exp.* 2018; 6(1): 5.

117. Liste-Calleja L, Lecina M, Lopez-Repullo J, Albiol J, Solà C, Cairó JJ. Lactate and glucose concomitant consumption as a self-regulated pH dctoxification mechanism in HEK293 cell cultures. *Appl Microbiol Biotechnol.* 2015; 99(23): 9951–9960.

118. Perico L, Morigi M, Rota C, Breno M, Mele C, Noris M, Introna M, et al. Human mesenchymal stromal cells transplanted into mice stimulate renal tubular cells and enhance mitochondrial function. *Nat Commun.* 2017; 8(1): 983.

119. Berridge MV, Tan AS. Characterization of the cellular reduction of 3-(4,5-dimethylthiazol-2-yl)-2,5-diphenyltetrazolium bromide (MTT): subcellular localization, substrate dependence, and involvement of mitochondrial electron transport in MTT reduction. *Arch Biochem Biophys.* 1993; 303(2): 474–482.

120. Hamid R, Rotshteyn Y, Rabadi L, Parikh R, Bullock P, et al. Comparison of AlamarBlue and MTT assays for high through-put screening. *Toxicol In Vitro*. 2004; 18(5): 703–710.

121. Hayden LJ, Pui AC, Roth SH. Human lung fibroblast cytotoxicity following acute sodium sulfide exposure. *Proc West Pharmacol Soc*. 1990; 33: 181–185.

122. Takahashi S, Abe T, Gotoh J, Fukuuchi, et al. Substrate-dependence of reduction of MTT: a tetrazolium dye differs in cultured astroglia and neurons. *Neurochem Int*. 2002; 40(5): 441–448.

123. Lewis DM, Blatchley MR, Park KM, Gerecht S. O2-controllable hydrogels for studying cellular responses to hypoxic gradients in three dimensions in vitro and in vivo. *Nat Protoc*. 2017; 12(8): 1620–1638.

124. O'Brien J, Wilson I, Orton T, Pognan F. Investigation of the Alamar Blue (resazurin) fluorescent dye for the assessment of mammalian cell cytotoxicity. *Eur J Biochem*. 2000; 267(17): 5421–5426.

125. Hudman DA, Sargentini NJ. Resazurin-based assay for screening bacteria for radiation sensitivity. *Springerplus*. 2013; 2(1): 55.

126. Mallick E, Scutt N, Scutt A, Rolf C. Passage and concentration-dependent effects of Indomethacin on tendon derived cells. *J Orthop Surg Res*. 2009; 4: 9.

127. Erikstein BS, Hagland HR, Nikolaisen J, Kulaweic M, Singh KK, Gjertsen BT, Tronstad KJ. Cellular stress induced by resazurin leads to autophagy and cell death via production of reactive oxygen species and mitochondrial impairment. *J Cell Biochem*. 2010; 111(3): 574–584.

128. White MJ, DiCaprio MJ, Greenberg DA. Assessment of neuronal viability with Alamar Blue in cortical and granule cell cultures. *J Neurosci Methods*. 1996; 70(2): 195–200.

129. Ahmed SA, Gogal RM Jr, Walsh JE. A new rapid and simple non-radioactive assay to monitor and determine the proliferation of lymphocytes: an alternative to [3H]thymidine incorporation assay. *J Immunol Methods*. 1994; 170(2): 211–224.

130. Al-Nasiry S, Geuses N, Hanssens M, Luyten C, Pijnenborg R, et al. The use of Alamar Blue assay for quantitative analysis of viability, migration and invasion of choriocarcinoma cells. *Human Rep*. 2007; 22(5): 1304–1309.

131. Lall N, Henley-Smith CJ, de Canha MN, Oozthuizen CB, Berrington D. Viability reagent, PrestoBlue, in comparison with other available reagents, utilized in cytotoxicity and antimicrobial assays. *Int J Microbiol*. 2013; 2013: 420601.

132. Xu M, McCanna DJ, Sivak JG. Use of the viability reagent PrestoBlue in comparison with AlamarBlue and MTT to assess the viability of human corneal epithelial cells. *J Pharmacol Toxicol Methods*. 2015; 71: 1–7.

133. Tang Z, Xie Y, Yang F, Huang Y, Wang C, Dai K, Zheng X, et al. Porous tantalum coatings prepared by vacuum plasma spraying enhance bmscs osteogenic differentiation and bone regeneration in vitro and in vivo. *PLoS One*. 2013; 8(6): e66263.

134. Davies DM, Snelling SJB, Quek L, Hakimi O, Yeh H, CArr A, Price AJ. Identifying the optimum source of mesenchymal stem cells for use in knee surgery. *J Orthop Res*. 2017; 35(9): 1868–1875.

135. Boncler M, Rozaiski M, Krajewska U, Posedek A, Watala C. Comparison of PrestoBlue and MTT assays of cellular viability in the assessment of anti-proliferative effects of plant extracts on human endothelial cells. *J Pharmacol Toxicol Methods*. 2014; 69(1): 9–16.

136. Fan X, White IM, Shopova SI, Zhu H, Suter JD, Sun Y. Sensitive optical biosensors for unlabeled targets: a review. *Anal Chim Acta*. 2008; 620(1–2): 8–26.

137. Becker W. Fluorescence lifetime imaging — techniques and applications. *J Microsc*. 2012; 247(2): 119–136.

138. Stringari C, Nourse JL, Flanagan LA, Gratton E. Phasor fluorescence lifetime microscopy of free and protein-bound NADH reveals neural stem cell differentiation potential. *PLoS One*. 2012; 7(11): e48014.

139. Guo HW, Chen CT, Wei YH, Lee OK, Gukassyan V, Kao FJ, Wang HW. Reduced nicotinamide adenine dinucleotide fluorescence lifetime separates human mesenchymal stem cells from differentiated progenies. *J Biomed Opt*. 2008; 13(5): 050505.

140. Wang HW, Gukassyan V, Chen CT, Wei YH, Guo HW, Yu JS, Kao FJ. Differentiation of apoptosis from necrosis by dynamic changes of reduced nicotinamide adenine dinucleotide fluorescence lifetime in live cells. *J Biomed Opt*. 2008; 13(5): 054011.

141. Evers M, Nunciada S, Osserian S, Casper M, Birngruber R, Evans CL, Manstein D. Enhanced quantification of metabolic activity for individual adipocytes by label-free FLIM. *Sci Rep*. 2018; 8(1): 8757.

142. Piston DW, Kremers GJ. Fluorescent protein FRET: the good, the bad and the ugly. *Trends Biochem Sci*. 2007; 32(9): 407–414.

143. Swain RJ, Stevens MM. Raman microspectroscopy for non-invasive biochemical analysis of single cells. *Biochem Soc Trans*. 2007; 35: 544–549.

144. Mallidis C, Sanchez V, Wistuba J, Wuebbeling F, Burger M, Fallnich C, Schlatt S. Raman microspectroscopy: shining a new light on reproductive medicine. *Hum Reprod Update*. 2014; 20: 403–414.

145. Duraipandian S, Zhang W, Ng J, Low JJ, Ilancheran A, Huang Z. In vivo diagnosis of cervical precancer using Raman spectroscopy and genetic algorithm techniques. *Analyst*. 2011; 136: 4328–4336.

146. Konorov SO, Schulze HG, Piret JM, Blades MW, Turner RF. Label-free determination of the cell cycle phase in human embryonic stem cells by Raman microspectroscopy. *Anal Chem*. 2013; 85: 8996–9002.

147. Bhai H, Li H, Han Z, Zhang C, Zhao J, Miao C, Yan S, *et al*. Label-free assessment of replicative senescence in mesenchymal stem cells by Raman microspectroscopy. *Biomed Opt Express*. 2015; 6: 4493–4500.

148. Brauchle E, Schenke-Layland K. Raman spectroscopy in biomedicine—non-invasive in vitro analysis of cells and extracellular matrix components in tissues. *Biotechnol J*. 2913; 8: 288–297.

149. Chen WW, Chen Ch, Wang CL, Wang HH, Wang LL, Ding ST, Lee TS, *et al*. Automated quantitative analysis of lipid accumulation and hydrolysis in living macrophages with label-free imaging. *Anal Bioanal Chem*. 2013; 405: 8549–8559.

150. Ghita A, Pascut FC, Mather M, Sottile V, Denning C, Notingher I. Cytoplasmic RNA in undifferentiated neural stem cells: a potential label-free Raman spectral marker for assessing the undifferentiated status. *Anal Chem*. 2012; 84: 3155–3162.

151. Gentleman E, Swain RJ, Evans ND, Boonrungsiman S, Jell G, Ball MD, Shean TA, et al. Comparative materials differences revealed in engineered bone as a function of cell-specific differentiation. *Nat Mater*. 2009; 8: 763–770.

152. Ghita A, Pascut FC, Sottile V, Notingher I. Monitoring the mineralisation of bone nodules in vitro by space- and time-resolved Raman micro-spectroscopy. *Analyst*. 2014; 139: 55–58.

153. Ghita A, Pascut FC, Sottile V, Denning C, Notingher I. Applications of Raman micro-spectroscopy to stem cell technology: label-free molecular discrimination and monitoring cell differentiation. *EPJ Tech Instrum*. 2015; 2: 6.

154. Homola J. Surface plasmon resonance sensors for detection of chemical and biological species. *Chem Rev*. 2008; 108: 462–493.

155. Brolo AG, Gordon R, Leathem B, Kavanagh KL. Surface plasmon sensor based on the enhanced light transmission through arrays of nanoholes in gold films. *Langmuir*. 2004; 20: 4813–4815.

156. Lertvachirapaiboon C, Baba A, Ekgaist S, Shinbo K, Kato K, Kaneko F. Transmission surface plasmon resonance techniques and their potential biosensor applications. *Biosens Bioelectron*. 2018; 99: 399–415.

157. Pasche S, Wenger B, Ischer R, Giazzon M, Angeloni S, Voirin G. Integrated optical biosensor for in-line monitoring of cell cultures. *Biosens Bioelectron*. 2010; 26: 1478–1485.

158. Mehrvar M, Abdi M. Recent developments, characteristics, and potential applications of electrochemical biosensors. *Anal Sci*. 2004; 20: 1113–1126.

159. Bard AJ, Faulkner LR. *Electrochemical Methods: Fundamentals and Applications*. New York, NY: Wiley New York, 2000.

160. Cho SK, Fan FR, Bard AJ. Electrochemical vapor deposition of semiconductors from gas phase with a solid membrane cell. *J Am Chem Soc*. 2015; 137: 6638–6642.

161. Salyk O, Vitecek J, Omasta L, Safarikova E, Stritesky S, Vala M, Weiter M. Organic electrochemical transistor microplate for real-time cell culture monitoring. *Applied Sci*. 2017; 7: 998.

162. Li X, Zhao L, Chen Z, Lin Y, Yu P, Mao L. Continuous electrochemical monitoring of extracellular lactate production from neonatal rat cardiomyocytes following myocardial hypoxia. *Anal Chem*. 2012; 84: 5285–5291.

163. Ges IA, Currie KP, Baudenbacher F. Electrochemical detection of catecholamine release using planar iridium oxide electrodes in nanoliter microfluidic cell culture volumes. *Biosens Bioelectron*. 2012; 34: 30–36.

164. Pitt JJ. Principles and applications of liquid chromatography-mass spectrometry in clinical biochemistry. *Clin Biochem Rev*. 2009; 30(1): 19–34.

165. Niessen WM. Advances in instrumentation in liquid chromatography-mass spectrometry and related liquid-introduction techniques. *J Chromatogr A*. 1998; 794(1–2): 407–435.

166. Gross JH. *SpringerLink (Online service), Mass Spectrometry A Textbook.* Berlin, Heidelberg: Springer Berlin, 2004, p. 1 online resource.

167. Parasuraman S, Anish R, Balamurugan S, Selvadurai Muralidharan S, Kumar KJ, Vijayan V. An overview of liquid chromatography-mass spectroscopy instrumentation. *Pharm Methods.* 2014; 5(2): 47–55.

168. Zhou D, Lin G, Zeng SC, Xiong B, Xie PY, Cheng DH, *et al.* Trace levels of mitomycin C disrupt genomic integrity and lead to DNA damage response defect in long-term-cultured human embryonic stem cells. *Arch Toxicol.* 2015; 89(1): 33–45.

169. Tsimbouri P, Gadegaard N, Burgess K, White K, Reynolds P, Herzyk P, Dalby MJ. Nanotopographical effects on mesenchymal stem cell morphology and phenotype. *J Cell Biochem.* 2014; 115(2): 380–390.

170. Surrati A, Linforth R, Fisk ID, Sottile V, Kim DH. Non-destructive characterisation of mesenchymal stem cell differentiation using LC-MS-based metabolite footprinting. *Analyst.* 2016; 141(12): 3776–3787.

171. Watanabe S, Yamada M, Ohtsu I, Makino K. Alpha-ketoglutaric semialdehyde dehydrogenase isozymes involved in metabolic pathways of D-glucarate, D-galactarate, and hydroxy-L-proline. Molecular and metabolic convergent evolution. *J Biol Chem.* 2007; 282(9): 6685–6695.

172. Sun Y, Chen D, Liu J, Xu Y, Shi X, Luo X, Pan Q, *et al.* Metabolic profiling associated with autophagy of human placenta-derived mesenchymal stem cells by chemical isotope labelling LC-MS. *Exp Cell Res.* 2018; 372(1): 52–60.

173. Wang D, Chen D, Yu J, Liu J, Shi X, Sun Y, Pan Q, *et al.* Impact of oxygen concentration on metabolic profile of human placenta-derived mesenchymal stem cells as determined by chemical isotope labeling LC-MS. *J Proteome Res.* 2018; 17(5): 1866–1878.

174. Risticevic S, Chen Y, Kudlejova L, Vatinno R, Baltensperger B, Stuff JR, Hein D, *et al.* Protocol for the development of automated high-throughput SPME-GC methods for the analysis of volatile and semi-volatile constituents in wine samples. *Nat Protoc.* 2010; 5(1): 162–176.

175. Dettmer K, Aronov PA, Hammock BD. Mass spectrometry-based metabolomics. *Mass Spectrom Rev.* 2007; 26(1): 51–78.

176. Amann A, Spanel P, Smith D. Breath analysis: the approach towards clinical applications. *Mini Rev Med Chem.* 2007; 7(2): 115–129.

177. Acevedo CA, Sanchez EY, Reyes JG, Young E. Volatile profiles of human skin cell cultures in different degrees of senescence. *J Chromatogr B Analyt Technol Biomed Life Sci.* 2010; 878(3–4): 449–455.

178. Wang C, Li P, Lian A, Sun B, Wang X, Guo L, Chi C, *et al.* Blood volatile compounds as biomarkers for colorectal cancer. *Cancer Biol Ther.* 2014; 15(2): 200–206.

179. Dent AG, Sutedja TG, Zimmerman PV. Exhaled breath analysis for lung cancer. *J Thorac Dis.* 2013; 5 Suppl 5: S540–550.

180. Cai G, Pauli GF, Wang Y, Jaki BU, Franzblau SG. Rapid determination of growth inhibition of mycobacterium tuberculosis by GC-MS/MS quantitation of tuberculostearic acid. *Tuberculosis (Edinb).* 2013; 93(3): 322–329.

181. Muñoz N, Kim J, Liu Y, Logan TM, Ma T. Gas chromatography-mass spectrometry analysis of human mesenchymal stem cell metabolism during proliferation and osteogenic differentiation under different oxygen tensions. *J Biotechnol.* 2014; 169: 95–102.

182. Roberts LD, Virtue S, Vidal-Puig A, Nicholis AW, Griffin JL. Metabolic phenotyping of a model of adipocyte differentiation. *Physiol Genomics.* 2009; 39(2): 109–119.

183. Klontzas ME, Vernardis S, Heliotis M, Tsiridis E, Mantalaris A. Metabolomics analysis of the osteogenic differentiation of umbilical cord blood mesenchymal stem cells reveals differential sensitivity to osteogenic agents. *Stem Cells Dev.* 2017; 26(10): 723–733.

184. Wang J, Alexander P, Wu L, Hammer R, Cleaver O, McKinght SL. Dependence of mouse embryonic stem cells on threonine catabolism. *Science.* 2009; 325(5939): 435–439.

185. Nagana Gowda GA, Raftery D. Recent advances in NMR-Based metabolomics. *Anal Chem.* 2017; 89(1): 490–510.

186. Jang MY, Chun SI, Hong KS, Shin J-W. Evaluation of metabolomic changes as a biomarker of chondrogenic differentiation in 3D-cultured human mesenchymal stem cells using proton (1H) nuclear magnetic resonance spectroscopy. *PLoS One.* 2013; 8(10): e78325.

187. Reynolds S, Ismail N, Calvert SJ, Pacey AA, Paley MNJ. Evidence for rapid oxidative phosphorylation and lactate fermentation in motile human sperm by hyperpolarized (13)C magnetic resonance spectroscopy. *Sci Rep.* 2017; 7(1): 4322.

188. Cuperlovic-Culf M, Barnett DA, Culf AS, Chute I. Cell culture metabolomics: applications and future directions. *Drug Discov Today.* 2010; 15(15–16): 610–621.

189. Dietmair S, Timmins NE, Gray PP, Nielsen LK, Krömer JO. Towards quantitative metabolomics of mammalian cells: development of a metabolite extraction protocol. *Anal Biochem.* 2010; 404(2): 155–164.

190. Ser Z, Liu X, Tang NN, Locasale JW. Extraction parameters for metabolomics from cultured cells. *Anal Biochem.* 2015; 475: 22–28.

191. Forsberg EM, Huan T, Rinehart D, Benton HP, Warth B, Hilmers B, Siuzdak G. Data processing, multi-omic pathway mapping, and metabolite activity analysis using XCMS Online. *Nat Protoc.* 2018; 13(4): 633–651.

192. Ren S, Hinzman AA, Kang EL, Szczesniak RD, Lu LJ. Computational and statistical analysis of metabolomics data. *Metabolomics.* 2015; 11(6): 1492–1513.

193. Hendriks MMWB, van Eeuwijk FA, Jellema RH, Westerhuis JA, Reijmers TH, Hoefsloot HCJ, Smilde AK. Data-processing strategies for metabolomics studies. *TrAC Trends in Analytical Chem.* 2011; 30(10): 1685–1698.

CHAPTER 4

Cell therapy for critical limb ischemia: Current progress and future prospects

Pavel Orekhov*, Mikhail Konoplyannikov*,†,‖, Vladimir Baklaushev*,
Peter Timashev†,‡,§, Anatoly Konoplyannikov⁶

*Federal Research Clinical Center of Specialized Medical Care and
Medical Technologies of the FMBA of Russia, Moscow, Russia
†Institute for Regenerative Medicine, Sechenov University, Moscow, Russia
‡N.N. Semenov Institute of Chemical Physics, Moscow, Russia
§Institute of Photonic Technologies, Research Center
"Crystallography and Photonics", Troitsk, Moscow, Russia
⁶A.Tsyb Medical Radiological Research Center of the Ministry of Health of Russia, Obninsk, Russia

ABSTRACT

In spite of the current notable advances in surgical management of critical limb ischemia (CLI), the most severe form of peripheral artery disease, it is still associated with the high frequency of amputations, lethality and low quality of life. Although the compensatory opportunities are mainly exhausted in the treatment of CLI, an efficient medical intervention remains possible. The purpose of this intervention is to eliminate a pronounced imbalance between the blood supply of the ischemic tissues and their metabolic needs. The physiological compensatory arteriogenesis, which actively proceeds at the initial stages of limb ischemia, almost ceases to the beginning of its transition into the final stages. Therefore, research efforts are focused on those technologies for tissue repair which are directed at the activation and expansion of the microvascular bed (angiogenesis) in the affected limb. Cell therapy, having been actively studied from the beginning of 2000s, is one of such approaches. This review discusses in-depth the advantages of different cell types for the CLI therapy, including peripheral bone marrow-derived mononuclear cells (BMMNCs) and mesenchymal stem cells (BMMSCs). The results of the most important pre-clinical and clinical studies, including the ongoing clinical trials, involving cell-based approach for CLI therapy have also been discussed besides optimization of the cell delivery techniques with or without the use of biomaterials as cell carriers.

‖ Corresponding author. Email: mkonopl@mail.ru

KEYWORDS

Bone marrow; Cell therapy; Clinical trials; EPCs; Ischemia; Mesenchymal; Mononuclear cells; Stem cells; Regenerative.

LIST OF ABBREVIATIONS

ABI	=	Ankle-brachial index
ASCs	=	Adipose tissue-derived stem cells
ASO	=	Arteriosclerosis obliterans
BM	=	Bone marrow
CLI	=	Critical limb ischemia
ECs	=	Endothelial cells
EPCs	=	Endothelial progenitor cells
EPC-MVs	=	EPCs-derived microvesicles
EVs	=	Extracellular vesicles
GFP	=	Green fluorescence protein
Hap	=	Hydroxyapatite
I/A	=	Intraarterial
I/C	=	Intracoronary
IL10	=	Interleukin-10
I/M	=	Intramuscular
I/V	=	Intravenous
miR	=	microRNA
MNCs	=	Mononuclear cells
MSCs	=	Mesenchymal stem cells
MSCs-EVs	=	MSC-derived extracellular vesicles
MVs	=	Microvesicles
OPGs	=	Objective performance goals
PAD	=	Peripheral artery disease
PB	=	Peripheral blood
PFWD	=	Pain-free walking distance
TAO	=	Thromboangitis obliterans
TcO2	=	Transcutaneous oxygen
TNFα	=	Tumor necrosis factor-α
VEGF	=	Vascular endothelial growth factor
VEGFR1	=	Vascular endothelial growth factor receptor-1
VEGFR2	=	Vascular endothelial growth factor receptor-2

4.1 INTRODUCTION

Among various manifestations of peripheral artery disease, critical limb ischemia (CLI) is considered the most severe one. Even with the modern treatment options such as endovascular surgery, CLI is characterized by a poor quality of life and poses threats of amputations and even death. Although CLI is generally associated with the exhaustion of the body's compensatory potential, there still exists the opportunity for efficient outside intervention. The term "critical limb ischemia", was introduced by Bell *et al.* to define the patients with clinical signs of an extreme degree of the disruption of blood circulation in the affected limb that leads to very high probability of its amputation.[1]

CLI is not a separate nosological form, but a syndrome belonging to a number of chronic obliterating diseases of peripheral arteries, diverging in their etiology and pathogenesis: obliterating atherosclerosis, thromboangitis obliterans (Buerger's disease), diabetic angiopathies and Raynaud's disease, with tissue ischemia as the main pathogenetic link. The clinical signs and characteristics features of CLI include intermittent claudication, rest pain in a limb, trophic ulcers and distal necroses, lowered blood pressure measured at an ankle or first toe, a progressive decrease of the pain-free walking distance (PFWD) and indices of a treadmill test, lowered skin temperature and transcutaneous partial oxygen pressure in limb tissues, occlusion of major blood vessels and depletion of microvascular bed, according to the data of X-ray contrast and ultrasound studies, laser Doppler imaging, etc.[2]

Chronic lower limb ischemia is diagnosed in 5–8% of persons older than 50, in about 30% of those who have risk factors (smoking, diabetes, hyperlipidemia, arterial hypertension),[3] and in general in 500–1000 per million population, according to the 1991 Second European Consensus Document on Chronic Leg Ischemia.[4] The number of major CLI-related limb amputations in the European Union and USA exceed 100,000 per year.[5] The incidence of amputations for this category of patients reaches 10–20%.[3] The lethality risk after this surgery reaches 15–30%, while the five-year survival rate does not exceed 30%.[3]

Surgical and endovascular limb revascularizations remain the main approaches of CLI therapy. Despite the impressive achievements in these fields,[6] such techniques cannot be applied in approximately in 20–40% of CLI patients due to peculiarities of the anatomical localization of major vessels, disease duration or concomitant pathology.[7] Besides, it is not always possible to achieve adequate perfusion of ischemic tissues. An important problem is early post-surgical thrombosis of stents and prostheses. Due to those reasons, limb amputation frequently appears as the single possibility to prolong a CLI patient's life. However, the existing conservative (non-surgical) treatment of CLI does not exhibit the required efficiency.[5]

4.2 CELL THERAPY OF CLI

The purpose of cell therapy in CLI is to eliminate the imbalance between the blood supply of ischemic tissues and their metabolic needs. Physiological compensatory arteriogenesis, which actively proceeds during the early stages of limb ischemia, is almost complete by the time of transition to the final stages, and the efficiency of any further stimulation of collateral growth of arterioles appears doubtful. Therefore, the focus of the current research is on the strategies based upon the molecular-cellular mechanisms of tissue and intracellular repair, which may lead to the growth and activation of the microvascular bloodstream (angiogenesis) of the affected limb.[8] Cell therapy is of such technology.

From the beginning of the 21st century, studies have been conducted on the application of regenerative medicine for the treatment of so-called "no option" CLI patients, i.e., those for whom all the standard therapy options including open and endovascular surgery have been exhausted.[9] Cell therapy offers multiple advantages in comparison with the use of specific growth factors. Firstly, cell therapy provides delivery of a series of cytokines instead of a single cytokine therapy. Secondly, cell therapy positively influences angiogenesis at the site of ischemia through resident stem cells homing and their endothelial differentiation and by the production of paracrine factors that promote ECs proliferation. The therapeutic benefits of cell therapy for the CLI patents in terms of enhancement of the regional blood supply and repair of ischemic tissues are reached in several ways:

1. Expanding ischemic tissue's vascular network and boosting its microperfusion at the level of the capillary bloodstream.
2. Remodeling of connective tissue that provides the growth of new capillaries and increase the interstitial fluid exchange.
3. Redirecting the response to the inflammatory process from tissue destruction to healing.

Realization of these benefits occurs via several inherent characteristics of all the stem cell types (to a certain extent) including the ability of migration, proliferation, tissue-specific differentiation, ability to synthesize and secrete a plethora of trophic/growth factors and other biologically active molecules (paracrine activity). The specificity of the stem cells effect is determined by their phenotypical belonging to a certain tissue type and the molecular-cellular characteristics of their microenvironment. Since a combination of these mechanisms may have a synergic character, combined use of different stem/ progenitor cells (MSCs, ECs, and anti-inflammatory) must play an important role in the enhancement of the therapeutic effects.[10]

Three types of cells are by far the most popular ones both in the research and in clinical trials — bone marrow-derived mononuclear cells (BMMNCs), peripheral blood mononuclear cells (PBMNCs) and mesenchymal stem (stromal) cells (MSCs).[11,12] Endothelial progenitor cells (EPCs) from the BM or peripheral blood are capable of incorporating into the existing vascular network to form new blood vessels.[13] Although no conclusive surface markers have been defined to identify EPCs, a combination of certain surface markers has been conventionally used for their identification, including CD34, CD133, and KDR (VEGF-/receptor 2). It is believed that during the structural rearrangement of the vascular network in the damaged tissue, which is based on the development of collateral arterioles and branching of capillaries, *de novo* formation of microvessels is possible with the participation of EPCs migrating from the bone marrow into the peripheral circulation and their homing into the ischemic regions.[13,14] An increase in the number of circulating EPCs in response tissue ischemia has been demonstrated.[15] There is mounting evidence regarding the possibility that EPCs get engrafted into the capillary wall with their subsequent differentiation. However, the fraction of bone marrow-derived cells integrated into the newly formed blood vessels of skeletal muscles has not been established as yet, although it may be essential in other tissues (e.g., in microvessels of the brain it comprises 26–42%). EPCs, as well as other CD34+ hematopoietic cells, are capable of activating resident ECs through the action of angiogenic factors.[16] Various studies have shown enhancement of neovascularization after transplantation of EPCs in pre-clinical experimental animal models of limb ischemia.[15,17]

MSCs, like other BM-derived cells, can promote angiogenesis through the establishment of new endothelial networks. MSCs are multipotent stromal cells derived from various body tissues.[18] Although the bone marrow remains the most frequently used source of MSCs, they can also be obtained from the skeletal muscle,[19] adipose tissue,[20] dental pulp[21], etc. MSCs have several peculiar immunomodulatory properties which make them favorable for the practical application in the treatment of various immune disorders. These include their anti-inflammatory action, suppression of dendritic cell maturation, proliferation and differentiation of T-lymphocytes, reduction of the natural killer cells' activity and activation of T-regulatory cells. MSCs reduce the secretion of pro-inflammatory

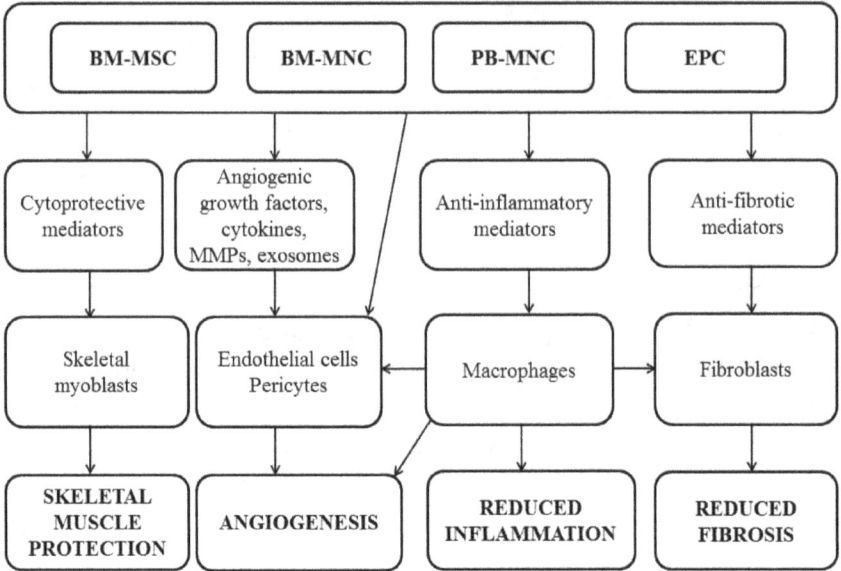

Figure-4.1.   Mechanisms of the cell therapy effects in CLI treatment.[27]

cytokines, i.e., interleukin-10 (IL)-10 and tumor necrosis factor-α (TNFα) by the T cells and augment the secretion of anti-inflammatory IL-4 [22]. Irrespective of their tissue of origin, MSCs do not express MHC class-II (HLA-DR) and possess immunosuppressive properties[22] and hence they are extremely promising for allogeneic transplantation.[23]

The revascularization of the ischemic limb after a direct MSCs injection into muscles has been demonstrated in a murine model of hind limb ischemia.[24] Similar neoangiogenic effects have been demonstrated after intramuscular (I/M) and intra-arterial (IA) administration of MSCs.[25] Both IA infusion of MSCs grown under 1% O_2 culture conditions and I/M implantation of MSCs pretreated with 2% O_2 have been reported to enhance revascularization of the hind limbs in a rodent model of hind limb ischemia.[26] Thus, all the above-mentioned cell types may either form new vessels or stimulate neoangiogenesis. Also, they may protect ischemic tissue, reduce inflammation and decrease tissue fibrosis (Figure-4.1).[27]

4.2.1 Clinical Studies

4.2.1.1 *Conducted clinical trials*

Different stem/progenitor cells and modes of cell delivery have been investigated for cell-based therapy to treat CLI. These include: direct injection of BMMNCs (I/M or I/A), direct injection of cytokine-mobilized PBMNCs after apheresis, direct injection of MSCs derived from BM or adipose tissue and direct injection of marker-selected or expanded cells derived from BM or PB. Several cell therapy-based clinical trials involving CLI patients have already been conducted (Table-4.1).

Most of these clinical trials have reported the safety and efficiency of cell therapy for CLI and have shown that I/M injections of BM-derived cells cause no serious adverse events, but reduce the rest pain and number of amputations, improve patients' quality of life, as well as enhance ankle-brachial index (ABI) and transcutaneous oxygen (TcO2) levels.[28-43] One of the most comprehensive data analyses on the cell-based therapy of CLI was published in 2018.[44] The meta-analysis was based on 23 randomized placebo-controlled clinical trials involving 962 patients. The data were sourced from PubMed, Embase, SinoMed, clinicaltrials.gov and the Cochrane databases. The meta-analysis showed a reduction of the amputation rates by 41%, increase in the probability of ulcer healing by 73%, increase of the ABI by 0.13, TcO2 by 12.22 mmHg, and PFWD by 144.84 meters, in the cell therapy group vs. those in the control group ($p<0.05$). The increased ABI and TcO2 level were consistent with improved wound healing and reduced amputation rate. The limb perfusion and the blood circulation in the toes essentially improved after the cell therapy versus the untreated group presumably due to enhanced angiogenesis. Indeed, there is evidence that BM-derived EPCs may promote the growth of microvasculature through paracrine as well as direct pathways.[21,22] The meta-analysis confirmed the safety and efficiency of cell therapy in CLI, with only non-serious and transient side effects present. Further clinical studies with a larger number of participants and better control remain necessary for the wide clinical application of cell therapy.

The USA Society for Vascular Surgery has suggested specific objective performance goals (OPGs) as tests for revascularization therapy in CLI.[45] The tests serve as both safety evaluation and assessment of the treatment efficiency and may play a key role in the strategy of cell therapy for CLI during the future clinical trials. They allow a direct comparison between different clinical studies and answer the question: at which stage of chronic arterial insufficiency would the use of cell therapy be optimal.

4.2.1.2 *Key factors determining the clinical efficiency cell-based therapy for CLI*

When analyzing the results of the various clinical studies, one can see a diversity of the key factors which determine the outcome of the study. Some of these factors include methods cell transplantation, cell dose, cell types, cell source and method of expansion in culture, pre-treatment of the cells before transplantation, selected

Table-4.1.   Results of cell-based clinical trials for CLI therapy.

Study location	Patients	Cells	Cell dose/Delivery	Follow-up (months)	Key findings
Prochazka et al. [28]; University Hospital Ostrava, Ostrava-Poruba, Czech Republic	N = 96 (42 BM MNCs & 54 std. care)	BM MNCs	40 ml of BM concentrate (BM MNCs) derived from 240 ml of patient's BM; I/M	4	Improved ABI and TcO2, rest pain in BM MNC group vs control; reduced amputation rate in BM MNC group vs control
Walter et al. [29] NCT00282646 (PROVASA); Div. of Cardiology and Vascular Medicine Frankfurt, Germany	N = 40 (21 BM MNCs & 19 placebo)	BM MNCs	10^8 single or double dose. I/A	6	Improved ulcer healing, rest pain in BM MNC vs placebo, no difference in amputation rate
Teraa et al. [30] NCT00371371 JUVENTAS; University Medical Center Utrecht, Utrecht, Netherlands	N = 160 (81 BM MNCs & 79 placebo)	BM MNCs	2×10^8 BM MNCs vs placebo; I/A (3X every three weeks)	6	No difference in all cause mortality rate, no difference in amputation rate vs placebo
Benoit et al. [31] NCT01245335 (BMAC); Tufts Medical Center, USA	N = 48 (34 BM MNCs & 14 placebo)	BM MNCs	10^9 BM MNCs vs placebo; I/M	6	Non-significant trend for reduced amputation rate vs placebo
Li et al. [32] Provincial Hospital affiliated to Shandong University, Shandong Province, China	N = 58 (29 BMMNCs & 29 placebo	BM MNCs	5×10^8–1.2×10^9 BM MNCs; I/M	6	Improved rest pain, ulcer healing, ABI in BM MNC group vs placebo. Major amputation rate 10% in BM MNC group vs 17% in placebo

Study	N	Cell type	Dose; Route	Follow-up (months)	Outcomes
Ponemone et al. [33] NCT01472289; Fortis Escorts Heart Institute & Research Centre, New Delhi, India	N = 17	BM MNCs	BM MNCs prepared using the Res-Q 60 tech. (0.5 cc/ injection for a total of 15–20 cc); I/M	12	Improved ABI and TcO2, mean rest pain and intermittent claudication pain scores, wound/ulcer healing, and 6-minute walking distance after BM MNC treatment. Major amputation-free survival (mAFS) rate and amputation-free rates (AFR) at 12 months were 70.6% and 82.3%, respectively
Osturk et al. [34] Gulhane Military Medical Academy Hematology Section, Turkey	N = 40 (20 PB MNCs & 20 std. care)	G-CSF Mobilized PB MNCs	10^9; I/M	3	Improved rest pain and ABI in PB MNC group vs control. Amputation rate: 15% in PB MNC group vs 25% in control
Huang et al. [35]; National Research Center for Stem Cell Engineering & Technology, State Key Laboratory of Experimental Hematology, Tianjin, China	N = 150 (76 PB MNCs vs 74 BM MNCs)	G-CSF mobilized PB MNCs vs BM MNCs	10^9 G-CSF Mobilized PB MNCs vs 10^8 BM MNCs; I/M	3	Improved rest pain and ABI with PB MNC vs BM MNC. Amputation rate: 5.3% in PB MNC group vs 8.1% in BM MNC group
Matoba et al. [36] NCT00145262 (TACT-NAGOYA); Nagoya University Graduate School of Medicine, Nagoya, Japan	N = 115 (74 atherosclerotic PAD & 41 patients with Buerger's disease	BM MNCs	10^9 BM MNCs; I/M	36	Improved rest pain, ulcer healing with BM MNC in both groups. Amputation rate: 40% in PAD vs 9% in TAO

(Continued)

Table-4.1.   *(Continued)*

Study location	Patients	Cells	Cell dose/Delivery	Follow-up (months)	Key findings
Lee *et al.* [37] NCT01663376; Pusan National University Hospital, Korea	N = 15 (12 patients with Buerger's disease, & 3 patients with diabetic foot)	Autologous ATSCs	300×10^6; I/M	6	Increased blood flow, substantial improvement of the Wong-Baker FACES pain rating score, claudication walking distance, and thermography results. No major amputations, minor amputation for 4 patients (26.7%)
Gupta *et al.* [38] NCT00883870; Stempeutics Research Pvt Ltd, Bangalore, India	N = 20 (10 BM MSCs & 10 placebo)	Allogeneic BM MSCs	2×10^6 cells/kg; I/M	6	Improved rest pain, ABI and ankle pressure, reduced adverse events in BM MSC vs placebo group. No difference in amputation rate
Lu *et al.* [39] NCT00955669; The Southwest Hospital Chongqing, Chongqing, China	N = 41 (21 BM MSCs & 20 BMMNCs, placebo) Rutherford Score 5-6, Ulcers	BM MSCs, BM MNCs	$5.0 \times 10^8 \sim 5.0 \times 10^9$ MSCs and MNCs; I/M	6	Improved ulcer healing, ABI and TcO2 in BM MSC group Amputation rate: No amputations in MSCs and MNCs groups; 16.2% in placebo group
Losordo *et al.* [40] NCT00616980 (*ACT34-CLI*); Northwestern Memorial Hospital, Chicago, USA	N = 28 (7 low-dose, 9 high-dose & 12 placebo)	Autologous CD34+ cells from G-CSF mobilized PB	10^5 (low-dose) or 10^6 (high-dose)/ kg; I/M	12	Improved ulcer healing and rest pain in CD34+ cells groups vs placebo group Amputation rate: 43% in low-dose group and 22% in high-dose group vs 75% in control group

Powell et al. [41] NCT00468000; (RESTORE-CLI) Cardiology, P.C. Alabama, USA; Arizona Heart Institute, Arizona, USA, and others (total 18 centers in USA)	N = 72 (48 Ixmyelocel-T & 24 placebo)	Ixmyelocel-T: autologous product including BM MSC (CD90+) and HSCs (CD45+)	$35\text{–}295 \times 10^6$ I/M	12	Improved rates of mortality, gangrene in Ixmyelocel-T group vs placebo. No significant difference in amputation rate (20.8% in Ixmyelocel-T group vs 25% in control group)
Szabo et al. [42] Semmelweis University Faculty of Budapest, Hungary	N = 20 (10 ACPs & 10 standard care)	PB-derived, expanded ACPs	$66.4 \pm 19.2 \times 10^6$ I/M	3 and 36	Improved pain score, wound healing and walking ability at 3 months in ACP group vs control. At 2 years: amputation rate 30% in ACP group vs 60% in control
Arici et al. [43] NCT01595776; Fondazione IRCCS Policlinico San Matteo, Pavia, Italy	8	EPCs (CD133+)	$16\text{–}40 \times 10^6$ cells per limb kg (mean number 25.2×10^6); I/M	12	75% (6 patients) had a complete healing of the wounds, rest pain cessation and walking recovery. Amputation rate: 25%

Abbreviations: ABI: Ankle-brachial index; ACPs: Angiogenic cell precursors; ATSCs: Adipose tissue derived stem cells; BM: Bone marrow; EPCs: Endothelial progenitor cells; G-CSF: Granulocyte colony stimulating factor; HSCs: Hematopoietic stem cells; I/A: Intra-arterial; I/M: Intramuscular; PB: Peripheral blood; MNCs: Mononuclear cells; MSCs: Mesenchymal stem cells; PB: Peripheral blood; PAD: Peripheral artery disease; TAO: Thromboangitis obliterans; TcO2: Transcutaneous Oxygen Tension.

cohorts of patients and their characteristics etc. These factors undoubtedly affect the efficacy of the conducted cell therapy and thus deserve a detailed consideration.

4.2.1.3 *Strategy of cell transplantation*

Discussing the ways of the cell delivery is of interest due to the two interrelated problems: a) the procedure's invasiveness and the simplicity of its application and b) the influence of the transplantation technique on its efficiency, that is directly associated with the "destiny" of the transplanted cells. The kinetics, distribution, and preservation of stem cells after their transplantation have been part of the detailed consideration in several experimental studies.[46,47]

4.2.1.3a *Intravenous transplantation*

Intravenous (I/V) administration is the simplest and the least invasive way of delivery which allows the cells to migrate from the central compartment into ischemic muscles.[48] However, it is generally considered that only a small number of stem cells are able to extravasate into the ischemic tissue after I/V administration *in vivo*, survive and undergo differentiation. In a study on a murine model of myocardial infarction, 83% of xenogenic MSCs were found in the lungs within one hour after I/V administration.[48] Due to this reason, the I/V route of cell delivery has only been rarely applied in the cell-based CLI therapy.

4.2.1.3b *Intra-arterial transplantation*

Intra-arterial (I/A) route of cell administration has been actively studied for CLI treatment.[30,50-52] I/A administration of cells may involve either transcutaneous contralateral femoral or antegrade femoral access for selective catheterization. The approach involves the use of a catheter of a small diameter (4F) which is advanced to the site of occlusion followed by a slow infusion.[52,53] It is generally believed that during a selective cell administration, the cells may reach the ischemic tissue borders with the blood flow, though the extent of their release from the microcirculation and grafting is unknown. Using intravital muscle microscopy, it was established that the majority of the cells after I/V or I/A administration get retained in the vascular network at the pre-capillary level (up to 92% for MSCs).[46] In spite of these data, the I/A administration has steadily provided positive clinical data in terms of reduction of the pain syndrome, ulcer healing or size reduction, decreased rate of amputation and improvement in the indices of perfusion, though the degree of efficiency differs between the.[46,52-54] It is important to note that all the studies report the absence or a very low frequency of various complications related to the procedure of I/A transplantation.[51,55]

4.2.1.3c *Repeated intra-arterial transplantation*

Repeated cell transplantation strategy has been rarely applied so far. In a non-randomized study, extremely high treatment efficiency has been reported which included 100% survival without amputations for a 12 months-long observation period, a very significant increase of pain-free walking distance, PFWD (up to 500–1000 meters), growth of ABI and several new capillaries by the data of video capillaroscopy.[56] In spite of such a phenomenal result, attentive reading of the article leads to serious critical considerations related to the criteria of patients' inclusion and methods of control. In 2012, a multicenter prospective placebo-controlled study was conducted which involved repeated I/A injections of cells.[57] The study results showed no fatal outcome and complications associated with the procedure (aspiration and infusion into the femoral artery), infectious complications and malignancies, and the survival without amputation in CLI patients showed a lower risk of lethality and/or amputations after cell therapy as compared to the natural course of the disease. These data showed high safety features of repeated cell delivery strategy. Almost all the patients noted pain alleviation, an increase of PFWD, improvement of the quality of life. The improvement of the perfusion indices by laser Doppler imaging and TcO2 remained unchanged for at least 12 months after the stem cell therapy. An 18 months follow-up showed a tendency to decrease in the mentioned indices that were explained by the disease progression. Such a decrease was noted also by other researchers.[58] Thus, in patients with a tendency to the perfusion reduction during 18 months follow-up, one may consider a possibility of a repeated course of treatment with BMMSCs.

To compare the efficiency of a placebo, single and double I/A infusions, PROVASA randomized double-blind study was conducted.[29] The significance of the study was that it diverged from the majority of randomized double-blind studies that investigated the efficiency of a single BMMSCs treatment to produce a limited duration of beneficial effects.[58] The PROVASA study showed high safety of SC aspiration and the I/A infusion procedure (only one hematoma and one pseudo-aneurysm, requiring a conservative treatment, were noted per 87 procedures); no complications and side-effects related to the cell therapy were observed. The PROVASA study has demonstrated that the repeated administration and a higher number of transplanted cells lead to a significantly better ulcer healing and alleviation of the pain syndrome than a single administration does. Thus, these results warrant a necessity of new randomized double-blind control studies with multiple courses of treatment.

The JUVENTAS study has been one of the largest RDBS which included 160 CLI patients randomized into a placebo-control group and a group of patients who underwent repeated I/A injections (3 times with a 3-week interval).[30] Like the PROVASA study, the JUVENTAS study also demonstrated the high level of safety of the technique, however, the basic results appeared disappointing as repeated I/A infusions of autogenic BMMSCs failed to significantly reduce the rate of major amputations in CLI. Nevertheless, the observed improvements in many parameters, i.e., quality of life, the intensity of rest pains, ABI, TcO2 in both groups at least make a basis for further studies.[30]

4.2.1.3d *Intramuscular transplantation*

The duration of the transplanted cells' presence in the ischemic area and specifics of the ischemic region itself affect the therapeutic efficiency of the cell-based therapy. For example, the rapid disappearance of cells from the myocardium was established after their I/M, I/C or I/V administration.[59] Contrary to the heart wherein the constant rhythmic contractility of the myocardium contributes to loss of cells besides other factors, the cells transplanted in the skeletal muscle are preserved for an essentially longer time. According to the data published by Kinnaird *et al.*,[60] large number of green fluorescence protein (GFP)-labelled MSCs injected in the adductor muscle of the murine model of hind limb ischemia were detected by day 7 after transplantation and the number of cells remained unchanged until day 14 of observation. They also observed the transplanted cells even by day 28. These observations provide a rationale for a preferred intramuscular route of cell transplantation under direct vision into a muscle. One of the advantages of intramuscular cell delivery is the creation of a depot of donor cells at the site of injection which locally releases pro-angiogenic/arteriogenic cytokines as a part of their paracrine activity in the ischemic tissue. In spite of a rather short life span of the transplanted cells, the induction of cell differentiation and direct cell-to-cell contact with resident cells contribute to the development of neovascular structures as well as an expansion of the existing vascular bed in the ischemic area.[5,52] Given the simplicity of I/M administration, it has emerged as the most preferred route of cell delivery in a majority of the clinical studies. The transplantation into the tibial muscles may be performed via the principle of a symmetric grid or in the projection of tibial arteries.[52,61] In rare cases, besides cell transplantation into the femur and lower leg, cell injections were performed into inter-metatarsal spaces of the foot.[62,63]

In a recently reported randomized controlled study on the comparative safety and efficacy of single and repeated intramuscular cell delivery strategies, 22 patients with thromboangitis obliterans (TAO) and diabetes mellitus (DM) were randomized into two equal groups. The patients received either a single or a four-fold (with a three-week interval) administration of autologous BMMNCs.[63] All the patients were followed-up for six months. No complications and side effects were registered, ABI increased by 82% of patients, the pain syndrome was reduced by 90% of cell transplanted patients with a notable improvement in the healing of ulcers in 90% while no significant differences in those indices were observed between the groups of single and repeated transplantation. PFWD was also essentially longer in both groups as compared to the initial values; however, a statistically significant increase towards the end of the observation period was noted in the group of repeated administration that was, by the authors' opinion, a reason for repeated courses of treatment.

Combined I/M and I/A administration has been rarely applied, although a desire to combine the advantages of both routes of cell delivery appears a logical reason to deliver the highest number of cells into the ischemic region.[50,64, 65] Franz *et al.* used the strategy of combined I/M and I/A treatment and observed that the rate of limb preservation reached 67% and a complete wound healing in those patients who had ulcers before the treatment and did not require amputation.[64] As was reported by Strauer *et al.* in one of their first studies on the comparative safety and efficiency of I/M and combined (I/M and I/A) routes of BMMNCs delivery in cardiac pathology, the improvement took place in both the treatment groups without any significant difference.[66] Van Tongeren *et al.* observed a substantial improvement in ABI and PFWD, as well as reduction of pain in both groups of the patients.[50] A major amputation was required in 2 patients out of 12 in the group of combined I/M and I/A treatment as compared to the 7 patients out of 15 in the group of I/M injection alone. It is pertinent to mention that in all the reported studies, no side-effects and complications were observed except for two patients with transient heart failure which occurred during BM aspiration in the study.[50]

4.2.2 Comparison of Different Ways of Transplantation

There are only a few studies which have directly compared the effect of multiple routes of administration for CLI therapy. Similar angiogenic effects after I/M and I/A cell delivery have been reported.[25] An extensive meta-analysis of clinical studies has shown that the ABI and PFWD improvement as well as the reduction of the rest pain and acceleration of ulcer healing at the affected limb do not differ statistically between I/M and I/A route of administration.[44] Klepanec *et al.* compared the efficiency of I/M and I/A routes of cell delivery in 41 patients and reported that both the procedures were well-tolerated and comparable in their therapeutic benefits.[52]

It is important to note that appreciable clinical effects after MSCs transplantation to the patients were independent of the route of administration (either I/M or I/A). The existing data do not allow an unambiguous answer to the question about the optimal route of cell transplantation. The known or suggested mechanisms of donor cell action, the high efficacy and a simply performed procedure makes the basis for a conclusion that I/M administration may be used preferentially in the future clinical trials and the routine clinical settings.

4.2.3 Dose of Transplanted Cells

To date, there is no convincing information allowing an objective calculation of the "required optimal dose" of transplanted cells. From our viewpoint, the definition of a "required optimal dose" requires an interpretation; it is assumed to be the lowest number of the cells reaching the target site to provide an optimal prognosis without a risk of side or adverse effects. Different numbers of cells have been used for delivery in the published studies with a general understanding that outcome of cell therapy is directly related to the number of the donor cells transplanted. The selection of an optimal dose is a complicated task since conventional pharmaceutical parameters (absorption, distribution, metabolism, and excretion) are difficult to measure or non-applicable in this case. The dose selection is currently performed by extrapolation of the results of animal studies or empirically, based on the efficiency observed during Phase-I or II clinical studies pertaining to incremental dose effect for comparison. Additional complexities with the cell dose selection are related to the indirect dose dependency of the results noted in different studies. The search of an optimal dose selection strategy remains one of the crucial issues of cell therapy.[67]

In many studies, a dose of 1.6×10^9 cells has been used, the same as used in the clinical study by Tateishi-Yuyama et al.,[68] which in turn was based on the experimental animal data. Among the conducted clinical studies, the cell doses vary from 0.1×10^9 to 50×10^9 with a positive effect observed even at the lowest concentration.[5] Klepanec et al. reportedly a statistically significant relationship between the numbers of CD34[+]-cells in the BM concentrate and the healing process. The tendency of increasing the dose of transplanted cells is indirectly reflected in the data presented in the meta-analysis of six randomized studies wherein autogenic cells of various origin were used with the doses from 1.36×10^6 to 3.23×10^9.[69] Nevertheless, the meta-analysis has shown insignificant improvement of the therapeutic effect with the increase in the number of transplanted cells. Thus, the issue of the lowest required and optimal therapeutic doses remains open for further investigation.

4.2.4 Cell Type

The origin/source of the donor cells is of major significance in terms of the outcome of cell therapy. From the clinical standpoint, the most frequently used cell type used in the randomized controlled trials is one of the following three types of cells: BMMNCs, PBMNCs, and BMMSCs.[70] In a study for the direct comparison of different cell types, Huang *et al.* randomly distributed 150 CLI patients between BMMNCs or PBMNCs therapies. The authors have reported that the amputation rate was low and statistically insignificant between the two groups (5.3% in the PBMNCs group vs. 8.1% in the BMMNCs group).[35] The improvement with ABI and rest pain was essentially better with the PBMNCs than with the BMMNCs. Similarly, Onodera *et al.* reported the absence of difference in the survival without amputation between the patients receiving BMMNCs and those receiving mobilized PBMNCs (20% vs 25.6% respectively).[71] Tateishi-Yuyama *et al.* grouped 22 patients with bilateral CLI to receive I/M injections of BMMNCs in one leg and PBMNCs implantation in the other leg, through a random distribution. The local BMMNCs were better than the PBMNCs therapy when compared for improvement of ABI, TcO2, rest pain and PFWD.[68] On the same note, a direct comparison of the efficacy of I/M injection of BMMNCs and BMMSCs for the treatment of diabetic CLI in 41 patients by Lu *et al.* showed that treatment with the latter was more efficient than the former, based on the improvement in ulcer healing rate, limb perfusion, PFWD, ABI, TcO2 and parameters of magnetic resonance angiography.[39] At the same time, no significant difference was observed in the pain relief, and no amputations were performed in both treatment groups. Besides, the researchers demonstrated that BMMSCs from diabetic patients secreted more VEGF, FGF-2, and angiopoietin-1 than BMMNCs in normoxic and hypoxic conditions. These growth factors facilitate differentiation of the transplanted cells into angioblasts, such as vascular ECs and vascular smooth muscle cells.[10]

It should be noted that the majority of studies comparing different types of cells, as well as meta-analyses by Xie *et al.*[43] and Rigato *et al.*,[70] include autologous cells. At the same time, it is generally believed that allogeneic MSCs may be preferred for CLI patients as compared to autologous MSCs since the former type of cells are available off-the-shelf and have been obtained from young healthy donors. This concept has been confirmed by several currently reported clinical trials.[38,72,73]

4.2.5 Characteristics of the Patient Population

CLI has several etiologies. Two basic diseases leading to CLI are arteriosclerosis obliterans (ASO) and thromboabgitis obliterans (TAO, or Buerger's disease).[74] The outcome of cell therapy in ASO patients has been somewhat inferior as compared to those observed in TAO due to lower functionality of BM cells in elderly patients who frequently have other systemic diseases as well.[58] Indeed, it has been established that the risk factors for atherosclerosis decrease the function of autologous stem and progenitor cells. The hypothesis about the relationship between the better clinical effect and the higher functionality of BMMNCs in patients with TAO as compared to ASO patients was confirmed in the clinical study by Idei et al.[58] TAO patients have a higher number of EPCs and those preserve their functionality. The authors believe that TAO patients in general have fewer risk factors by definition since TAO starts before the age of 50 and age-associated typical risk factors. The important data were obtained in respect to the stability of the achieved result. For the average observation period of 4.8 years, a different efficacy of autologous BM cell transplantation was demonstrated for ASO and TAO patients. While the primary ABI and TcO2 increase was observed in both groups of the patients, only TAO patients showed stable unchanged indices during three-year follow-up as compared to the ASO patients who suffered a gradual decline and return to the initial level.

One should also take into account the degree of ischemia in patients being included in the clinical trials for cell therapy. For example, Walter et al. demonstrated that patients with Rutherford scale Stage 6, (patients with extensive gangrene and impending amputation) did not respond to cell therapy (I/A administration of BMMNCs) as compared to the patients with Rutherford scale Stages 4–5. As Rigato et al. pointed out in their meta-analysis that up to 50% of CLI patients are not candidates for revascularization, cell therapy demonstrated its safety for all patients and the arising adverse effects were transient and mild.[70] Hence, cell therapy is the only medical option available for these patients.

4.2.6 New Technologies to Improve the CLI Therapy Efficiency

As mentioned above, the positive effects of stem and progenitor cells are mainly mediated by their paracrine action rather than by their ability to graft and differentiate post engraftment. Extracellular vesicles (EV) that are actively released from such cells contain a great number of bioactive molecules, thus playing a crucial role in the intercellular communication and performing a lot of functions in the damaged tissue regeneration,[75] in particular, in the repair of vascular damage resulting from lower limb ischemia.[76] For example, Gangadaran et al. demonstrated that EVs, obtained from murine MSCs (MSCs-EVs) carried a pronounced pro-angiogenic potential, activating VEGF (VEGFR1 and VEGFR2) receptors in murine SVEC-4 ECs after their treatment with MSCs-EV *in vitro*.[77] The downstream pathways involved in angiogenesis (SRC, AKT и ERK) are also activated. MSCs-EV administration with the addition of Matrigel *in vivo* enhanced the reperfusion and formation of new blood vessels in an ischemic limb of an experimental animal model. Similarly, microvesicles (MVs), obtained from murine EPCs (EPCs-MVs) were capable of activating the angiogenic signaling in resting ECs, via a horizontal RNA transfer and of enhancing the restoration of the ischemic hind limb in a murine model of hind limb ischemia.[78] There was substantially increased regional limb perfusion and capillary density at the site of the cell graft. The EPCs-MVs have been shown to contain pro-angiogenic microRNAs: miR-126 and miR-296.

Another novel technology for the enhancement of the CLI cell therapy efficiency is based on the use of biomaterials. Some biomaterials, primarily the injected hydrogels, have properties which not only enhance physical retention of cells at the site of the cell graft in the ischemic region but also provide an additional protective microenvironment to improve their survival. Young *et al.* used an injectable *in situ* gelatin hydrogel for I/M delivery of human adipose-derived stromal cells (ASCs).[79] After an I/M injection of hydrogel containing ASCs in a murine model of hind limb ischemia, ASCs demonstrated significantly enhanced survival for 28 days of observation and enhanced vascular density, as compared to the cells delivered without the hydrogel. Similarly, PEGylated fibrin gel was used for I/M delivery of MSCs in a rodent model of hind limb ischemia.[80] The authors demonstrated that MSCs delivery in PEGylated fibrin gel resulted in essentially enhanced formation of mature blood vessels and better functional restoration of the skeletal muscle.

Takeda *et al.* have reported nano-scaled hydroxyapatite (HAp)-coated polymer microsphere as an injectable cell scaffold (ICS) for I/M delivery of BMMNCs in a murine model of hind limb ischemia.[81] The mice used had streptozotocin-induced diabetes and mice without the streptozotocin treatment (non-diabetic) were used as controls.[81] When BMMNCs were injected alone, they were inefficient in limb preservation in diabetic mice. On the contrary, BMMNCs injection together with ICS revealed significant limb preservation in the diabetic mice as much as in mice without diabetes. The authors concluded that ICS significantly enhanced the therapeutic benefits in terms of angiogenesis in the diabetic animals due to local accumulation of proangiogenic and anti-apoptotic factors, i.e., vascular endothelial growth factor (VEGF) and fibroblast growth factor-2 (FGF2).

The results of the above-mentioned studies help us to infer the clinical potential of EVs obtained from the stem and progenitor cells for CLI therapy. As for the appropriate biomaterials, including hydrogels, they are already being used in some clinical studies as discussed in the following section.

4.3 ACTIVE OR ONGOING CLINICAL TRIALS

To date, there are 21 active or ongoing clinical trials registered at clinicaltrials.gov (as of March 2019) using cell therapy for the treatment of CLI (Table-4.2).

Table-4.2 enlists the use of autologous MNCs in clinical trials (eight clinical trials with MNCs of which six clinical trials used BMMNCs and two clinical trials used PBMNCs). The next popular type of stem cells are MSCs used in a total of six clinical trials (either autologous or allogeneic) (four clinical trials using BMMSCs, one clinical trial with Wharton's jelly-derived MSCs and one clinical trial with ASC MSCs). Pro-angiogenic progenitor cells have been used in a total of four enlisted clinical trials, and the three remaining studies used cells modified with angiogenic genes and cells in combination with biomaterials.

4.4 CONCLUSION

The cell-based therapy approach for CLI is an attractive treatment option and a promising field of medical research. Numerous randomized clinical studies have demonstrated the safety and efficacy of cell therapy which results in improvement of distal perfusion, an increase of pain-free walking distance, pain alleviation, as well as in boosting the ABI and TcO2 indices in CLI patients. However, in spite of the achieved positive results, the limb loss and lethality remain very high in CLI. In the majority of randomized control trials wherein the follow-up period reaches one year and more, the rate of amputation does not decrease as compared to a control group. Currently, there exists no established optimal protocol for cell transplantation in the CLI patients. There is a clear dearth of information regarding the optimal cell type, cell dose, route of administration and frequency of the treatment for a certain CLI stage. The criteria for the selection of the most suitable candidate CLI patients for cell therapy remain yet to be clarified. Traditional methods, mainly related to the local application of BMMSCs or PBMNCs, MSCs, or EPCs are still used in active clinical trials. Unfortunately, the standard delivery for cells of any type does not result in a significant reduction of the amputation rate in CLI. Obviously, one needs to apply principally new technologies. The use of biomaterials has significantly improved the donor cell survival and is being viewed as the most promising in the new clinical trials. Put together, these advancements warrant further multicenter randomized placebo-controlled clinical trials for their assessment in terms of safety and efficacy.

Table 4.2. Active or ongoing clinical trials on the CLI cell therapy (clinicaltrials.gov March, 2019).

No.	Title, ClinicalTrials.gov Identifier	Interventions	Delivery	Locations
Autologous Mononuclear cells (MNCs)				
1	Safety and Efficacy Study of Autologous Concentrated Bone Marrow Aspirate (cBMA) for Critical Limb Ischemia (CLI) NCT01049919	Biological: BMMNCs Other: Placebo	I/M	Central Arkansas Veterans Healthcare System, Little Rock, Arkansas, United States; University of California-Davis Medical Center, Sacramento, California, United States; University of Miami, Miami, Florida, United States and 22 more.
2	CLI Rapid Delivery by SurgWerks-CLI Kit and VXP System (CLIRST III); CT02538978	Biological: BMMNCs Other: Placebo (diluted autologous PB)	I/M	Cesca Therapeutics, Inc., California, USA
3	Use of Autologous Concentrated BM Aspirate in Preventing Wound Complications in Below Knee Amputation (MarrowCHAMP); NTC02863926	Biological: BMMNCs concentrated via the MarrowStim device (cBMA)	I/M	Indiana University School of Medicine, Indianapolis, USA
4	The Efficacy and Safety of REX-001 to Treat Ischemic Rest Pain in Subjects with CLI Rutherford Category 4 and DM; NCT03111238	Biological: REX-001 (autologous BMMNCs) Other: Placebo (diluted suspension of RBCs)	I/M	First site: Hospital Reina Sofia de Córdoba, Spain
5	The Efficacy and Safety of REX-001 to Treat Ischemic Ulcers in Subjects with CLI Rutherford Category 5 and DM; NCT03174522	Biological: REX-001 (autologous BMMNCs) Other: Placebo (diluted suspension of RBCs)	I/A	First site: Hospital Reina Sofia de Córdoba, Spain

(*Continued*)

Table-4.2.   (*Continued*)

No.	Title, ClinicalTrials.gov Identifier	Interventions	Delivery	Locations
6	Safety and Efficacy Study of Autologous BM Aspirate Concentrate for No-Option Critical Limb Ischemia (DIALEG) NCT01818310	Biological: Autologous BM Aspirate Concentrate having BM MNCs), using 3 types of applications Other: Standard treatment (surgical endovascular treatment with maximum drug therapy Group D)	I/M I/A I/V	University Hospital Ostrava, Ostrava, Czechia
7	HemaTrate in the Treatment of CLI; NCT03809494	Biological: Autologous concentrated total nucleated cells (TNCs), derived using HemaTrate Blood Filtration system Other: Saline	I/M	University Hospital Ostrava, Ostrava, Czechia; Upper Silesian Medical Centre, Katowice, Poland; St Thomas' Hospital, London, UK; Manchester Royal Infirmary, Manchester, UK; Newcastle Upon Tyne Hospital NHS, Newcastle, UK
8	Treatment of No-option CLI by G-CSF-mobilized PB-MNC; NCT03686228	Biological: PBMNCs mobilized by G-CSF Other: Standard of care	I/M	Vascular Surgery, Siriraj Hospital, Mahidol University, Bangkoknoi, Bangkok, Thailand

Mesenchymal stem cells (MSCs)

No.	Title, ClinicalTrials.gov Identifier	Interventions	Delivery	Locations
9	A Safety and Efficacy Study of Autologous BM- MSCs in CLI; NCT02477540	Biological: Cellgram-CLI (Autologous BMMSCs)	I/M	Gachon University Gil Medical Center, Incheon, South Korea
10	Allogeneic MSCs for the CLI Therapy NCT03239535	Biological: allogeneic BM MSCs Other: Normal saline	I/M	Federal Research Clinical Center of Federal Medical & Biological Agency, Moscow, Russian Federation
11	Evaluation of Tissue Genesis Icellator Cell Isolation System to Treat (CLI-DI) NCT02234778	Biological: TGI SVF material (Autologous ADS Vascular Fraction Cells)	I/M	Indiana University Hospital, Indianapolis, Indiana, USA

12	Autologous Stem Cells for the Treatment of No Option CLI; NCT03455335	Biological: 20 million hMSCs; Biological: 40 million hMSCs; Biological: 80 million hMSCs (autologous BM MSC)	I/M	Galway University Hospital, Galway City, Galway, Ireland
13	Cardiovascular Clinical Project to Evaluate the Regenerative Capacity of CardioCell in Patients with No-option CLI; NCT03423732	Drug: CardioCell (MSCs from Wharton's Jelly) Other: Placebo	I/M	The John Paul II Hospital, Cracovia, Poland
14	Allogeneic MSCs for Angiogenesis and Neovascularization in No-option Ischemic Limbs (SAIL); NCT03042572	Drug: Allogeneic BM MSCs Other: Placebo	I/M	University Medical Center Utrecht, Utrecht, Netherlands
Marker-selected or expanded cells				
15	Study to Assess the Efficacy and Safety of CLBS12 in Patients with CLI; NCT02501018	Biological: CLBS12 (autologous G-CSF-mobilized PB-derived CD34+ cells) Other: Standard of care (SOC)	I/M	Asahikawa Medical University Hospital — 1-1-1 Higashi-2jou, Midorigaoka, Asahikawa-shi, Japan; Fukuoka Sanno Hospital, Japan; Shonan Kamakura General Hospital, Japan; Kobe City Medical Center General Hospital, Japan; Shinsuma General Hospital, Kobe, Japan; Osaka Saiseikai Nakatsu Hospital, Japan; Nippon Medical School Hospital, Tokyo, Japan. Toho University Medical Center Ohashi Hospital, Japan; Tokyo Medical University Hospital, Japan; Tokyo Women's Medical University, Japan

(Continued)

Table-4.2. (*Continued*)

No.	Title, ClinicalTrials.gov Identifier	Interventions	Delivery	Locations
16	Assessment of Blood-Borne Autologous APCs Therapy in Patients with CLI; NCT02140931	Biological: APCs Other: Placebo	I/M	Vancouver General Hospital, Vancouver, British Columbia, Canada; Toronto General Hospital, Toronto, Ontario, Canada
17	Blood-Derived Autologous Angiogenic Cell Precursor Therapy in Patients with CLI; NCT02551679	Biological: APCs Other: Cell culture medium	I/M	Tibor Rubin VA Medical Center, Long Beach, California, USA; University of Florida, Gainesville, Florida, USA; Clinovation Research, LLC, Miami, Florida, USA; Clinical Research of Central Florida, Winter Haven, Florida, USA; Novant Health Heart and Vascular Institute, Charlotte, North Carolina, USA; Temple University Hospital, Philadelphia, Pennsylvania, USA; Houston Methodist DeBakey Heart & Vascular Center, Houston, Texas, USA; Clinical Trials of Texas, Inc. (CTT), San Antonio, Texas, USA; Vancouver General Hospital, Vancouver, British Columbia, Canada; Toronto General Hospital, Ontario, Canada
18	Assessing the Feasibility of BGC101 in the Treatment of PAD & CLI NCT02805023	Biological: BGC101 (autologous cell preparation composed of a mixture of cells enriched for EnEPCs) and adult HSPCs; Other: Control medium	I/M	Laniado Hospital, Netanya, Israel

Modified cells and cells combined with biomaterials

19	Safety Study of MultiGeneAngio in Patients with Chronic CLI; NCT00956332	Biological: MultiGeneAngio (ECs and smooth muscle cells harvested from the patient's vein segment, and gene modified by transfer of angiogenic genes)	I/A	Barzilai Medical Center, Ashkelon, Israel; Soroka Medical Center, Israel; Rambam Medical Center, Israel; Shaare Zedek Medical Center, Israel; Hadassah University Hospital, Ein Kerem, Israel; Kaplan Medical Center, Rehovot, Israel; Chaim Sheba Medical Center, Tel-Hashomer, Israel
20	Autologous Transplantation of BM-ECs With PRPE for Treatment of CLI; NCT02993809	Biological: BMECs and PRPE; Biological: BMECs	I/M	South China Research Center for Stem Cell and Regenerative Medicine, Guangzhou, China
21	Autologous BM-MNC Combined with Hyaluronan Therapy for PAOD; NCT03214887	Biological: RV-P1501: hyaluronan combined with autologous BMMNCs	I/M	National Cheng Kung University Hospital, Tainan, Taiwan National Taiwan University Hospital, Taipei, Taiwan

Abbreviations: ABI: Ankle-brachial index; ASCs: Adipose tissue-derived stromal cells; APCs: Angiogenic precursor cells; BMMSCs: BM-derived Mesenchymal stem cells; BM: Bone marrow; BMMNCs: bone marrow mononuclear cells; CLI: Critical limb ischemia; ECs: Endothelial cells; EPCs: Endothelial progenitor cells; HSPCs: Hematopoietic stem/progenitor cells; I/A: Intra-arterial; IM: Intramuscular; I/V: Intravenous; PBMNCs: Peripheral Blood mononuclear cells; PRPE: platelet rich plasma extract; PAD: Peripheral artery disease; RBCs: red blood cells; TAO: Thromboangitis obliterans; TcO2: Transcutaneous Oxygen Tension.

Moreover, such issues as the duration of the follow-up period for estimation of the distant results, reason for repeated clinical trial courses, assessment of risk factors and concomitant diseases must be addressed.

REFERENCES

1. Bell PF, Charlesworth D, DePalma RG. The difinition of critical ischemia of a limb. *Brit J Surg*. 1982; 69: 2.
2. Williams DT. The physiological evaluation of critical lower limb ischemia. In: *Critical Limb Ischemia* (eds.) Bosiers M, Schneider PA. New York, NY: Informa Healthcare USA, Inc., 2009, pp. 27–51.
3. Hirsch AT, Haskal ZJ, Hertzer NR, Curtis W, Bakal CW, Creager MA, Halperin JL, Hiratzka F, *et al*. Practice guidelines for the management of patients with peripheral arterial disease (lower extremity, renal, mesenteric, and abdominal aortic): a collaborative report. *Circulation*. 2006; 113: 463–654.
4. Second European Consensus Document on chronic critical leg ischemia. *Circulation*. 1991; 84(4): 16–26.
5. Lawall H, Bramlage P, Amann B. Stem cell and progenitor cell therapy in peripheral artery disease. A critical appraisal. *Thromb Haemost*. 2010; 103: 696–709.
6. Norgren L, Hiatt WR, Dormandy JA, Nehler MR, Harris KA, Fowkes FG. TASC II Working Group. Inter-society consensus for the management of peripheral arterial disease (TASC II). *Eur J Vasc Endovasc Surg*. 2007; 33(1): 1–75.
7. Guidelines for percutaneous transluminal angioplasty. Standards of practice committee of the society of cardiovascular and interventional radiology. *Radiology*. 1990; 177: 619–626.
8. Cooke JP, Losordo DW. Modulating the vascular response to limb ischemia: angiogenic and cell therapies. *Circ Res*. 2015; 116: 1561–1578.
9. Davies MG. Critical limb ischemia: cell and molecular therapies for limb salvage. *Methodist Debakey Cardiovasc J*. 2012; 8: 20–27.
10. Hart CA, Tsui J, Khanna A, *et al*. Stem cells of the lower limb: their role and potential in management of critical limb ischemia. *Exp Biol Med (Maywood)*. 2013; 238(10): 1118–1126.
11. Barc P, Skora J, Pupka A, Turkiewicz D, Dorobisz AT, Garcarek J, *et al*. Bone-marrow cells in therapy of critical limb ischaemia of lower extremities down experience. *Acta Angiol*. 2006; 12: 155–166.
12. Dash NR, Dash SN, Routray P, *et al*. Targeting nonhealing ulcers of lower extremity in human through autologous bone marrow-derived mesenchymal stem cells. *Rejuvenation Res*. 2009; 12: 359–366.
13. Asahara T, Murohara T, Sullivan A, *et al*. Isolation of putative progenitor endothelial cells for angiogenesis. *Science*. 1997; 275: 964–967.
14. Takahashi T, Kalka C, Masuda H, *et al*. Ischemia- and cytokine-induced mobilization of bone marrow-derived endo-thelial progenitor cells for neovascularization. *Nat Med*. 1999; 5: 434–438.

15. Kalka C, Masuda H, Takahashi T, *et al.* Transplantation of ex vivo expanded endothelial progenitor cells for therapeutic neovascularization. *Proc Natl Acad Sci USA.* 2000; 97: 3422–3427.

16. Kamihata H, Matsubara H, Nishiue T, *et al.* Implantation of bone marrow mononuclear cells into ischemic myocardium enhances collateral perfusion and regional function via side supply of angioblasts, angiogenic ligands, and cytokines. *Circulation.* 2001; 104: 1046–1052.

17. Murohara T, Ikeda H, Duan J, *et al.* Transplanted cord blood-derived endothelial precursor cells augment postnatal neovascularization. *J Clin Invest.* 2000; 105: 1527–1536.

18. Tolar J, Le Blanc K, Keating A, *et al.* Concise review: hitting the right spot with mesenchymal stromal cells. *Stem Cells.* 2010; 28: 1446–1455.

19. Williams JT, Southerland SS, Souza J, *et al.* Cells isolated from adult human skeletal muscle capable of differentiating into multiple mesodermal phenotypes. *Am Surg.* 1999; 65: 22–26.

20. Zuk PA, Zhu M, Mizuno H, *et al.* Multilineage cells from human adipose tissue: implication for cell-based therapies. *Tissue Engl.* 2001; 7: 211–228.

21. Gronthos S, Arthur A, Bartold PM, *et al.* A method to isolate and culture expand human dental pulp stem cells. *J Methods Mol Biol.* 2011; 698: 107–121.

22. Puissant B, Barreau C, Bourin P, *et al.* Immunomodulatory effect of human adi-pose tissue-derived adult stem cell: comparison with bone marrow mesenchymal stem cells. *Brit J Haematol.* 2005; 129(1): 118–129.

23. Prockop D. Repair of tissues by adult stem/progenitor cells [MSCs]: controversies, myths, and changing paradigms. *Mol Ther.* 2009; 17: 939–946.

24. Moon MH, Kim SY, Kim YJ, *et al.* Human adipose tissue-derived mesenchymal stem cells improve postnatal neovascularization in a mouse model of hind limb ischemia. *Cell Physiol Biochem.* 2006; 17(5–6): 279–290.

25. Yoshida M, Horimoto H, Mieno S. Intra-arterial bone marrow cell transplantation induces angiogenesis in rat hind limb ischemia. *Eur Surg Res.* 2003; 35: 86–91.

26. Rosova I. Hypoxic preconditioning results in increased motility and improved therapeutic potential of human mesenchymal stem cells. *Stem Cells.* 2008; 26: 2173–2182.

27. Frangogiannis NG. Cell therapy for peripheral artery disease. *Curr Opin Pharmacol.* 2018; 39: 27–34.

28. Prochazka V, Gumulec J, Jaluvka F, *et al.* Cell therapy, a new standard in management of chronic critical limb ischemia and foot ulcer. *Cell Transplant.* 2010; 19: 1413–1424.

29. Walter DH, Krankenberg H, Balzer JO, *et al.* Intraarterial administration of bone marrow mononuclear cells in patients with critical limb ischemia: a randomized-start, placebo-controlled pilot trial (PROVASA). *Circ Cardiovasc Interv.* 2011; 4: 26–37.

30. Teraa M, Sprengers RW, Schutgens RE, *et al.* Effect of repetitive intra-arterial infusion of bone marrow mononuclear cells in patients with no-option limb ischemia: the randomized, double-blind, placebo-controlled rejuvenating endothelial progenitor cells via transcutaneous intra-arterial supplementation (JUVENTAS) trial. *Circulation.* 2015; 131: 851–860.

31. Benoit E, O'Donnell TF Jr, Iafrati MD, et al. The role of amputation as an outcome measure in cellular therapy for critical limb ischemia: implications for clinical trial design. *J Transl Med*. 2011; 9: 165.

32. Li M, Zhou H, Jin X, et al. Autologous bone marrow mononuclear cells transplant in patients with critical leg ischemia: preliminary clinical results. *Exp Clin Transplant*. 2013; 11: 435–439.

33. Ponemone V, Gupta S, Sethi D, et al. Safety and effectiveness of bone marrow cell concentrate in the treatment of chronic critical limb ischemia utilizing a rapid point-of-care system. *Stem Cells Int*. 2017; 2017: 4137626.

34. Ozturk A, Kucukardali Y, Tangi F, et al. Therapeutical potential of autologous peripheral blood mononuclear cell transplantation in patients with type 2 diabetic critical limb ischemia. *J Diabetes Complications*. 2012; 26: 2933.

35. Huang PP, Yang XF, Li SZ, et al. Randomised comparison of G-CSF-mobilized peripheral blood mononuclear cells versus bone marrow-mononuclear cells for the treatment of patients with lower limb arteriosclerosis obliterans. *Thromb Haemost*. 2007; 98: 1335–1342.

36. Matoba S, Tatsumi T, Murohara T, et al. Long-term clinical outcome after intramuscular implantation of bone marrow mononuclear cells (Therapeutic Angiogenesis by Cell Transplantation [TACT] trial) in patients with chronic limb ischemia. *Am Heart J*. 2008; 156: 1010–1018.

37. Lee HC, An SG, Lee HW, et al. Safety and effect of adipose tissue-derived stem cell implantation in patients with critical limb ischemia: a pilot study. *Circ J*. 2012; 76(7): 1750–1760.

38. Gupta PK, Chullikana A, Parakh R, et al. A double blind randomized placebo controlled phase I/II study assessing the safety and efficacy of allogeneic bone marrow derived mesenchymal stem cell in critical limb ischemia. *J Transl Med*. 2013; 11: 143.

39. Lu D, Chen B, Liang Z, et al. Comparison of bone marrow mesenchymal stem cells with bone marrow-derived mononuclear cells for treatment of diabetic critical limb ischemia and foot ulcer: a double-blind, randomized, controlled trial. *Diabetes Res Clin Pract*. 2011; 92: 26–36.

40. Losordo DW, Kibbe MR, Mendelsohn F, et al. A randomized, controlled pilot study of autologous CD34+ cell therapy for critical limb ischemia. *Circ Cardiovasc Interv*. 2012; 5: 821–830.

41. Powell RJ, Marston WA, Berceli SA, et al. Cellular therapy with Ixmyelocel-T to treat critical limb ischemia: the randomized, double-blind, placebo-controlled RESTORE-CLI trial. *Mol Ther*. 2012; 20: 1280–1286.

42. Szabo GV, Kovesd Z, Cserepes J, et al. Peripheral blood-derived autologous stem cell therapy for the treatment of patients with late-stage peripheral artery disease-results of the short- and long-term follow-up. *Cytotherapy*. 2013;15:1245–1252.

43. Arici V, Perotti C, Fabrizio C, et al. Autologous immuno magnetically selected CD133+ stem cells in the treatment of no-option critical limb ischemia: clinical and contrast enhanced ultrasound assessed results in eight patients. *J Transl Med*. 2015; 13: 342.

44. Xie B, Luo H, Zhang Y, et al. Autologous stem cell therapy in critical limb ischemia: a meta-analysis of randomized controlled trials. *Stem Cells Int*. 2018; 2018: 7528464.

45. Goodney PP, Schanzer A, Demartino RR, *et al.* Validation of the Society for Vascular Surgery's objective performance goals for critical limb ischemia in everyday vascular surgery practice. *J Vasc Surg.* 2011; 54(1): 100–108.e4.

46. Toma C, Wagner WR, Bowry S, *et al.* Fate of cultured-expanded mesenchymal stem cells in the microvasculature: in vivo observations of cell kinetics. *Circ. Res.* 2009; 104(3): 398–402.

47. Furlani D, Ugurlucan M, Ong L, *et al.* Is the intravascular administration of mesenchymal stem cells safe? Mesenchymal stem cells and intravital microscopy. *Microvasc. Res.* 2009; 77(3): 370–376.

48. Gruenloh W, Kambal A, Sondergaard C, *et al.* Characterization and in vivo testing of mesenchymal stem cells derived from human embryonic stem cells. *Tissue Eng. Part A* 2011; 17: 1517–1525.

49. Lee RH, Pulin AA, Seo MJ, *et al.* Intravenous hMSCs improve myocardial infarction in mice because cells embolized in lung are activated to secrete the anti-inflammatory protein TSG-6. *Cell Stem Cell.* 2009; 5: 54–63.

50. Van Tongeren RB, Hamming JF, Fibbe WE, *et al.* Intramuscular or combined intramuscular/intra-arterial administration of bone marrow mononuclear cells: a clinical trial in patients with advanced limb ischemia. *J Cardiovasc Surg (Torino).* 2008; 49: 51–58.

51. Sprengers RW, Lips DJ, Moll FL, *et al.* Progenitor cell therapy in patients with critical limb ischemia without surgical options. *Ann Surg.* 2008; 247: 411–420.

52. Klepanec A, Mistrik M, Altaner C, *et al.* No difference in intra-arterial and intramuscular delivery of autologous bone marrow cells in patients with advanced critical limb ischemia. *Cell Transplant.* 2012; 21: 1909–1918.

53. Chochola M, Pytlik R, Kobylka P, *et al.* Autologous intra-arterial infusion of bone marrow mononuclear cells in patients with critical limb ischemia. *Int Angiol.* 2008; 27: 281–290.

54. Ruiz-Salmeron R, de la Cuesta-Diaz A, Constantino-Bermejo M, *et al.* Angiographic demonstration of neoangiogenesis after intraarterial infusion of autologous bone marrow mononuclear cells in diabetic patients with critical limb ischemia. *Cell Transplant.* 2011; 20: 1629–1639.

55. Bartsch T, Brehm M, Zeus T, *et al.* Autologous mononuclear stem cell transplantation in patients with peripheral occlusive arterial disease. *J Cardiovasc Nurs.* 2006; 21: 430–432.

56. Cobellis G, Silvestroni A, Lillo S, *et al.* Long-term effects of repeated autologous transplantation of bone marrow cells in patients affected by peripheral arterial disease. *Bone Marrow Transplant.* 2008; 42: 667–672.

57. Schiavetta A, Maione C, Botti C, *et al.* A phase II trial of autologous transplantation of bone marrow stem cells for critical limb ischemia: results of the Naples and Pietra Ligure Evaluation of Stem Cells study. *Stem Cells J Transl Med.* 2012; 1: 572–578.

58. Idei N, Soga J, Hata T, *et al.* Autologous bone-marrow mononuclear cell implantation reduces long-term major amputation risk in patients with critical limb ischemia: a comparison of atherosclerotic peripheral arterial disease and Buerger disease. *Circ Cardiovasc Interv.* 2011; 4: 15–25.

59. Li SH, Lai TY. Tracking cardiac engraftment and distribution of implanted bone marrow cells: comparing intra-aortic, intravenous and intramyocardial delivery. *J Thorac Cardiovasc Surg*. 2009; 137(5): 1225–1233.

60. Kinnaird T, Stabile E, Burnett MS, *et al*. Local delivery of marrow-derived stromal cells augments collateral perfusion through paracrine mechanisms. *Circulation*. 2004; 109(12): 1543–1549.

61. Amann B, Luedemann C, Ratei R, *et al*. Autologous bone marrow cell transplantation increases leg perfusion and reduces amputations in patients with advanced critical limb ischemia due to peripheral artery disease. *Cell Transplant*. 2009; 18: 371–380.

62. Powell RJ, Comerota AJ, Berceli SA, *et al*. Interim analysis results from the RESTORE-CLI, a randomized, double-blind multicenter phase II trial comparing expanded autologous bone marrow-derived tissue repair cells and placebo in patients with critical limb ischemia. *J Vasc Surg*. 2011; 54: 1032–1041.

63. Molavi B, Zafarghandi MR, Aminizadeh E, *et al*. Safety and efficacy of repeated bone marrow mononuclear cell therapy in patients with critical limb ischemia in a pilot randomized controlled trial. *Arch Iran Med*. 2016; 19(6): 388–396.

64. Franz RW, Parks A, Shah KJ, *et al*. Use of autologous bone marrow mononuclear cell implantation therapy as a limb salvage procedure in patients with severe peripheral arterial disease. *J Vasc Surg*. 2009; 50: 1378–1390.

65. Franz RW, Shah KJ, Johnson JD, *et al*. Short to mid-term results using autologous bone-marrow mononuclear cell implantation therapy as a limb salvage procedure in patients with severe peripheral arterial disease. *Vasc Endovasc Surg*. 2011; 45: 398–406.

66. Strauer BE, Brehm M, Zeus T, *et al*. Repair of infarcted myocardium by autologous intracoronary mononuclear bone marrow cell transplantation in humans. *Circulation*. 2002; 106: 1913–1938.

67. Gupta R, Losordo DW. Cell therapy for critical limb ischemia. moving forward one step at a time. *Circulation: Cardiovasc Interventions*. 2011; 4: 2–5.

68. Tateishi-Yuyama E, Matsubara H, Murohara T, *et al*. Therapeutic angiogenesis for patients with limb ischaemia by autologous transplantation of bone-marrow cells: a pilot study and a randomized controlled trial. *Lancet*. 2002; 360: 427–435.

69. Ai M, Yan CF, Xia FC, *et al*. Safety and efficacy of cell-based therapy on critical limb ischemia: a meta-analysis. *Cytotherapy*. 2016; 18(6): 712–724.

70. Rigato M, Monami M, Fadini GP. Autologous cell therapy for peripheral arterial disease: systematic review and meta-analysis of randomized, nonrandomized, and non-controlled studies. *Circ Res*. 2017; 120(8): 1326–1340.

71. Onodera R, Teramukai S, Tanaka S, *et al*. Bone marrow mononuclear cells versus G-CSF-mobilized peripheral blood mononuclear cells for treatment of lower limb ASO: pooled analysis for long-term prognosis. *Bone Marrow Transplant*. 2011; 46: 278–284.

72. Wijnand JGJ, Teraa M, Gremmels H, *et al*. Rationale and design of the SAIL trial for intramuscular injection of allogeneic mesenchymal stromal cells in no-option critical limb ischemia. *J Vasc Surg*. 2018; 67(2): 656–661.

73. Orekhov PY, Chupin AV, Konoplyannikov MA, *et al*. Transplantation of allogeneic mesenchymal stem cells for the treatment of critical lower limb ischemia. *Eur Cell Mater*. 2016; 31(1): 289.

74. Weinberg I, Jaff MR. Nonatherosclerotic arterial disorders of the lower extremities. *Circulation*. 2012; 126: 213–222.

75. Chen B, Li Q, Zhao B, *et al*. Stem cell-derived extracellular vesicles as a novel potential therapeutic tool for tissue repair. *Stem Cells Transl Med*. 2017; 6(9): 1753–1758.

76. Todorova D, Simoncini S, Lacroix R, *et al*. Extracellular vesicles in angiogenesis. *Circ Res*. 2017; 120: 1658–1673.

77. Gangadaran P, Rajendran RL, Lee HW, *et al*. Extracellular vesicles from mesenchymal stem cells activates VEGF receptors and accelerates recovery of hind limb ischemia. *J Control Release*. 2017; 264: 112–126.

78. Ranghino A, Cantaluppi V, Grange C, *et al*. Endothelial progenitor cell-derived microvesicles improve neovascularization in a murine model of hind limb ischemia. *Int J Immunopathol Pharmacol*. 2012; 25(1): 75–85.

79. Young SA, Sherman SE, Cooper TT, *et al*. Mechanically resilient injectable scaffolds for intramuscular stem cell delivery and cytokine release. *Biomaterials*. 2018; 159: 146–160.

80. Ricles LM, Hsieh PL, Dana N, *et al*. Therapeutic assessment of mesenchymal stem cells delivered within a PEGylated fibrin gel following an ischemic injury. *Biomaterials*. 2016; 102: 9–19.

81. Takeda K, Fukumoto S, Motoyama K, *et al*. Injectable cell scaffold restores impaired cell-based therapeutic angiogenesis in diabetic mice with hind limb ischemia. *Biochem Biophys Res Commun*. 2014; 454(1): 119–124.

CHAPTER 5

Stem cell-derived paracrine factors modulate cardiac repair

Sadia Mohsin*, Mohsin Khan†
*Center for Cardiovascular Research
†Center for Metabolic Disease Research
Lewis Katz School of Medicine, Temple University,
Philadelphia, USA

ABSTRACT

Heart disease is the primary cause of mortality and morbidity in the world. Existing therapies limit the extent of injury without structural restoration of the lost myocardial tissue. Consequently, injured myocardial tissue continues to remodel that ultimately leads to cardiac failure. Cell therapy provides a promising alternative for enhancement of cardiac structure and function, yet the mechanism explaining salutary effects remain elusive. Ability of the transplanted stem cells to adopt cardiac cell morphology dubbed as the "Transdifferentiation Hypothesis" is widely believed to be the mechanism for cell therapy. Recently, however multiple studies provide evidence that challenges the transdifferentiation hypothesis, thereby questioning the ability of transplanted stem cells to differentiate into tissue cell types. Alternatively, stem cells secrete growth factors, proteins and extracellular vesicles including exosomes that possess cardioprotective and regenerative properties, thereby forming the "Paracrine Hypothesis". This chapter aims to summarize the cell therapy and its applications for cardiac repair and regeneration including mechanisms explaining beneficial effect of the transplanted stem cells. A particular focus will be on the emerging importance of the paracrine hypothesis and its future implication for cardiac tissue repair after injury.

KEYWORDS

Cardiomyocytes; Embryonic; Exosome; miRNA; MSCs; Paracrine; Stem cells.

---

*Corresponding author. Email: sadia.mohsin@temple.edu

LIST OF ABBREVIATIONS

ANP	=	Atrial natriuretic peptide
ASCs	=	Adult stem cells
CPCs	=	Cardiac progenitor cells
CSCs	=	Cardiac stem cells
ESCs	=	Embryonic stem cells
ESCRT	=	Endosomal sorting complex required for transport
GCSF	=	Granulocyte colony-stimulating factor
GSK3β	=	Glycogen synthase kinase 3β
HIF	=	Hypoxia inducible factor-1
hnRNPA2B1	=	Heterogeneous nuclear ribonucleoprotein (A2B1)
HSC70	=	Heat shock cognate 70
iPSCs	=	Induced pluripotent stem cells
IVF	=	In vitro fertilization
LBPA	=	Lysobisphosphatidic acid
LIF	=	Leukemia inhibitory factor
MHC	=	Major histocompatibility complex
miRNA	=	microRNA
MSCs	=	Mesenchymal cells
MVBs	=	Multivesicular bodies
nSMase 2	=	Neutral sphingomyelinase 2
TNF	=	Tumor necrosis factor

5.1 INTRODUCTION

The concept of stem cell-based research and medical applications has existed for a relatively short time, but has experienced an exciting journey that has been a popular point of discourse in public forums. First characterized in mice in the early 1980's, stem cells quickly became a heavily debated field by 1998 when James Thomson, funded by the Geron Corporation, proved that it was possible to derive stem cells from human embryonic blastocysts By 2006, Shinya Yamanaka and his team in Japan discovered that it was possible to reprogram cells that have already differentiated into adult fibroblasts into a pluripotent embryonic-like state by introducing four transcription factors, Oct3/4, Sox2, c-Myc, and Klf4. Newspapers and optimists heralded this discovery as the dawn of a new form of medicine that in theory could provide quick and easy regeneration of any of the body's two hundred cell types through simple transplantation of stem cells. In this chapter, we will realize that this concept: transplanted stem cells naturally differentiating into the host organ's cells, known as the "Transdifferentiation Hypothesis", is more challenging than previously envisioned. Alternatively, stem cells possess the ability to secrete regenerative paracrine factors mediating their repair, thereby constituting the "Paracrine Hypothesis".

By definition, stem cells are classified as the cells possessing the ability to self-renew and undergo an indefinite number of cell divisions while still maintaining an undifferentiated state. Secondly, stem cells exhibit some degree of potency or the capacity to differentiate into a specialized cell type. Totipotent stem cells such as the morula and blastocyst are considered totipotent because they can form all cell types, including extraembryonic cells like the umbilical cord and placenta. During embryogenesis the embryoblast forms, containing inner mass cells that are pluripotent and give rise to all cell types composing the eutherian. These pluripotent stem cells further divide during embryogenesis into three germ layers ectoderm, mesoderm, and endoderm. Pluripotent stem cells of the mesoderm, the middle layer of the embryo, will eventually give rise to muscle, bone, and organ tissue including the heart.

Following embryonic development into fully developed organs and tissue systems, the body retains multipotent adult stem cells in dedicated areas known as stem cell niches. Multipotent stem cells such as hematopoietic bone marrow stem cells have been shown to further produce oligopotent stem cells like myeloid and lymphoid lineages of blood cells, where there are even more restrictions to the potency of daughter cells. For example, an oligopotent lymphoid cell can give rise to new B and T cells but there is no evidence that it is able to differentiate into a red blood cell. In summary, the overall theme of the chapter is to develop an in-depth understanding of stem cell biology, its application for tissue regeneration including limitations and future challenges. A particular focus will rest on mechanisms explaining stem cell salutary effects in the context of cardiac regeneration.

5.2 EMBRYONIC STEM CELLS

Embryonic stem cells (ESCs) are derived from the inner cell mass of the blastocyst from a developing embryo. ESCs are pluripotent i.e. that can grow and differentiate into cell types that are derived from all three primary germ layers i.e. ectoderm, endoderm, and mesoderm. This feature distinguishes them from adult stem cell types. They can essentially form all 220-cell types in the adult body. Adult stem cells compared to ESCs are multipotent meaning they can produce only a limited number of cell types. Another important characteristic of stem cells is self-renewal. ESCs are capable of self-propagating under certain conditions. Because of their plasticity and unlimited self-renewal, they are an important tool to study different disease conditions. Some of the diseases that could potentially be treated with ESCs include several blood and immune disorders, diabetes, blindness, and spinal cord injuries. Other potential uses of ESCs include investigating embryo development, different genetic disorders and *in vitro* toxicology testing. ESCs from human embryos are developed from eggs that have been fertilized (*in vitro*) in an IVF clinic and then donated for research after obtaining informed donor consent.

5.2.1 Pathway from ESCs to Cardiac Cells

Human ESCs originate from the inner cell mass of the blastocyst within the first week of post-coitum embryo. Specialized transcription factors control the pluripotency and differentiation of these ESCs. *Sox2* and *Oct4* function together to maintain the pluripotent state throughout early development.[1] *Sox2* is required to maintain cells of the epiblast in an undifferentiated state. Reduction of *Sox2* leads to epiblast cells differentiating into the trophoectoderm and extraembryonic endoderm, while embryonic cells lacking *Oct4* can only form trophoectoderm derivatives.[2] Similarly, another transcription factor *Nanog* persists through inner cell mass formation into the formation of the primitive ectoderm and ensures proliferation in a pluripotent state.[3] The ectoderm expresses other pluripotent cell markers such as SSEA1, but the expression of *Fgf5, Gbx2,* and *Rex1* is lost, modifying differentiation capabilities. *In vitro* experiments have shown that these ectoderm cells will rapidly revert to primitive inner cell mass cells when leukemia inhibitory factor (LIF) is introduced to their growth media.[4] This shows that early differentiation events are not terminal and hence, pluripotent cells can adopt distinct cell states in response to specific extracellular factors, which is an important argument for the Paracrine Hypothesis, something we will discuss later in this chapter.

Shortly after endoderm cell differentiation from ESCs, gastrulation initiates. The mesoderm begins rapid production of three major types of mesoderm cells: organizer, embryonic mesoderm, and extraembryonic mesoderm. The mesoderm arises in two major waves of differentiation. The first wave generates *Flk-1+ Pdgfra+* lateral plate mesoderm population, which contributes primarily to hematopoietic and endothelial cell lines. The second wave will give rise to *Flk-1+ Pdgfra–* cells, which eventually form the cardiac mesoderm cell line. Mesoderm Posterior BHLH Transcription Factor 1, *Mesp1,* is an essential factor in the development of cardiac cells from mesoderm precursors, and is expressed in almost all precursors of the cardiovascular system.[5] *Mesp1* plays a key role in the early specification for cardiac precursor cells, and is critical in identifying cardiac cell progenitors, but Mesp1 expression is transient, and primarily restricted to cells within and immediately egressing from the primitive streak.[6] We have discussed earlier how the presence or absence of essential transcription factors defines the proper differentiation and progression into other pluripotent cell lines. In the case of Mesp1, expression occurs in both *Pdgfra+* and *Pdgfra–* cells, but expression at different times during development causes the formation of different cell lines. Therefore, it is not the overall presence or lack thereof of Mesp1 in these cells, but the temporal expression that dictates further differentiation.[5]

In an *in vitro* experiment where *Mesp1* expression is controlled by an doxycycline-inducible promoter in a mouse embryoid body model, scientists were able to show that the timing of expression for *Mesp1* dictates the type of cell produced. Two days after embryoid body formation, inducing *Mesp1* will not result in cardiomyocyte production but increases *Flk-1+ Pdgfra+* hematopoietic progenitor populations. But by waiting just one more day to induce Mesp1, *Flk-1+ PDGFRα–* cardiovascular mesoderm cells appear within 24 hours, and *cTnT+* (cardiac troponin T) cardiomyocytes are observed by the eighth day. Significant upregulation of hematopoietic-specific genes, such as *Gata1, Sfpi1, Runx1*, and *Hbb-bh1*, occur by early induction of Mesp1 but not by the later induction. In contrast, late induction upregulated cardiac-specific markers, including *Myl7, Myl2, cTnT, Nkx-5, Tnni3*, and *sarcomeric α-actinin*.[6] Embryos that lack *Mesp1* expression entirely will have defective mesodermal migratory activity, defective primitive streak formation, head fold formation, and aberrant heart morphogenesis resulting in cardia bifida.[5,7]

The lateral plate mesoderm will give rise to the parietal body wall, the cells that differentiate to form the heart. As this heart-forming region produces the cardiac crescent, a plate of pro-myocardial cells intermingles with endothelial strands. Pro-myocardial cells can be determined by the expression of *Nkx-2.5*, a master cardiac development gene in the *Bmp-Nkx-2.5-Mef2c* pathway, located in the cardiac crescent.[8,9] By the second month after fertilization, the developing heart is now known as the primary linear tube and resembles an inverted Y where the two arms form the atrial chambers and the stem becomes the left ventricle.[8] Recently it has been found that the LIM homeodomain transcription factor islet-1, *Isl1*, is essential for cardiac progenitor cell line differentiation and heart development. *Isl1* is a marker for a specific set of cardiac progenitors that contribute to the general area of right heart development, but note that there is a second population of cardiac progenitor cells that do not require *Isl1* for proper signaling and differentiation.[10] *Isl1* promotes the demethylation of histone H3K27me3 and promotes a signaling cascade involving core cardiac transcription factors, *Myocd*, and myocyte enhancer factor-2, *Mef2c*.

Prior to the fifth week in humans, cells from the second cardiac crescent migrate into the heart tube where they contribute to the development of the cranial pole and will contribute to the outflow tract and early development of the right ventricle. *Mef2c* expression is directly required for proper looping morphogenesis of the linear tube in the primitive heart; leading to the failure to form the right ventricular region.[11] After the completion of looping, the primary heart tube within the pericardial cavity can be divided into atrial and ventricular components with an outflow tract. As development continues, pronounced changes occur in all parts of the tube to help create separate left and right components. The left and right ventricles will originate from the ventricular loop of the primary heart tube[12]. The new myocardium can be distinguished from the primary myocardium by its expression of atrial natriuretic peptide (ANP), which is an important secreted protein regulator in the secretome, which we will discuss later on. This peptide is a very important vasodilator in the fully developed human heart. Ventricles undergo a more asymmetric development process in comparison to the development of the atria.

5.3 INDUCED PLURIPOTENT STEM CELLS

5.3.1 What is an iPSC? / How to Make an iPSC

Yamanaka and his team were able to prove their induced somatic cells entered a pluripotent state through resistance to G418 reporter system. Each of Yamanaka's twenty-four candidate genes were introduced into this cell line through retroviral transduction. What they discovered was that somatic cells with induced expression of *Oct3/4, Sox2, c-Myc,* and *Klf4* will produce G418 resistant cells. The first induced pluripotent stem cell was discovered.[13] Just a year later, it was shown that while *Oct3/4* and Sox2 were essential to induce a pluripotent state, the induction of *c-Myc* and *Klf4* could be replaced with *Nanog* and *Lin28* expression. Both induction schemas produced pluripotent cells that have normal karyotypes, exhibited morphology similar to ESCs, including the phenotypic round shape with large nucleoli, scant cytoplasm, telomerase activity, expression of cell surface markers, and proliferation properties that maintain embryoid or teratoma formation potential, the developmental potential to differentiate into advanced derivatives of all three primary germ layers.[13,14] By 2008, it was shown that scientists could induce the pluripotent cell state without genome integrating viruses and induction of either four transcription factors using transiently expressing adenoviruses.[15]

Induced pluripotent stem cells appear to be indistinguishable from ESCs based on the pluripotent ability to form embryoid bodies and teratomas. While induced pluripotent cells are quite similar to their embryonic counterparts, regardless of origin, they possess a recurrent gene expression signature different from embryonic cells that extends to miRNA expression and epigenetic factors. Current research suggests that signature gene expression differences are due to the reprogramming factors differentially binding to promoter regions.[16] The conversion of somatic cells into pluripotent stem cells via overexpression of reprogramming factors involves epigenetic remodeling. Furthermore, induced pluripotent cells have DNA methylation patterns that are substantially different at CpG islands in all known cell lines to date compared to ESCs and cancer lines. Differentially methylated regions in pluripotent cells are significantly enriched in tissue-specific and cancer-specific regions and the change in methylation of CpG islands in pluripotent cells occur in areas broadly involved in tissue differentiation, epigenetic reprogramming, and cancer. Whether different sets of reprogramming factors influence the type and extent of aberrant DNA methylation in induced pluripotent cells differently remains unknown. DNA methylation at a significant proportion of CpG sites in induced pluripotent stem cells differs from that of ESCs, and amongst different induced pluripotent lines. For instance, reprogramming with Yamanaka factors mainly failed to demethylate CpGs, but reprogramming with Thomson factors, resulted in failure to methylate CpGs. Differences in the level of transcripts encoding *DNMT3b* and *TET3* between Yamanaka's and Thomson's pluripotent cells may contribute partially to the distinct types of aberrations. For example, aberrantly methylated genes in Yamanaka induced pluripotent stem cells were enriched for *Nanog* targets that are also aberrantly methylated in some cancers, which were absent in Thomson lines. Therefore, the method of reprogramming influences the level, loci, and type of DNA methylation aberrations in induced pluripotent cells and pluripotent cells should be considered a unique subtype of pluripotent cells rather than a facsimile of its embryonic counterpart.[17]

5.3.2 Murine Model

With the advent of induced pluripotent stem cells (iPSCs) came a new field of potential regenerative medicine. For example, it has been shown that retrovirally induced pluripotent cells from adult fibroblasts could be cultured and transplanted into the eyes of immune-compromised retinal degenerative mice. These cells would proceed to form teratomas containing tissue comprising of all three germ layers. Within a month after sub-retinal implantation, a large portion of said cells implanted in the retinal outer nuclear layer and gave rise to increased electro retinal-function.[18] Another example would be in spinal cord injury mouse models, induced pluripotent cells can give rise to neurospheres that could be grafted and promote functional recovery accompanied by stimulated synapse formation just months after transplant. However, long-term observation of these mice reveals deterioration in function accompanied by tumor formation. These tumors usually consist of undifferentiated neural cells and exhibited activation of *OCT4*.[19] Similarly, experiments in the myocardial infracted rat and mouse model showed that induced pluripotent stem cells differentiate into functional cardiomyocytes, but long-term studies show the auto-generation of heterogeneous tumors at both intra-cardiac and extra-cardiac sites. Tumorigenesis is independent of cell dose, transplant duration, and the presence or absence of myocardial infarction. Further studies have shown that teratoma formation across long term studies is unavoidable, even in autologous stem cell lines.[20,21] Undifferentiated ESCs also consistently formed cardiac teratomas in immunocompetent mice, regardless of ESCs transplanted into infarcted hearts or non-infarcted hearts. This further indicates that myocardial injury per se does not regulate signaling with direct effect on proper differentiation.[22]

5.3.3 Human Model

The very transcription factors that are upregulated to induce pluripotency from somatic cells are upregulated in many human cancer lines. In this section, we will briefly describe some of the transcription factors that are important for pluripotency but are known modulators of cancers as well. Normally, differentiated cells lose expression of *Oct3/4*, and it is expressed in only a few cells found in the basal layer of the epidermis. But ESCs and ASCs, as well as cancer stem cells maintain expression of *Oct3/4*.[23] Oct4 is one of the highly upregulated factors mediating self-renewal and differentiation found in cancer stem cell lines as well as other malignant cells of the bladder, prostate, colon, and cervix.[24-26] *Sox2*, another pluripotent marker, is the most upregulated transcription factor in cancer stem cells of squamous skin tumors in mice. While *Sox2* is absent in healthy epidermal cells, it is expressed in the majority of pre-neoplastic skin tumors and invasive mouse and human skin squamous cell carcinoma. Ablation of the *Oct3/4* or *Sox2* expressing cells within primary benign or malignant carcinomas leads to tumor regression or conditional deletion of either transcription factor will decrease the ability of cancer cells to propagate if transplanted, showing the critical role for *Oct3/4* and *Sox2* expression to maintain tumorigenesis.[27,28]

While *Oct3/4* and *Sox2* overexpression is essential to induce pluripotency in somatic cells it must be coupled with either *c-Myc* and *Klf4*, or *Nanog* and *Lin28*. *c-Myc* occupies a large number of genomic target genes that are involved predominantly in cellular metabolism, cell cycle, and protein synthesis pathways, whereas the targets of core factors relate toward developmental and transcription-associated processes. Inactivation of *c-Myc* is sufficient to induce sustained regression of some invasive cancers. *c-Myc* functions as an oncogene in cancer lines and its inactivation has the potential to reverse tumorigenesis. This displays the pluripotent capacity of tumors to differentiate into normal cellular lineages and tissue structures, while retaining the potential to become cancerous, in a similar fashion to that of the reversal of somatic cells to a pluripotent stem cell state.[29]

The role of *Klf4* and its other family members in oncogenesis and pluripotency has not been fully parsed, and there have been many incidences that indicate opposite roles for *Klf4* by using different cell models.[30] *Klf4* in some cases exerts anti-apoptotic effects, while in myeloid leukemia cells, *Klf4* promotes hydrogen peroxide-induced apoptosis. Using *in vivo* colon cancer models, scientists found that overexpression of *Klf4* induces G1/S cell cycle arrest, thereby reducing the tumor formation and cell migration.[31] This puzzlement may be due to *Klf4* possessing roles in multiple parallel pathways.

The homeobox domain transcription factor Nanog is a key regulator of embryonic development, and is broadly expressed in many human cancers as well due to its pro-tumorigenic properties. Like other transcription factors involved in maintaining potency, Nanog is found at low levels in differentiated somatic cells and silencing in stem cells induces differentiation.[32] Expression of Nanog has been reported in many human cancer lines, including brain carcinomas, breast, cervix, colon, gastric, head and neck, liver, lung, kidney, oral cavity, ovary, pancreas, and prostate. Like the other stem cell transcription factors we mentioned, the expression level of Nanog is positively correlated with treatment resistance and poor survivability. Scientists have shown that inducing Nanog expression enhances tumorigenicity *in vivo* and *in vitro* but ablation or inhibition of Nanog has anti-tumorigenic properties.[33]

Lin28, a conserved RNA binding protein has also been found in primary human cancers and cancer stem cells. *Lin28* is a Let-7 microRNA antagonist and is well documented for controlling the timing of embryonic development. *Lin28* is expressed at high levels throughout the embryo at early developmental stages and declines through development and differentiation. *Lin28* functions as an oncogene that drives tumorigenesis by suppressing Let-7. Knocking out *Lin28a* and *Lin28b* prevents *c-myc*-driven hepato-carcinogenesis, indicating that these transcription factors work together to induce pluripotency and oncogenesis, tying cancer and stem cell potency even closer together.[34] The tight link between the transcription factors responsible for inducing pluripotency as well as oncogenesis have created formidable barricades for stem cells to be used in transplant-based therapies for organ damage due to incomplete differentiation of the reprogrammed cells into the target tissue and eventual tumorigenesis.

5.4 ADULT STEM CELLS

The adult stem cells (ASCs) are undifferentiated cells that are found in a niche among differentiated cells in a tissue. The ASCs, ESCs, can self-renew and differentiate into major specialized tissue types. The primary role of ASCs is to maintain and/or repair the resident tissue. ASCs are also known as somatic stem cells, referring to their origin from cells of the body instead of germ cells. ASCs have opened a new era of research as they have now been discovered in many organs. Additionally, there are no major ethical concerns raised with the use of adult stem cells as opposed to ESCs. Research on adult stem cells began in the 1960s when two types of stem cells were discovered within the bone marrow known as hematopoietic stem cells and mesenchymal stem cells. Now there is a growing consensus that ASCs are also present in organs that were previously thought to be post-mitotic, including the heart and brain. This is one of the major discoveries of the era as it brings hope to treat patients with cardiac and brain injuries.

Despite the major advances in ASCs research there are still questions left to be answered. How many kinds of ASCs are present, and what are their relative numbers in the tissue of residence? How are stem cells maintained during organ development? Are they the left-over stem cells from organ development? What triggers their differentiation during the injury? Finding answers to some of the key questions can lead to much needed breakthrough discoveries in the field.

5.4.1 Differences between ESCs and ASCs

As discussed previously, epiblast cells differentiate into specialized populations of fetal stem cells during embryonic development. Further into development these fetal stem cells are replaced by ASCs with even more limited potency. Not only do ASCs have higher attenuated potency compared to their ESCs counterparts, but they are mostly quiescent and have a much slower proliferative rate. However, ASCs will generate transient populations of progenitor cells that regain the rapid proliferation rate of ESCs.

The coordination of *Oct4, Nanog, Sox2*, and *Bmi1*, as well as other transcription factors, directly determines the epigenetic state that activates or represses differentiation into the three germ layers or target cells in ESCs. Aberrant activation and high levels of expression of these transcription factors are found in cancer cell lines and have been used as markers for the identification of tumors. Contrary to ESCs, expression of many of these known transcription factors have only been observed at very low levels in ASCs lines like cardiac progenitor cells.[35] This observation indicates that maintaining pluripotency in adult stem cells is controlled via separate pathways and remains an active field of research.

Nearly all stem cell purification strategies involve purification for cell surface markers c-Kit and Sca-1 and selecting against markers of differentiated lineages that are indicative of somatic cells originating from the given cell source. For example, hematopoietic stem cells can be selected against markers for mature hematopoietic cell lineages (B220, CD4, CD8, Gr-1, Mac-1, and Ter-119). A purified c-Kit+Lin-Sca-1+(KLS) culture is enriched for pluripotent activity in hematopoietic cells, but the bone marrow compartment contains progenitor cells in addition to self-renewing hematopoietic stem cells.[36] Note, only around 10% of KLS cells will be true hematopoietic stem cells. There exists a variety of strategies to enrich bone marrow and other organ samples for ASCs, with or without the KLS surface marker.[37]

5.5 TRANSDIFFERENTIATION AS BASIS FOR CELLULAR TURNOVER IN THE HEART

The adult heart is largely dormant and injury to the myocardium is followed by fibrosis and scar tissue. Cell-based therapies have been proposed as a promising treatment strategy for cardiac repair and regeneration. Three main pathways explain the salutary effects of stem cell transplantation that lead to new cardiomyocyte generation: differentiation of ASCs or progenitor cells, division of preexisting mature cardiomyocytes, and amplification of dedifferentiated cardiac myocytes. Evidence, to an extent, exists for all three processes. It may be possible that more than one of these mechanisms can operate simultaneously or at different developmental and pathological stages. But, the overall turnover rate of cardiac myocytes in the normal human adult heart though is vanishingly low. The approximate turnover rate of cardiomyocytes in the young adult is only 1% and drops to just 0.3% by the time someone reaches 70 years of age. This was proven through an elegant carbon dating assay, made available by nuclear armament and testing. The genomic DNA of cells generated during the pulse or the chase reflects the earth's atmospheric 14C concentration at that given point in time. Scientists determined the age of cardiac myocytes by measuring the concentration of 14C in their nuclei and showed that the adult heart contains a modest number of cardiomyocytes generated during the human adulthood.[38]

Cardiomyocytes frequently synthesize genomic DNA during S-phase without passing through mitosis. This results in a quarter to half of all post-natal human myocytes to become bi-nucleated. Most myocytes are tetraploid or octoploid by human adulthood. Polyploidization is not correlated to age per se but is positively correlated with hypertrophy. Hypertrophic hearts, and even more so if coupled with interstitial fibrosis, arteriosclerosis, hypertension, or other cardiac-related ailments, can have over twice as much polyploidization when compared to the normal adult heart.[39] Interestingly, mitotic checkpoint kinase, *Mps1*, in zebrafish is involved in cardiomyocyte regeneration and the elimination of scar formation. Following removal of 20% of the heart through transection of the ventricular apex in zebrafish, pre-existing cardiac myocytes adjacent to the site appear to undergo a process of dedifferentiation. Observable dissolution of sarcomeric structures and newly generated cardiac myocytes are functionally integrated with the preexisting myocardium and the heart is left with little residual evidence of the injury; it has been shown that this is done through an evolutionarily conserved *Mps1*-based manner. Downregulation of *Mps1* in zebrafish resulted in increased scarring and fibrosis.[40]

Studies have demonstrated that mature human cardiomyocytes also reenter the cell cycle. Cultures of cardiac tissue from human biopsies have shown that cell dedifferentiation characterized by the loss of sarcomeric structure, condensation, and extrusion of sarcomeric proteins is possible in myocytes with two distinct morphologies. Type I cells are larger and have extensive sarcolemma spreading, with stress fibers and nascent myofibrils. Type II cells appeared smaller, with more mature myofibril organization and focal adhesions. These cells did not require viral induction but morphogenesis was completely dependent upon the extracellular culture.[41] Non-coding RNA, microRNA, and large intergenic non-coding, lncRNA, modulates the transcriptional network of stem cells and induction of reprogramming.[42] The miR302/367 clusters can rapidly induce mouse and human fibroblasts pluripotent stem cells and avoid the use of transcription factors by targeting hundreds of ESCs related mRNAs directly. The mechanism underlying reprogramming by miRNA differs from that of transcription factor induced reprogramming in that there is no requirement for protein translation or viral induction of *Oct4, Sox2, Klf4,* and *Myc*.[43] Differentiation is not a terminal process and this indicates that differentiated cells can alter their cell fate without reverting to a pluripotent state and these morphological changes can be induced without forced overexpression of genes through viral induction but simply by exploiting the innate cellular ability to respond to the extracellular environment.

It has been found that the adult heart maintains a subpopulation of Lin⁻c-kit⁺ cardiac stem cells (CSCs) supporting cellular turnover under physiological and pathological states. CSCs are self-renewing, clonogenic, and multipotent. When injected into a heart with myocardial ischemia, CSCs give rise to myocytes, smooth muscle, and endothelial cells. CSCs and their clonal progeny integrate with existing myocardial cells and form new blood vessels.[44] Moreover, heart transplantation in male patients from female donors shows chimerism through observation of the male Y-chromosome, providing further evidence of cellular turnover in the adult heart. Undifferentiated cells expressing surface stem cell related antigens like c-kit, MDR1, and Sca-1 were Y-chromosome positive in the transplanted female heart in less than a month post-transplant. Therefore, primitive cells translocated from the male host into the atria and ventricles. Loss of stem-cell markers and further proliferation coupled with differentiation into adult cardiac cells followed colonization to form new myocytes, coronary arterioles, and capillaries. Within the first year, female transplanted hearts into male hosts possess anywhere between 10–20% of myocytes, arterioles, and capillaries of male origin.[45] Similarly, following infarction, Lin−c-kit+ CD34+ adult stem cells and hematopoietic stem cells from bone marrow can migrate to the damaged areas of the heart in response to secreted stem cell factors and granulocyte colony-stimulating factor. This cytokine-mediated translocation of bone marrow stem cells has been observed in the human heart within the first month following myocardial infractions.[46] Factors involved in fetal cardiac development like Mef2D, Gata4, nestin, and Flk1 have been found in the migrated progenitor cells.[46] Further long-term studies were unable to show a correlated increase in ejection fraction following bone marrow stem cell injection therapy scheme when treating myocardial infarction, questioning the efficiency of differentiated stem cells in the heart.

5.6 PARACRINE HYPOTHESIS FOR CARDIAC REPAIR

While the transdifferentiation hypothesis focused on the stem cells themselves to differentiate into the targeted tissue to mediate organ repair, the paracrine hypothesis includes a growing amount of data suggesting that cells, including stem cells, can mediate repair via excretion of exosomes containing various factors like cytokines, chemokines, growth factors, various proteins, microRNAs, and messenger RNAs that indirectly regulate mechanisms involved in recruiting endogenous progenitors, inducing angiogenesis, downregulating fibrosis and inflammation (Figure-5.1).

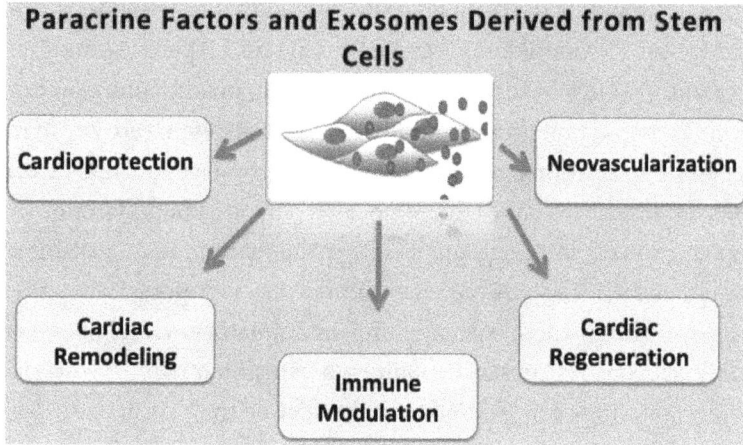

Figure-5.1.   Paracrine factors and exosomes derived from stem cells target multiple biological processess to augment repair.

5.6.1 Cytokines

Cytokines are a broad-based category of ~5–20 kDa immunomodulating proteins that have a specific effect on inter- and intra-cellular communication. The first cytokine was described in 1972 as a soluble mitogenic protein found in cultured lymphocyte cell media able to stimulate human T-cell growth.[47] The field has quickly grown since then and now involves chemokines, interferons, interleukins, lymphokines, and tumor necrosis factors regulating autocrine signaling, paracrine signaling and endocrine signaling. The distinction between cytokines to hormones and growth factors is that a cytokine may be produced by more than one type of cell rather than a specialized or discrete cell type, unlike hormone counterparts. Secondly, hormones circulate in nanomolar concentrations, while cytokines circulate in picomolar concentrations and potentially increase a 1000-fold during trauma or infection.

Important to note, in human HIV-1 models, it was shown that cytokines are not free-floating in the blood plasma but are associated with exosomes and are distinct from virions.[48] Hematopoietic cytokines are known to influence cellular proliferation, differentiation, maturation, and lineage commitment in bone marrow, such as granulocyte colony-stimulating factor (G-CSF), granulocyte-macrophage colony-stimulating factor, stem cell factor, Flt-3 ligand, and erythropoietin have been found to also confer cardiovascular benefits in the myocardial infarction and heart failure model. This cytoprotective and the cardioprotective effect is due to the cytokines directing mobilization and homing of bone marrow-derived stem cells into the infarcted heart and subsequently inducing myocardial repair.[49]

Secretion of angiotensin II and atherogenic factors have been found to interact with transcription factors like nuclear factor-kappaB or peroxisome proliferator-activated receptors. This, in turn, induces circulating pro-inflammatory cytokines, monocyte chemoattractant protein-1 and interleukin-8, which are key in regulating atherosclerosis by recruiting monocytes and macrophages to the cardiac vessel wall. Levels of these cytokines increase after myocardial infarction and after percutaneous coronary intervention, leading to acute restenosis post-intervention. Interleukin-10 and HMG-CoA reductase inhibitors, with potentially other statins, can effectively inhibit these specific pro-inflammatory vessel factors.[50] Novel gene therapy approaches or statin regiments can potentially downregulate pro-inflammatory cytokines and in the future prevent or treat cardiovascular diseases and trauma.

5.6.2 Exosomes

All cell types secrete vesicles of varying sizes into the extracellular space in response to changes in the physiological or pathological state. Membrane vesicles are composed of a lipid bilayer enclosing soluble cytosolic material and nuclear components and include apoptotic bodies, exosomes, and microvesicles (microparticles). Originating from different subcellular compartments, extracellular vesicles act as vehicles for the transfer of biological information, acting locally and remotely.[51] Exosomes contain a variety of signaling factors and can be generally identified based on the unique signaling content reminiscent of the parent cell.[52,53]

Exosomes are microvesicles that are released by cells to help regulate cell communication and respond to different cell physiological states.[54] Exosomes modulate normal cell functioning and changes that occur during the development of pathologies. On average, exosomes are 30–150 nm in size and carry material that assists in intercellular signaling and waste management. The generation of exosomes involves a series of processes starting with the endosomes. Endosomes are membrane bound compartments in eukaryotic cells that can form into multivesicular bodies (MVBs). These MVBs fuse with the plasma membrane and release as luminal vesicles. At this point, the vesicles are called exosomes. If direct budding from the plasma membrane occurs to surround cytoplasmic contents, then the vesicle is known as microvesicles. All mammals and almost all examined cells have been found to some degree exhibit exosome secretion.[48] They are found in all biological fluids, are recoverable, and therefore a potential tool as a biomarker for diseases or homeostasis. Within the myocardium, vesicles can influence cardiomyocytes and immune cells to help facilitate a proper environmental change to maintain myocardial homeostasis.

5.6.2.1 Exosome Cargo
5.6.2.1a Lipids

An early observation found that lipids are sorted along routes of membrane traffic and are directly involved in the regulation of protein sorting and membrane flow in exosomes. Exosomes contain lipid molecules involved in exosome biogenesis such as lysobisphosphatidic acid (LBPA) and mediate cellular signaling such as prostaglandins activation.[55]

Characteristically, signaling factors are released in the extracellular medium via their association with lipid raft domains like cholesterol, sphingomyelin, phosphatidylcholine, sphingolipids ceramide, and glycerophospholipid on the exosomal membrane. Raft-associated molecules can be detected by immunoblot and in Triton X-100-insoluble fractions isolated from exosomes.[56,57]

Intraluminal vesicles of multivesicular endosomes are either sorted for cargo degradation into lysosomes or secreted as exosomes into the extracellular milieu. This segregation process into distinct subdomains on the endosomal membrane and the transfer of exosome associated domains into the lumen of the endosome depends on ceramide via sphingomyelinases. A neutral sphingomyelinase 2 (nSMase 2)-dependent pathway is suggested to be responsible for the enrichment of miRNA in the exosome compared to the intracellular content of the parent cell.[58]

Compared to the parental cells, exosomes are enriched in cholesterol and sphingomyelin and their accumulation in cells potentially modulates target cell homeostasis through interactions with receptors on the target cell periphery, which in turn causes them to be internalized into the endosome.[59]

5.6.2.1b Proteins

An overlapping field of research is emerging regarding the cellular secretome and the cellular proteome. It has been found that the proteomes secreted by cardiac stem cells and myocytes potentially act synergistically to regulate cardiac repair processes after injury. Due to their endosomal origin, exosomes contain numerous membrane transport and fusion proteins like Rab GTPases, annexins, flotillins, integrins, and tetraspanins. Many proteins specific to the plasma membrane and cytoplasm are present in the exosomes, and aid in the transport across the target cell.[60] Regardless of cellular origin, exosomes do contain conserved proteins as well, such as heat shock cognate 70 (HSC70) and tetraspanin CD63[61] along with proteins for cytoskeleton and metabolism (GAPDH). Cell signaling proteins like β-catenin, Wnt5, Notch ligand, delta-like 4, TNFα, TGF-β are known to shuttle within many different species of exosomes, and in some cases exosomes can carry major histocompatibility complex (MHC) class I and II molecules albeit at relatively low levels.[62]

Recent efforts have focused on characterizing pathological biomarkers expressed by exosomes thereby providing ways to detect disease progression and pathology. Interestingly, exosomes direct immunoregulation through influencing transportation of immune response proteins, presence of antigens, and transfer of infectious agents. The relationship between the immune system and exosomes was first discovered in B-lymphocytes that secrete exosomes containing MHC class II molecules presenting them to T-lymphocytes.[63] Research later revealed that dendritic cells secrete exosomes as well that contain MHC class I molecules and influence antigen presentation to T-cells. Dendritic cells under stress or infection increase exosome secretion of pro-inflammatory and immunosuppressive signals. Similarly, research studies have shown a relationship between exosomes and cancer. In malignant cells, exosomes stimulate angiogenesis and assist tumor growth.[64] Exosomes carry cancer signaling factors and tumor-promoting mRNAs that will activate oncogenic pathways. Analyzing exosomes content may help in early diagnosis of malignant tumors; exosomes of tumor cells have been shown to secrete factors that inhibit the natural defense immune mechanisms and promote anti-apoptotic features, which allows growth of malignant tumors. Exosome-based cancer treatments are non-toxic, non-immunogenic, and can be engineered to have the robust delivery capacity and target specificity, thus acting as a powerful nano-carrier to deliver anti-cancer drugs or genes for cancer stem cell targeting therapy.[65]

5.6.2.1c *Sorting proteins*

The mechanisms involved in protein sorting and loading into exosomes are still under active research. The endosomal sorting complex required for transport (ESCRT-0, -1, and -2) has been found to recognize and sequester ubiquitinated proteins to the endosomal membrane where the ESCRT-3 complex promotes membrane budding. Analysis of purified exosomes from various cell types show enrichment of different ESCRT components and ubiquitinated proteins.[66] Potential methods of treatment for the ischemic heart using a purely proteomic-based approach have been emerging in recent years.

In the myocardium, there seems to be an increased role for cellular secretome in modulating cardiomyocytes and immune cells to maintain myocardial homeostasis.

Exosomes have been identified from almost every cell within the myocardium and are secreted in response to various physiological and pathological stimuli. Cardiomyocytes in the ischemic rat myocardium exposed to secreted frizzled related protein 2 (*Sfrp2*) up-regulate the expression of antiapoptotic genes, such as cellular β-catenin.[67] Grafting bone marrow-derived mesenchymal cells (MSCs) overexpressing Akt1, a secreted serine-threonine kinase, resulted in a 100-fold increase of Sfrp2 expression in cardiomyocytes. Moreover, conditioned media from hypoxic *Akt1* MSCs inhibits hypoxia-induced apoptosis and triggers vigorous spontaneous contraction of adult rat cardiomyocytes *in vitro*. When injected into infarcted hearts, *Akt* MSC conditioned media significantly limits infarct size and improves ventricular function relative to controls.[53] Similarly, overexpression of *Sfrp2* inhibited cardiac remodeling through reduction of intramyocardial inflammation, collagen deposition, and cardiomyocyte hypertrophy. The study observed regeneration of 90% of lost myocardial volume and completely normalized systolic and diastolic cardiac function. While normally Wnt3a is upregulated in ischemic cardiomyocytes and induces apoptosis, *Sfrp2* overexpression was capable of blocking the proapoptotic effect of *Wnt3a*, showing that *Akt* and *Sfrp2* can modulate Wnt/β-catenin signaling pathway in the heart.[68]

In other studies, β-catenin, a target protein for glycogen synthase kinase 3β (*GSK3β*), sensitizes neonatal cardiomyocytes to hypoxia/reoxygenation-induced apoptosis. Stabilization or upregulation of β-catenin protects neonatal rat cardiomyocytes against apoptosis with a mechanism similar to the pharmacological action of statins. Statins are known to inhibit cardiomyocyte apoptosis through *GSK3β* inactivation and stabilization of β-catenin levels in cardiomyocytes.[67,69]

Intercellular protein transfer is another feature of exosomal proteins regulating different cell signaling pathways. Exosome lipid layer encompassing the exosome has protective proteins that help when traveling through the extracellular matrix. Tissue development and regeneration has been correlated with exosomes shuttling of *Wnt* protein in Drosophila.[70] Additionally, studies of breast cancer cell lines have shown that exosomes contain many oncogenic proteins like HER-2,[71] tumor necrosis factor (TNF), and TNF receptors 1 and 2.[72]

In a similar approach, anti-apoptotic oncogene *Bcl-2* overexpressed rat MSCs were injected after myocardial infarction that led to reduced MSCs apoptosis and enhanced vascular endothelial growth factor secretion under hypoxic conditions. *Bcl-2* overexpression in MSCs enhanced the salutary effect of stem cell therapy by increasing cell survival and left ventricle function and decreasing infarction size. The same study demonstrated increased *VEGF* expression in *Bcl-2* modified MSC in response to hypoxic conditions.[73] Persistent *VEGF* overexpression has been linked to *Hsp90* mediated induction of *Bcl-2* expression that promotes anti-apoptotic function and enhances the survival of leukemia cells and tumorigenesis.[74] Importantly, no vascular tumor formation was observed in animals transplanted with *Bcl2* overexpressing MSCs, implying transient enhancement of *VEGF* and *Bcl2* as a mechanism for cardiac repair without the risk of tumorigenesis.[73]

5.6.2.1d *RNAs*

Exosomes can carry messenger and microRNA cargo that is able to mediate functional responses in recipient cells, thereby providing another mechanism for intercellular communication. MicroRNAs (miRNAs) have been found within exosomes in a cell type-dependent manner, are preferentially packaged and enriched in concentrations that are higher than the intracellular content of the parent cell.[75]

Similar to that of ubiquitination, small ubiquitin-like modifier (SUMO) mediates sumoylation of miRNAs by heterogeneous nuclear ribonucleoprotein A2B1 (*hnRNPA2B1*) that binds to exosomal miRNAs promoting specific miRN sorting.[76] Other research indicates preferential distribution and enrichment of miRNA in exosomes are naturally achieved in the cells through 3′ end posttranscriptional modification. Interestingly, 3' adenylated miRNAs are relatively enriched in cells, whereas 3′ end uridylated isoforms appear overrepresented in exosomes.[77] The exact functional significance of 3′-end modifications of miRNAs and sumoylation remains to be determined, but post-transcriptionally tagging of miRNAs promotes sorting and recruitment into exosomes. The lack of any apparent motif upon global analysis of miRNAs enriched in exosomes, besides the fact that not all exported miRNAs are 3′end modified, suggests multiple strategies for loading of miRNAs into EVs, and that not all exosomes contain identical cargo thus further proving evidence that different cell types secrete a heterogeneous population of vesicles.

Therapies involving the modulation of active exosome specific miRNAs are the most promising route of treatment. The first evidence showing that exosomes can be used to deliver desired RNA products was exhibited by Matthew J Wood and his team in 2011. He and his team were successfully able to inject modified exosomes with siRNA to specifically knockdown GAPDH production in neurons, microglia, oligodendrocytes in the brain.[78] Approaches have shown that it is possible to generate exosomes from CD34+ stem cells that carry Cy3 labeled pre-miRNA (miRNA precursors) that rigorously enter into recipient cells and deliver the precursors to regulate expression.[79]

Beneficial effects of stem cell-derived exosomes for ischemic tissue repair and regeneration are dependent upon exosome-mediated transfer of miRNAs as shown by several preclinical studies. For instance, hypoxia-inducible factor-1 (*HIF-1*) is a transcription factor that mediates adaptive responses to ischemia. Co-delivery of cardiac progenitor cells (CPCs) with a non-viral minicircle plasmid carrying HIF-1 into the ischemic heart improves survival of transplanted CPCs.[80] Furthermore, expression analysis showed that exosomes from hypoxic CPCs contain a different subset of miRNAs than normoxic CPCs. Similarly, CPC exosomes show an enrichment of several miRNAs including miR-210 and miR-132. Both these miRNAs are known to protect myocytes against apoptosis, and miR-132 possesses additional properties for promoting endothelial tube formation.[81]

5.7 STEM CELL-DERIVED EXOSOMES FOR CARDIAC REPAIR

Stem cell therapy has been widely used to enhance cardiac function after injury, but the functional benefit is limited. Transplanted cells do not persist very long in the heart and are unable to differentiate into fully functional cardiac myocytes. With the advent of the paracrine hypothesis, recent work has uncovered the cardiac repairability of stem cell-derived exosomes. The following section summarizes the recent work in this context.

5.7.1 Pluripotent Stem Cell Exosomes

ESCs and induced pluripotent stem cells (iPS) represent the most regenerative cell types. However, direct transplantation of both ESCs and iPSs is associated with tumor formation limiting their efficacy for cardiac repair. Recently, Khan and colleagues showed that exosomes derived from ESCs possess incredible cardiac repairability. ESCs exosomes were transplanted in the heart after myocardial infarction and the animals were followed for eight weeks. Results showed increased cardiac function including activation of cardiomyocyte proliferation, angiogenesis, and cardiac progenitor cell mobilization. The salutary effects were linked to the delivery of miR-294 enriched ESC exosomes to the injured myocardium.[82] In another study, iPS-derived exosomes were used for cardiac repair in a myocardial infarction model in mice. Authors showed that iPS exosomes promoted cardiomyocyte survival and angiogenesis thereby augmenting cardiac function through the transfer of miR-210.

5.7.2 Endothelial Progenitor Cell (EPCs) Exosomes

Endothelial progenitor cells (EPCs) possess the ability to secrete paracrine growth factors with multiple functions, including enhancement of migratory ability of mature endothelial cells and activation of tissue resident cardiac progenitor cells

(CPCs). A microarray study of cytokine expression found distinct expression patterns for angiogenic growth factors across endothelial progenitor cells (EPCs), human umbilical vein endothelial cells (HUVECs), and human microvascular endothelial cells (HMVEC) derived from human peripheral blood. *VEGF-A, VEGF-B, SDF-1,* and *IGF-1* mRNA levels were higher in the EPCs as compared to HUVECs or HMVECs, which positively correlated to the secretome of cells. Furthermore, the presence of *VEGF* and *SDF-1* in EPC conditioned media was found to induce migration of c-kit$^+$ CPCs, further showing that secretome content can directly affect the behavior of other cells.[83] Studies performed with bone marrow-derived CD34+ cells have shown that these cells release exosomes that are able to enhance cardiac function via promoting neovascularization in the injured heart. Further analysis revealed that several proangiogenic miRNAs such as miR-126 and miR-130a are enriched within EPC exosomes. Similarly, human EPCs-derived microvesicles carry cell surface receptors such as CD44 and CD29 along with mRNAs for genes involved in angiogenesis, survival, and proliferation.

5.7.3 Mesenchymal Stromal/Stem Cells (MSC) Exosomes

Mesenchymal stromal/stem cells are non-hematopoietic stem cells in the stroma of bone marrow and comprise around 0.001%–0.01% of the total nucleated bone marrow cells.[84] While mainly in bone marrow, mesenchymal stem cells reside in virtually all post-natal organs and tissues and can be isolated from human adipose tissue, liver, spleen, thymus, umbilical cord blood, placenta, Wharton's jelly, brain, lung, dental pulp, palatine tonsils, peripheral blood and almost any other source.[85] Due to originating in different tissue conditions, MSCs do not exhibit identical surface markers and possess functional differences, but not phenotype.[86] MSCs are broadly characterized and identified by their plastic adherence properties in standard culture conditions, expression of CD105, CD90, and CD73, and lack of expression of CD34, CD45, CD14 or CD11b, CD79a or CD19 and HLA-DR markers, and differentiation potency to adipocytes, chondroblasts, and osteoblasts *in vitro*.[87] MSCs are of great research interest due to their ability to migrate to injured tissue and differentiate into various cell lineages while also secreting exosomes that can recruit other cell types for tissue repair. MSCs are capable of producing exosomes enriched with pro-angiogenic factors, anti-fibrotic factors, neuroprotective factors, and immunosuppressive factors.[88] Apart from common surface exosome markers like CD9 and CD81, MSC-derived exosomes express several molecules that are generally specific to MSCs, such as CD29, CD44, CD90 and CD73.[89]

Intravenous injection of MSC derived exosomes from human umbilical cord was found to be tolerable in animals and have beneficial effects on weight loss with no harmful effects on renal or liver function.[90] These MSC exosomes were found to have an important role in maintaining tissue homeostasis and possess cardioprotective effects through preventing apoptosis, oxidative stress suppression, anti-inflammatory effects, cardiac regeneration, neovascularization, and anti-vascular remodeling effects.[91] Similarly, MSC derived exosomes have been found to increase ATP levels and promote myocardial viability and function post-infarction through activation of the PI3K/Akt pathway.[92]

Several miRNAs found in adult MSC derived exosomes have been identified as regulating cell proliferation, including miRNA-21, miRNA-191, miRNA-222, and let-7a. Additionally, miR-21, miR-222, and let-7f induce angiogenesis while miR-6087 promotes endothelial cell differentiation.[93] Muscle regeneration through promoting myogenesis and angiogenesis was found to be mediated through miRNA-494 dependent on the effect of cytokines present in exosomes.[94] In contrast, MSC exosomal miRNAs may also cause deleterious effects as well. An *in vivo* study showed that tumor growth was promoted through MSC-derived exosomes via increasing VEGF expression that activates extracellular signal-regulated kinase1/2 (ERK1/2).[95] Therefore it is important to note here that simple administration of MSC-derived exosomes may not be an effective treatment, rather it is necessary to deliver a combination of various signaling factors within each exosome type working synergistically.

5.7.4 Cardiac Progenitor Cells (CPCs) Exosomes

Cardiac progenitor cells (CPCs) reside within different regions of the heart regulating cardiac homeostasis in response to physiological states. Cell-based therapy using CPCs has shown cells to be safe and augment cardiac structure and function. CPCs are clonogenic and differentiate into all three cell lineages in the heart. Yet, CPC transdifferentiation potential has been challenged recently as several studies have shown inability of the cells to form functional cardiac myocytes. Alternatively, CPCs possess the ability to release paracrine factors including exosomes with cardiac regenerative ability. In a recent study, a total of fifteen proteins were identified in cultured rat CPC media and from cultured neonatal rat ventricular myocytes to be potential paracrine factors that promote cardiomyocyte proliferation. The rat CPCs were found to secrete fibrinogen beta chain, hemopexin, histidine-rich glycoprotein, interleukin-1 receptor-like1, osteopontin, and SPARC. Specifically, connective tissue growth factor (CTGF) enhanced cardiac stem cell proliferation whereas atrial natriuretic factor (ANP) and interleukin-1 receptor-like 1 (IL1RL1) were found to down-regulate proliferation in cardiac stem cells.[86] Similarly, characterization of microvesicles derived from human cardiac progenitor cells revealed enrichment of several cardioprotective miRNAs including miR-210, miR-132 and miR-146.[81] Changes in cell physiological state impacts vesicle release and this concept was tested recently. Authors used characterized CPC exosomes released under normoxic and hypoxic conditions and determined hypoxic exosomes to be more cardioprotective compared to normoxic CPC exosomes together with a unique microRNA expression profile. Collectively, CPC-derived exosomes represent a promising choice for cardiac repair and regeneration, allowing for a treatment strategy that circumvents issues associated with direct cell administration in the heart.

5.8 LIMITATIONS AND FUTURE POTENTIAL

Stem cells have created excitement in the regeneration therapy field. However, the mechanisms by which stem cells work are open for debate and need to be researched further. Many studies to date have provided large data sets that indicate stem cell transplantation is safe and effective in small and large animal models and humans (in some cases). However, more recently the transdifferentiation hypothesis has been challenged as an explanation of stem cells' therapeutic effects. ESCs and induced pluripotent-derived stem cells have shown transdifferentiation into cardiac-derived lineages, however adult cardiac-derived stem cells have shown very limited differentiation capacity. However, the beneficial effect of cell therapy is documented widely in small and large animals plus clinical trials, suggesting involvement of many other biological processes that potentially linked to the beneficial effects seen after cell therapy. The secretion of paracrine factors by cells, as suggested by the paracrine hypothesis, is one of the most likely mechanisms and warrants further investigation. However, recent excitement over paracrine secretion by stem cells must be channeled into meaningful research efforts toward fully understanding the role of stem cells and their secreted factors in progressing the field of reparative biology. Cellular uptake of paracrine factors and exosome is quick, resulting in rapid dissemination of the vesicular contents to the target cells. Therefore, an important area for consideration is how long the beneficial effect of paracrine factors including exosome therapy lasts after delivery? Unlike stem cells which have short half-life, most of the studies involving exosomes administration have shown activation of endogenous repair response thus leading to augmented function after injury even after the exosomes have cleared. Concurrently, systemic administration via intravenous infusion may carry the risk of off-target effects in other organs besides the target organ. Optimal paracrine factor-based treatment would warrant tailored therapy designed for specific cell types within an organ to minimize adverse effects, but this approach has not been tested yet and certainly represents a limitation.

Stem cell treatment resulted in a significant reduction of symptoms after heart injury, laying the foundation for future studies. Nevertheless, it is important to assess the safety of paracrine factors and exosome-based therapy in clinical settings, including their biodistribution, toxicity, and half-life studies. Moreover, careful dose estimation must be performed with the development of procedures for repeated paracrine factor administration.

ACKNOWLEDGEMENTS

We would like to thank our colleagues for preparation of this book chapter. The work presented here was supported by NIH and AHA grants to both Dr. Khan and Dr. Mohsin.

CONFLICT OF INTEREST

The authors confirm that this chapter contents have no conflict of interest.

REFERENCES

1. Thomson JA, Itskovitz-Eldor J, Shapiro SS, Waknitz MA, Swiergiel JJ, Marshall VS, Jones JM. Embryonic stem cell lines derived from human blastocysts. *Science*. 1998; 282(5391): 1145–1147.
2. Niwa H, Miyazaki J, Smith AG. Quantitative expression of Oct-3/4 defines differentiation, dedifferentiation or self-renewal of ES cells. *Nat Genet*. 2000; 24(4): 372–376.
3. Darr H, Mayshar Y, Benvenisty N. Overexpression of NANOG in human ES cells enables feeder-free growth while inducing primitive ectoderm features. *Development*. 2006; 133(6): 1193–1201.
4. Rathjen J, Lake JA, Bettess MD, Washington JM, Chapman G, Rathjen PD. Formation of a primitive ectoderm like cell population, EPL cells, from ES cells in response to biologically derived factors. *J Cell Sci*. 1999; 112(Pt 5): 601–612.
5. Chan SS, Shi X, Toyama A, Arpke RW, Dandapat A, Iacovino M, Kang J, *et al*. Mesp1 patterns mesoderm into cardiac, hematopoietic, or skeletal myogenic progenitors in a context-dependent manner. *Cell Stem Cell*. 2013; 12(5): 587–601.
6. Saga Y, Kitajima S, Miyagawa-Tomita S. Mesp1 expression is the earliest sign of cardiovascular development. *Trends Cardiovasc Med*. 2000; 10(8): 345–352.

7. Buckingham M, Meilhac S, Zaffran S. Building the mammalian heart from two sources of myocardial cells. *Nat Rev Genet.* 2005; 6(11): 826–835.

8. Moorman A, Webb S, Brown NA, Lamers W, Anderson RH. Development of the heart: (1) formation of the cardiac chambers and arterial trunks. *Heart.* 2003; 89(7): 806–814.

9. Lints TJ, Parsons LM, Hartley L, Lyons I, Harvey RP. Nkx-2.5: a novel murine homeobox gene expressed in early heart progenitor cells and their myogenic descendants. *Development.* 1993; 119(3): 969.

10. Wang Y, Li Y, Guo C, Lu Q, Wang W, Jia Z, Chen P, *et al.* ISL1 and JMJD3 synergistically control cardiac differentiation of embryonic stem cells. *Nucleic Acids Res.* 2016; 44(14): 6741–6755.

11. Lin Q, Schwarz J, Bucana C, Olson EN. Control of mouse cardiac morphogenesis and myogenesis by transcription factor MEF2C. *Science.* 1997; 276(5317): 1404–1407.

12. Christoffels VM, Habets PE, Franco D, Campione M, de Jong F, Lamers WH, Bao ZZ, *et al.* Chamber formation and morphogenesis in the developing mammalian heart. *Dev Biol.* 2000; 223(2): 266–278.

13. Takahashi K, Yamanaka S. Induction of pluripotent stem cells from mouse embryonic and adult fibroblast cultures by defined factors. *Cell.* 2006; 126(4): 663–676.

14. Yu J, Vodyanik MA, Smuga-Otto K, Antosiewicz-Bourget J, Frane JL, Tian S, Nie J, *et al.* Induced pluripotent stem cell lines derived from human somatic cells. *Science.* 2007; 318(5858): 1917–1920.

15. Stadtfeld M, Nagaya M, Utikal J, Weir G, Hochedlinger K. Induced pluripotent stem cells generated without viral integration. *Science.* 2008; 322(5903): 945–949.

16. Chin MH, Mason MJ, Xie W, Volinia S, Singer M, Peterson C, Ambartsumyan G, *et al.* Induced pluripotent stem cells and embryonic stem cells are distinguished by gene expression signatures. *Cell Stem Cell.* 2009; 5(1): 111–123.

17. Planello AC, Ji J, Sharma V, Singhania R, Mbabaali F, Müller F, Alfaro JA, *et al.* Aberrant DNA methylation reprogramming during induced pluripotent stem cell generation is dependent on the choice of reprogramming factors. *Cell Regen (Lond).* 2014; 3(1): 4.

18. Tucker BA, *et al.* Transplantation of adult mouse iPS cell-derived photoreceptor precursors restores retinal structure and function in degenerative mice. *PLoS One.* 2011; 6(4): e18992.

19. Nori S, Okada Y, Nishimura S, Sasaki T, Itakura G, Kobayashi Y, Renault-Mihara F, *et al.* Long-term safety issues of iPSC-based cell therapy in a spinal cord injury model: oncogenic transformation with epithelial-mesenchymal transition. *Stem Cell Reports.* 2015; 4(3): 360–373.

20. Zhang Y, *et al.* Intramyocardial transplantation of undifferentiated rat induced pluripotent stem cells causes tumorigenesis in the heart. *PLoS One.* 2011; 6(4): e19012.

21. Mauritz C, Schwanke K, Reppel M, Neef S, Katsirntaki K, Maier LS, Nguemo F, *et al.* Generation of functional murine cardiac myocytes from induced pluripotent stem cells. *Circulation.* 2008; 118(5): 507–517.

22. Nussbaum J, Minami E, Laflamme MA, Virag JA, Ware CB, Masino A, Muskheli V, *et al.* Transplantation of undifferentiated murine embryonic stem cells in the heart: teratoma formation and immune response. *FASEB J.* 2007; 21(7): 1345–1357.

23. Trosko JE. From adult stem cells to cancer stem cells: Oct-4 Gene, cell-cell communication, and hormones during tumor promotion. *Ann N Y Acad Sci.* 2006; 1089: 36–58.

24. Linn DE, Yang X, Sun F, Xie Y, Chen H, Jiang R, Chen H, *et al.* A role for OCT4 in tumor initiation of drug-resistant prostate cancer cells. *Genes Cancer.* 2010; 1(9): 908–916.

25. Hatefi N, Nouraee N, Parvin M, Ziaee SAM, Mowla SJ. Evaluating the expression of oct4 as a prognostic tumor marker in bladder cancer. *Iran J Basic Med Sci.* 2012; 15(6): 1154–1161.

26. Wang YD, Cai N, Wu X-L, Cao HZ, Xie LL, Zheng PS. OCT4 promotes tumorigenesis and inhibits apoptosis of cervical cancer cells by miR-125b/BAK1 pathway. *Cell Death Dis.* 2013; 4: e760.

27. Talebi A, Kianersi K, Beiraghdar M. Comparison of gene expression of SOX2 and OCT4 in normal tissue, polyps, and colon adenocarcinoma using immunohistochemical staining. *Adv Biomed Res.* 2015; 4: 234.

28. Boumahdi S, Driessens G, Lapouge G, Rorive S, Nassar D, Le Mercier M, Delatte B, Caauwe A, *et al.* SOX2 controls tumour initiation and cancer stem-cell functions in squamous-cell carcinoma. *Nature.* 2014; 511(7508): 246–250.

29. Shachaf CM, *et al.* MYC inactivation uncovers pluripotent differentiation and tumour dormancy in hepatocellular cancer. *Nature.* 2004; 431(7012): 1112–1117.

30. Ding B, Liu P, Liu W, Sun P, Wang CL. Emerging roles of Kruppel-like factor 4 in cancer and cancer stem cells. *Asian Pac J Cancer Prev.* 2015; 16(9): 3629–3633.

31. Dang DT, Chne X, Feng J, Torbenson M, Dang LH, Yang VW. Overexpression of Kruppel-like factor 4 in the human colon cancer cell line RKO leads to reduced tumorigenecity. *Oncogene.* 2003; 22(22): 3424–3430.

32. Hanna J, Saha K, Pando B, van Zon J, Lenger CJ, Creyghton MP, Oudennarden A, *et al.* Direct cell reprogramming is a stochastic process amenable to acceleration. *Nature.* 2009; 462(7273): 595–601.

33. Bourguignon LY, Peyrollier K, Gilad E. Hyaluronan-CD44 interaction activates stem cell marker Nanog, Stat-3-mediated MDR1 gene expression, and ankyrin-regulated multidrug efflux in breast and ovarian tumor cells. *J Biol Chem.* 2008; 283(25): 17635–17651.

34. Nguyen LH, Robinton DA, Seligson MT, Wu L, Li L, Rakheja D, Comerford SA, *et al.* Lin28b is sufficient to drive liver cancer and necessary for its maintenance in murine models. *Cancer Cell.* 2014; 26(2): 248–261.

35. Amini S, Fathi F, Mobalegi J, Sofimajidpour H, Ghadimi T. The expressions of stem cell markers: Oct4, Nanog, Sox2, nucleostemin, Bmi, Zfx, Tcl1, Tbx3, Dppa4, and Esrrb in bladder, colon, and prostate cancer, and certain cancer cell lines. *Anat Cell Biol.* 2014; 47(1): 1–11.

36. Adolfsson J, Borge OJ, Bryder D, Theilgaard-Mönch K, Astrand-Grundström I, Sitnicka E, Sasaki Y, *et al.* Upregulation of Flt3 expression within the bone marrow Lin(-)Sca1(+)c-kit(+) stem cell compartment is accompanied by loss of self-renewal capacity. *Immunity.* 2001; 15(4): 659–669.

37. Kiel MJ, Radice GL, Morrison SJ. Lack of evidence that hematopoietic stem cells depend on N-cadherin-mediated adhesion to osteoblasts for their maintenance. *Cell Stem Cell.* 2007; 1(2): 204–217.

38. Bergmann O, Bhardwaj RD, Bernard S, Zdunek S, Barnabé-Heider F, Walsh S, Zupicich J, *et al.* Evidence for cardiomyocyte renewal in humans. *Science.* 2009; 324(5923): 98–102.

39. Adler CP, Friedburg H. Myocardial DNA content, ploidy level and cell number in geriatric hearts: post-mortem examinations of human myocardium in old age. *J Mol Cell Cardiol.* 1986; 18(1): 39–53.

40. Poss KD, Wilson LG, Keating MT. Heart regeneration in zebrafish. *Science.* 2002; 298(5601): 2188–2190.

41. Bird SD, Doevendans PA, van Rooijen MA, Brutel de la Riviere A, Hassink RJ, Passier R, Mummery CL. The human adult cardiomyocyte phenotype. *Cardiovasc Res.* 2003; 58(2): 423–434.

42. Judson RL, Babiarz JE, Venere M, Blelloch R. Embryonic stem cell-specific microRNAs promote induced pluripotency. *Nat Biotechnol.* 2009; 27(5): 459–461.

43. Anokye-Danso F, Trivedi CM, Juhr D, Gupta M, Cui Z, Tian Y, Zhang Y, Yang W, *et al.* Highly efficient miRNA-mediated reprogramming of mouse and human somatic cells to pluripotency. *Cell Stem Cell.* 2011; 8(4): 376–388.

44. Beltrami AP, Barlucchi L, Torella D, Baker M, Limana F, Chimenti S, Kasahara H, Rota M, *et al.* Adult cardiac stem cells are multipotent and support myocardial regeneration. *Cell.* 2003; 114(6): 763–776.

45. Quaini F, Urbanek K, Beltrami AP, Finato N, Beltrami CA, Nadal-Ginard B, Kajstura J, *et al.* Chimerism of the transplanted heart. *N Engl J Med.* 2002; 346(1): 5–15.

46. Orlic D, Kajstura J, Chimenti S, Limana F, Jakoniuk I, Quaini F, Nadal-Ginard B, *et al.* Mobilized bone marrow cells repair the infarcted heart, improving function and survival. *Proc Natl Acad Sci USA.* 2001; 98(18): 10344–10349.

47. Gery I, Gherson RK, Waksman BH. Potentiation of the T-lymphocyte response to mitogens. II. The cellular source of potentiating mediator(s). *J Exp Med.* 1972; 136(1): 143–155.

48. Konadu KA, Chu J, Huang MB, Amancha PK, Armstrong W, Powell MD, Villinger F, *et al.* Association of cytokines with exosomes in the plasma of HIV-1-seropositive individuals. *J Infect Dis.* 2015; 211(11): 1712–1716.

49. Sanganalmath SK, Abdel-Latif A, Bolli R, Xuan Y-T, Dawn B. Hematopoietic cytokines for cardiac repair: mobilization of bone marrow cells and beyond. *Basic Res Cardiol.* 2011; 106(5): 709–733.

50. Ito T, Ikeda U. Inflammatory cytokines and cardiovascular disease. *Curr Drug Targets Inflamm Allergy.* 2003; 2(3): 257–265.

51. Boulanger CM, Loyer X, Rautou PE, Amabile N. Extracellular vesicles in coronary artery disease. *Nat Rev Cardiol.* 2017; 14(5): 259–272.

52. Tang XL, Rokosh G, Sanganalmath SK, Yuan F, Sato H, Mu J, Dai S, *et al.* Intracoronary administration of cardiac progenitor cells alleviates left ventricular dysfunction in rats with a 30-day-old infarction. *Circulation.* 2010; 121(2): 293–305.

53. Gnecchi M, He H, Noiseux N, Liang OD, Zhang L, Morello F, Mu H, Melo LG, *et al.* Evidence supporting paracrine hypothesis for Akt-modified mesenchymal stem cell-mediated cardiac protection and functional improvement. *FASEB J.* 2006; 20(6): 661–669.

54. Ratajczak J, Miekus K, Kucia M, Zhang J, Reca R, Dvorak P, Ratajczak MZ. Embryonic stem cell-derived microvesicles reprogram hematopoietic progenitors: evidence for horizontal transfer of mRNA and protein delivery. *Leukemia.* 2006; 20(5): 847–856.

55. Kobayashi T, Gu F, Gruenberg J. Lipids, lipid domains and lipid-protein interactions in endocytic membrane traffic. *Semin Cell Dev Biol.* 1998; 9(5): 517–526.

56. Lai RC, Arslan F, Lee MM, Sze NS, Choo A, Chen TS, Salto-Tellez M, *et al.* Exosome secreted by MSC reduces myocardial ischemia/reperfusion injury. *Stem Cell Res.* 2010; 4(3): 214–222.

57. de Gassart A, Geminard C, Fevrier B, Raposo G, Vidal M. Lipid raft-associated protein sorting in exosomes. *Blood.* 2003; 102(13): 4336–4344.

58. Trajkovic K, Hsu C, Chiantia S, Rajendran L, Wenzel D, Wieland F, Schwille P, *et al.* Ceramide triggers budding of exosome vesicles into multivesicular endosomes. *Science.* 2008; 319(5867): 1244–1247.

59. Record M, Carayon K, Poirot M, Silvente-Poirot S. Exosomes as new vesicular lipid transporters involved in cell-cell communication and various pathophysiologies. *Biochim Biophys Acta.* 2014; 1841(1): 108–120.

60. Simons M, Raposo G. Exosomes — vesicular carriers for intercellular communication. *Curr Opin Cell Biol.* 2009; 21(4): 575–581.

61. Mathivanan S, Ji H, Simpson RJ. Exosomes: extracellular organelles important in intercellular communication. *J Proteomics.* 2010; 73(10): 1907–1920.

62. Simpson RJ, Lim JW, Moritz RL, Mathivanan S. Exosomes: proteomic insights and diagnostic potential. *Expert Rev Proteomics.* 2009; 6(3): 267–283.

63. Peters PJ, Geuze HJ, Van der Donk HA, Slot JW, Griffith JM, Stam NJ, Clevers HC, *et al.* Molecules relevant for T cell-target cell interaction are present in cytolytic granules of human T lymphocytes. *Eur J Immunol.* 1989; 19(8): 1469–1475.

64. Hood JL, Pan H, Lanza GM, Wickline SA, *et al.* Paracrine induction of endothelium by tumor exosomes. *Lab Invest.* 2009; 89(11): 1317–1328.

65. Azmi AS, Bao B, Sarkar FH. Exosomes in cancer development, metastasis, and drug resistance: a comprehensive review. *Cancer Metastasis Rev.* 2013; 32(3–4): 623–642.

66. Buschow SI, *et al.* Exosomes contain ubiquitinated proteins. *Blood Cells Mol Dis.* 2005; 35(3): 398–403.

67. Mirotsou M, Zhang Z, Deb A, Zhang L, Gnecchi M, Noiseux N, Mu H, *et al.* Secreted frizzled related protein 2 (Sfrp2) is the key Akt-mesenchymal stem cell-released paracrine factor mediating myocardial survival and repair. *Proc Natl Acad Sci USA.* 2007; 104(5): 1643–1648.

68. Mangi AA, Noiseux N, Kong D, He H, Rezvani M, Ingwall JS, Dzau VJ. Mesenchymal stem cells modified with Akt prevent remodeling and restore performance of infarcted hearts. *Nat Med.* 2003; 9(9): 1195–1201.

69. Bergmann MW, Rechner C, Freund C, Baurand A, El Jamali A, Dietz R, *et al.* Statins inhibit reoxygenation-induced cardiomyocyte apoptosis: role for glycogen synthase kinase 3beta and transcription factor beta-catenin. *J Mol Cell Cardiol.* 2004; 37(3): 681–690.

70. Korkut C, Ataman B, Ramachandran P, Ashley J, Barria R, Gherbesi N, Budnik V. Trans-synaptic transmission of vesicular Wnt signals through Evi/Wntless. *Cell.* 2009; 139(2): 393–404.

71. Chairoungdua A, Smith DL, Pochard P, Hull M, Caplan MJ. Exosome release of beta-catenin: a novel mechanism that antagonizes Wnt signaling. *J Cell Biol*, 2010. 190(6): 1079–1091.

72. Soderberg A, *et al.* Redox-signaling transmitted in trans to neighboring cells by melanoma-derived TNF-containing exosomes. *Free Radic Biol Med.* 2007; 43(1): 90–99.

73. Li W, Ma N, Ong LL, Nesselmann C, Klopsch C, Ladilov Y, Furlani D, *et al.* Bcl-2 engineered MSCs inhibited apoptosis and improved heart function. *Stem Cells.* 2007; 25(8): 2118–2127.

74. Dias S, Shmelkov SV, Lam G, Rafii S. VEGF(165) promotes survival of leukemic cells by Hsp90-mediated induction of Bcl-2 expression and apoptosis inhibition. *Blood.* 2002; 99(7): 2532–2540.

75. Goldie BJ, Dun MD, Lin M, Smith ND, Verrills NM, Dayas CV, Cairns MJ. Activity-associated miRNA are packaged in Map1b-enriched exosomes released from depolarized neurons. *Nucleic Acids Res.* 2014; 42(14): 9195–9208.

76. Villarroya-Beltri C, Gutiérrez-Vázquez C, Sánchez-Cabo F, Pérez-Hernández D, Vázquez J, Martin-Cofreces N, Martinez-Herrera DJ, *et al.* Sumoylated hnRNPA2B1 controls the sorting of miRNAs into exosomes through binding to specific motifs. *Nat Commun.* 2013; 4: 2980.

77. Koppers-Lalic D, *et al.* Nontemplated nucleotide additions distinguish the small RNA composition in cells from exosomes. *Cell Rep.* 2014; 8(6): 1649–1658.

78. Alvarez-Erviti L, *et al.* Delivery of siRNA to the mouse brain by systemic injection of targeted exosomes. *Nat Biotechnol.* 2011; 29(4): 341–345.

79. Mathiyalagan P, Sahoo S. Exosomes-based gene therapy for MicroRNA delivery. *Methods Mol Biol.* 2017; 1521: 139–152.

80. Ong SG, *et al.* Cross talk of combined gene and cell therapy in ischemic heart disease: role of exosomal microRNA transfer. *Circulation.* 2014; 130(11 Suppl 1): S60–69.

81. Barile L, *et al.* Extracellular vesicles from human cardiac progenitor cells inhibit cardiomyocyte apoptosis and improve cardiac function after myocardial infarction. *Cardiovasc Res.* 2014; 103(4): 530–541.

82. Khan M, Nickoloff E, Abramova T, Johnson J, Verma SK, Krishnamurthy P, Mackie AR, *et al.* Embryonic stem cell-derived exosomes promote endogenous repair mechanisms

and enhance cardiac function following myocardial infarction. *Circ Res.* 2015; 117(1): 52–64.

83. Urbich C, Aicher A, Heeschen C, Dernbach E, Hofmann WK, Zeiher AM, Dimmeler S. Soluble factors released by endothelial progenitor cells promote migration of endothelial cells and cardiac resident progenitor cells. *J Mol Cell Cardiol.* 2005; 39(5): 733–742.

84. Pittenger MF, Mackay AM, Beck SC, Jaiswal RK, Douglas R, Mosca JD, Moorman MA, *et al.* Multilineage potential of adult human mesenchymal stem cells. *Science.* 1999; 284(5411): 143–147.

85. da Silva Meirelles L, Chagastelles PC, Nardi NB. Mesenchymal stem cells reside in virtually all post-natal organs and tissues. *J Cell Sci.* 2006; 119(Pt 11): 2204–2213.

86. Kellner J, Sivajothi S, McNiece I. Differential properties of human stromal cells from bone marrow, adipose, liver and cardiac tissues. *Cytotherapy.* 2015; 17(11): 1514–1523.

87. Dominici M, Le Blanc K, Mueller I, Slaper-Cortenbach I, Marini F, Krause D, Deans R, *et al.* Minimal criteria for defining multipotent mesenchymal stromal cells. The International Society for Cellular Therapy position statement. *Cytotherapy.* 2006; 8(4): 315–317.

88. Aggarwal S, Pittenger MF. Human mesenchymal stem cells modulate allogeneic immune cell responses. *Blood.* 2005; 105(4): 1815–1822.

89. Yang Y, Bucan V, Baehre H, von der Ohe J, Otte A, Hass R. Acquisition of new tumor cell properties by MSC-derived exosomes. *Int J Oncol.* 2015; 47(1): 244–252.

90. Sun L, Xu R, Sun X, Duan Y, Han Y, Zhao Y, Qian H, *et al.* Safety evaluation of exosomes derived from human umbilical cord mesenchymal stromal cell. *Cytotherapy.* 2016; 18(3): 413–422.

91. Gallina C, Turinetto V, Giachino C. A new paradigm in cardiac regeneration: the mesenchymal stem cell secretome. *Stem Cells Int.* 2015; 2015: 765846.

92. Arslan F, Lai RC, Smeets MB, Akeroyd L, Choo A, Aguor EN, Timmers L, *et al.* Mesenchymal stem cell-derived exosomes increase ATP levels, decrease oxidative stress and activate PI3K/Akt pathway to enhance myocardial viability and prevent adverse remodeling after myocardial ischemia/reperfusion injury. *Stem Cell Res.* 2013; 10(3): 301–312.

93. Merino-Gonzalez C, Zuñiga FA, Escudero C, Ormazabal V, Reyes C, Nova-Lamperti E, Salomón C, *et al.* Mesenchymal stem cell-derived extracellular vesicles promote angiogenesis: potencial clinical application. *Front Physiol.* 2016. 7: 24.

94. Nakamura Y, Miyaki S, Ishitobi H, Matsuyama S, Nakasa T, Kamei N, Akimoto T, *et al.* Mesenchymal-stem-cell-derived exosomes accelerate skeletal muscle regeneration. *FEBS Lett.* 2015; 589(11): 1257–1265.

95. Zhu W, Huang L, Li Y, Zhang X, Gu J, Yan Y, Xu X, *et al.* Exosomes derived from human bone marrow mesenchymal stem cells promote tumor growth in vivo. *Cancer Lett.* 2012. 315(1): 28–37.

CHAPTER 6

Cancer stem cells and their microenvironment

**Valentina Masciale*, Giulia Grisendi†, Federico Banchelli‡,
Roberto D'Amico‡, Uliano Morandi*, Massimo Dominici†,
Khawaja Husnain Haider§, Beatrice Aramini*,¶**

*Division of Thoracic Surgery
†Division of Oncology
‡Center of Statistics
Department of Medical and Surgical Sciences for Children and Adults
University of Modena and Reggio Emilia, Modena, Italy
§Sulaiman AlRajhi Medical School, AlQaseem, Kingdom of Saudi Arabia

ABSTRACT

Cancer Stem Cells (CSCs) are a small population of cells within tumors holding stemness properties that sustain cancer progression, such as enhanced capacities for self-renewal, growing, metastasizing, homing, and reproliferating. CSCs show remarkable organizing capacities as they can educate neighboring cells to provide nutrients and collaborate in the elusion from the immune system, creating an environment favorable for tumor growth. In particular, tumor-specific microenvironments comprise stromal cells, immune cells, networks of cytokines and growth factors, hypoxic regions, and the extracellular matrix. The contribution of the microenvironment in this picture is crucial: it is now accepted that the "cancer" scenario is not simply composed of transformed cells working together in isolated and strictly autonomous machinery. Tumor microenvironment actively collaborates with neoplastic cells at different levels: promoting proliferation while evading growth suppression and immune surveillance, overcoming cell death, modulating cell metabolism, activating angiogenesis and invasion/metastasis programs. Also, the interactions between CSC and microenvironment help in their survival of common anti-cancer therapies thus being partly responsible for disease recurrence. Further studies regarding CSC/microenvironment seem to be promising for new CSC-targeting therapies, which may represent an innovative strategy for the cure of lung cancer.

¶Corresponding author. Email: beatrice.aramini@unimore.it

KEYWORDS

Cancer; CSCs; Carcinogenesis; Differentiation; Signaling; Stem cell; Tumor.

LIST OF ABBREVIATIONS

APC	=	Antigen presenting cells
CAFs	=	Cancer-associated fibroblasts
cCAFs	=	Circulating CAFs
CCL18	=	C-C motif ligand 18
CSCs	=	Cancer stem cells
CSF-1	=	Colony stimulating factor-1
CSF-1R	=	Colony stimulating factor-1receptor
CTC	=	Circulating tumor cells
CTLA-4	=	Cytotoxic T lymphocyte-associated protein 4
CXCR4	=	C-X-C-chemokine receptor 4
ECM	=	Extracellular matrix
EGF	=	Epidermal growth factor
EMT	=	Endothelial-to-Mesenchymal transition
Glu-GNPs	=	Glucose-coated gold nanoparticles
HA	=	Hyaluronan
HGF-1	=	Hepatocyte growth factor-1
ITH	=	Inter-tumoral heterogeneity
Mang-NPs	=	Mangostin-encapsulated Poly (lactic-co-glycolic acid) nanoparticles
miRNAs	=	MicroRNAs
MMPs	=	Matrix metalloproteinases
PD-1	=	Programmed cell death 1
PDGF	=	Platelet-derived growth factor
PDGFR	=	Platelet-derived growth factor receptor
SCC	=	Squamous cell carcinoma
SDF-1	=	Strromal cell-derived factor-1
Shh	=	Sonic hedgehog
SPARC	=	Secreted protein acidic and rich in cysteine
TAMs	=	Tumor associated macrophages
TCR	=	T-cell receptor
TIL	=	Tumor infiltrating lymphocytes
TME	=	Tumor microenvironment
VEGF	=	Vascular endothelial growth factor

VEGFR1 = Vascular endothelial growth factor receptor-1
VEGFR2 = Vascular endothelial growth factor receptor-2
VM   = Vascular mimicry

6.1 INTRODUCTION

6.1.1 Heterogeneity of the Tumor

Given the assortment of cell and tissue types that are known to exist within tumors, recent research has more focused on tumor heterogeneity.[1] Among the early investigators during 1800s, Virchow and Cohnheim postulated the existence of cancer stem cells (CSCs) that arise from what they believed to be "activation of dormant embryonic tissue remnants".[1,2] CSCs are defined as a subpopulation of cancer cells with the capability to auto-regenerate, proliferate and differentiate into multiple cancer cell lineages through symmetric and asymmetric cell division, with tumorigenic potential and specific surface markers.[3-7] The unique characteristics of CSCs include the requirement of a small number of CSCs to initiate new tumor, self-renewal and differentiation potential, possession of specific and distinguishing surface markers that help in their identification and isolation, and resistance to conventional chemotherapy and radiotherapy.[3,4,7,8] Despite these well-known characteristics, the definition of the cellular components of the tumor mass remains contentious.[3]

The differentiation hierarchy that underpins the development of all cellular compartments is indispensable for understanding the origin of tumor cells (Figure-6.1). However, for many tissues, the cellular hierarchies have not been adequately refined thus making it difficult to confirm the cellular origins of cancer.[9] The scientific community has marked the existence of tumor heterogeneity that can be classified as inter-tumoral heterogeneity and intra-tumoral heterogeneity. Inter-tumoral heterogeneity (ITH) is currently defined as multiple interactions as variations between tumors of different tissue and cell types, between tumors of the same tissue type from different patients, and between different tumors within the same individual. On the other hand, intra-tumoral heterogeneity refers to variations observed within a single tumor.[1,10] Different tumor cell populations differ in surface marker expression, genetic or epigenetic changes, genetic stability, resistance or susceptibility to therapy and growth rates.[11]

The inter-tumoral heterogeneity provides the basis of cancer classification into different types and sub-types based on divergence in their histological features, genetic profile, protein signatures, and surface markers expression profile. Many of these variables provide clinically relevant prognostic features and/or predictive

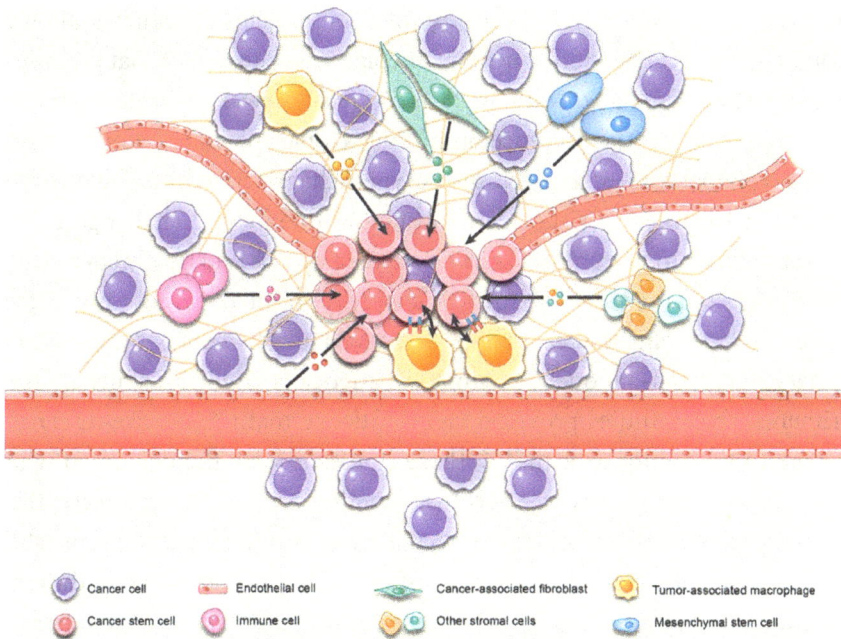

Figure-6.1. The cellular components of the tumor microenvironment.

information.[12] On the other hand, intra-tumoral heterogeneity complicates cancer prognosis and treatment.[13] Currently, clinical evaluation of tumor heterogeneity is an emerging issue to improve clinical oncology while intra-tumor heterogeneity is closely related to cancer progression, resistance to therapy, and the probability of recurrence.[14] It is inter-connected with complex molecular mechanisms including spatial and temporal phenomena, which are often peculiar for every single patient.[14]

Because of ITH in the primary tumors as well as metastases, and due to the wide clinical heterogeneity amongst cancer patients, it is imperative to apply clinical research methods directly to patients' material in contemporary clinical practice to ensure specific, optimal and effective treatment.[14] For any tumor type, only few molecular biomarkers are being currently employed for diagnosis and an only minor part of the available treatment targets is being exploited. It is now anticipated that clinical research, directly performed on cancer patients, will be increasingly diffused as a requirement to obtain more efficient and personalized tumor therapy protocols.[15] Moreover, phase III clinical trials in oncology have recently encountered wide criticisms,[16–18] because of the long duration required, the high cost, and the less than expected results.

The concept of CSCs has been included in the definition of the intra-tumor heterogeneity which is postulated to develop over time as CSCs divide and differentiate asymmetrically.[19,20] The loss of normal cellular controls allows the development and propagation of genetic or epigenetic alterations that give the cells novel properties associated with metastasis, self-renewal, treatment resistance, and recurrence.[19,21] As the presence of multiple clonal sub-populations within the same tumor imparts divergent cell phenotypes, characterized by obvious growth advantage or treatment resistance, a substantial therapeutic challenge exists, as only some cells within a tumor would be affected by any one treatment.[22] The first clear evidence to support a role for CSCs activity in intact tumors was provided by three independent studies carried out using experimental brain, skin and intestinal tumor mouse models.[23] Using the genetically engineered lineage-tracing experiments, these studies provided clear evidence that CSCs arise *de novo* and contribute to the tumor growth.[24-26] These studies resolve the debate on whether CSCs do exist or are merely a xenotransplantation artifact. Nevertheless, the key question remains whether targeting of CSCs alone would be sufficient or whether non-CSCs could take their place after de-differentiation. Unfortunately, the efficacy of CSCs targeting and the capacity to revert to the CSC state has been difficult to study due to the limited characterization of CSC-specific markers. Several markers including CD133, CD44, CD166, CD24, and ALDH1 activity, have proven useful for prospective isolation of CSCs in multiple solid tumors.[27] However, CSC-specific marker expression profile differs between tumor types. For instance, while CD133 has been used as a marker to identify CSCs in glioblastoma[28] and CRC,[23] it is not a reliable marker in breast cancer where CD44+ CD24– is commonly used to enrich CSCs.[29] CSC-specific marker expression also varies between cancer sub-types and even, between patients in the same subtype.[30] For instance, CD44high-CD24low fails to efficiently enrich CSCs in triple-negative breast cancer[31] and CD133 has been debated in colon cancer. Furthermore, the lack of consistency has generated confusion in the identification of CSCs and questioned the importance of CSC-specific markers.[32-34] These observations indicate that the phenotype of CSCs is not as well defined as would be required for optimal detection in clinical material.

CSCs also exhibit several genetic and cellular adaptations that confer resistance towards classical therapeutic strategies. These include relative dormancy/slow cell cycle kinetics, efficient DNA repair, high expression of multidrug-resistance-type membrane transporters, and resistance to apoptosis. Cancer often acquires resistance to chemotherapy or radiotherapy after non-lethal exposure.[35,36] This process most likely represents the natural selection of resistant CSCs. Radiotherapy and most types of chemotherapy protocols exert their antineoplastic function by disrupting cancer cell DNA integrity and hence, it is possible that the oncogenic resistance of CSCs results from increased expression of DNA integrity-maintenance systems.[36] Besides, the elevated expression of drug efflux pumps may promote oncogenic resistance against chemotherapeutic agents.[37,38] Logically, combination of therapeutic regimens targeting both tumor cells as well as CSCs could be a more effective strategy to improve long-term prognosis.[39]

There are two models about the origin, maintenance, progression, and heterogeneity of tumors.[12,13,39] These models include the stochastic or clonal evolution (CE) model, and the hierarchy or CSC model.[12,13,39] According to the stochastic model, malignancy constitutes a homogeneous population of cells which generates their heterogeneity in response to some unique combinations of endogenous and exogenous factors.[40] While endogenously, these would include gene dosage effects, transcriptional and translational control mechanisms, exogenously it includes cytokine concentrations, cell–cell interactions and niche environment (Figure-6.2).[40]

On the contrary, the hierarchy model predicts malignancy in a manner analogous to the normal tissue hierarchy wherein cancer/tissue stem cells are able to produce identical daughter stem cells with self-renewal capacity, and committed progenitor daughter cells with limited, although potentially still significant, potential to divide.[40,21] The limitation of the stochastic model is that it is based on

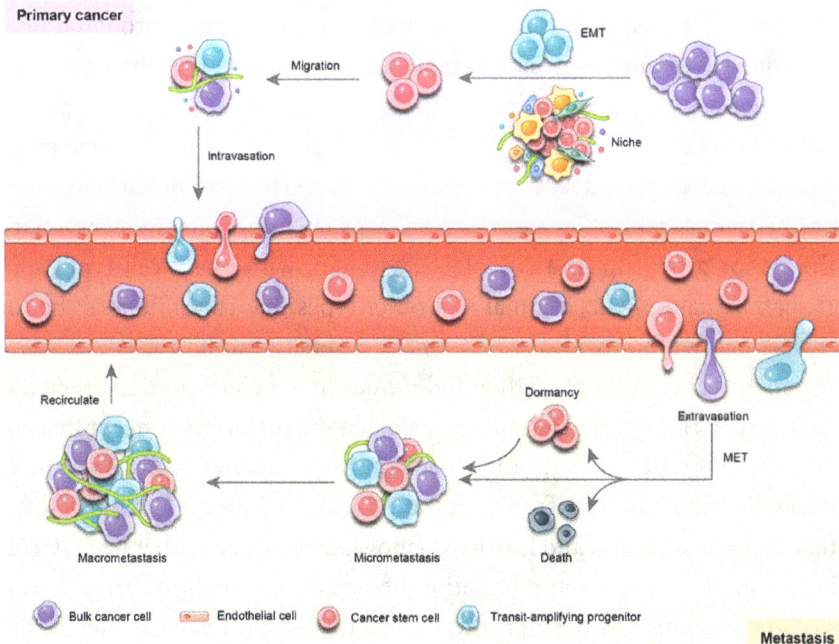

Figure-6.2.   Cells and tumor microenviroment interactions.

the unpredictable capacity to understand whether stemness is found truly within each population, or whether the cells first undergo a process of de-differentiation to a more tissue-specific stem cell-like phenotype before re-acquiring stemness in the process (Figure-6.3).[1,24,40] There is convincing evidence in the published data that cancer cells, as well as stem cells, are subject to clonal evolution during which new clones continuously develop with new genetic, and potentially epigenetic, changes.[21,24,40] Environmental factors result in constantly adapting cancer cell populations with altered characteristics in terms of rate of proliferation, metastatic potential, and drug resistance.[42] These processes could be accommodated by the CSC model as well as the hierarchical and stochastic models of heterogeneity.[1,21,40] Nevertheless, a scientifically sound and globally accepted definition regarding the CSC model is still warranted and necessitates future research to clarify the origin and the development of CSCs.

6.1.2. Role of CSCs in Carcinogenesis

Cancer is a group of diseases involving abnormal cell growth with the potential to invade or spread to other parts of the body with a high mortality rate and without sustainable treatment.[43] Peter Nowell first described the concept of clonal cell cancer evolution in 1976.[44] It has been applied to try to understand tumor growth, aggressiveness, and resistance to treatment, migration, proliferation, and mediatization. Aberrant cell division initiates cancer that also gains the augmentation ability throughout the body by some complex biochemical and signalling processes. To elucidate cancer progression, the CSC model gives a comprehensive proposal about cancer development and progression.[43] Nevertheless, much about cancer is yet to be understood. We cannot solely focus on tumor heterogeneity but also that the tumor grows up in a complex ecosystem, with many cell types such as endothelial, hematopoietic, stromal and other types that can influence the tumor main driver pathway to survival.[11,45,46] Genetic diversity, tumor micro-environment and epigenetics are coming together and influence the concept of maintenance of stem cell state.[11] This revolutionary idea changed the historical concept that tumor cells harbor stem cells, and with these active pro-normal stem cells, are rare intra-organ cells with the capacity of self-renewal, which can generate all kind of different cells that make up an organ and lead to organogenesis.[3] On the other hand, CSCs are rare intra-tumoral cells, a sub-population of cancer cells with un-bridled renewal capacity; they generate phenotypically diverse tumor cell lineages thus leading to tumorigenesis. These cells are considered highly malignant, fundamental for the growth of neoplasia, for recurrence, for drug resistance and metastasis.[47] Many signalling pathways have been shown to get dysregulated in CSCs.[48] The most well-studied and established pathways include Wnt/β-catenin, Hedgehog (Shh), Notch, and JAK/STAT3 pathways.

6.1.2.1 Wnt/β-catenin

The Wnt family of proteins transduce signals through the Frizzled (FZD) and LRP5/6 receptors to the Wnt/β-catenin and Wnt/STOP (stabilization of proteins) signalling cascades, also known as the canonical Wnt signalling cascade, and through the FZD and/or ROR1/ROR2/RYK receptors to the Wnt/PCP (planar cell polarity), Wnt/RTK (receptor tyrosine kinase) and Wnt/Ca^{2+} signalling cascades (also known as the non-canonical Wnt signalling cascades).[49–55] The canonical Wnt/β-catenin signalling cascade is involved in self-renewal of stem cells and proliferation or differentiation of progenitor cells,[56,57] whereas non-canonical Wnt signalling cascades are involved in the maintenance of stem cells, directional cell movement or inhibition of the canonical Wnt signalling cascade.[58–61] Both canonical and non-canonical Wnt signalling cascades are instrumental in the development and evolution of CSCs. Wnt activating mutations occur early during colon tumorigenesis whereas the progression of the disease is often accompanied by other genetic alterations, most commonly seen in *KRAS, BRAF, TP53,* and *SMAD4.*[34] Although these alterations are recurrently described as driver mutations in various cancers, it is still unknown which of these are required to maintain established tumors and whether interfering with Wnt signalling might be a viable therapeutic target in the background of additional drivers.

6.1.2.2 Sonic Hegdehog/GLI (Shh/GLI)

The Sonic Hegdehog/GLI (Shh/GLI) pathway has been extensively studied for its role in both developmental biology as well as cancer biology.[62] The Shh pathway is involved mainly in pattern formation during early embryonic development, while in latter stages its function in stem/progenitor cell proliferation becomes increasingly relevant.[62] During postnatal development and in adult tissues, Shh/GLI promotes cell homeostasis by actively regulating gene transcription, recapitulating the function observed during normal tissue growth. The fundamental importance of Shh/GLI in tumor growth and cancer evolution and insights into a novel mechanism of Shh action in cancer through autophagy modulation in cancer stem cells have been previously described.[62,63] In a recent study focussed on autophagy it was observed that the disruption of autophagy accelerates tumor progression in both cancer cells and the stroma that harbors tumorigenesis.[63]

6.1.2.3 *Notch*

CSCs self-renew and generate more differentiated cancer progenitor cells upon replication which possesses the capacity to dedifferentiate and acquire a stem-like phenotype by following a series of signalling pathways, molecular circuitries and epigenetic modifications.[64] Notch is one of the highly conserved pivotal signalling pathways that regulate cell proliferation, maintenance of stemness, cell fate specification, differentiation, and homeostasis of multicellular organism in general.[65] Notch also plays a key role in embryonic vasculature development.[66] Given its significance in various cellular processes, Notch signalling is one of the most activated pathways in cancer cells and their metastasis. Taking into consideration the critical participation of Notch pathway in both CSCs self-renewal and tumor angiogenesis, it has been extensively studied as a target to eliminate CSCs. The inhibition of Notch signalling has been reported as an emerging therapeutic strategy to cure cancer and eliminate CSCs.[67]

6.1.2.4 *JAK/STAT3 pathways*

STAT3 is an important regulator of cell proliferation and survival; it has a major role in the maintenance of stem cells and their differentiation and is involved in the cancerous potential of many cell types. STAT3 acts through regulation of oncogenes and tumor suppressor genes, as well as influencing tumor microenvironments.[68-72] It exerts various but sometimes contrasting functions in the normal as well as transformed cells. As STAT3 expression and activation are regulated by multiple signals and it has a role in many signalling pathways, STAT3 is considered as a flexible and adaptable regulator of cell function in different types of cells under different conditions and regulate gene expression either directly or indirectly through interaction with other transcription factors.[73]

Many novel small molecules are now being developed and tested in clinical trials to block the above-mentioned signalling pathways, which otherwise become dysregulated in CSCs. Some of these small molecules block the self-renewal and induction of apoptosis in CSCs.[45] Although not recognized as kinase inhibitors, they act by inhibiting the Wnt/β-catenin pathway, STAT3 pathway, NOTCH pathway and the Shh pathway. The STAT3 pathway is critical for the self-renewal and survival of CSCs in various neoplasms. Inhibition of STAT3 pathway inhibits cell proliferation *in vitro* and reduces tumor growth *in vivo*.[74,75] CSCs are also involved in tumor relapse and mediastization possibly due to mutations or epigenetic modifications in the daughter CSCs that exhibit more aggressive growth to become the driver of tumor formation thereafter.[76]

Genetic signatures in CSCs are thought to predict tumor recurrence and metastases, providing support for the concept that CSCs are the metastatic precursors.[77] For example, expression of the CSCs marker CD133 in glioblastoma and lung adenocarcinomas has been correlated with both the expression of cell proliferation marker Ki67 and poorer clinical outcomes.[78,79] CD133 expression has also been correlated with patient survival in high-grade oligodendroglial tumors,[80] rectal cancer,[81] gastric adenocarcinomas,[82] and non-small cell lung cancer.[83] Additionally, in patients with colorectal carcinoma, combined expression of CD133, CD44, and CD166 successfully identify the patients at low-, intermediate-, and high-risk of recurrence and metastasis.[84] Likewise, methylation of Wnt target-gene promoter is a strong predictor of colorectal cancer recurrence thus suggesting that CSC gene signatures, rather than reflecting CSC number, reflect the differentiation status of the malignant tissue and the risk for dissemination.[85] One of the key steps in the metastatic cascade is the migration of tumor cells to the distant tissues and organs away from the primary tumor that is facilitated by CSCs migration. The emigrational potential of cells is a physiological process in development, and tumor cells appear to capitalize on these physiologic mechanisms. Most adult tissues maintain some aspect of this emigrational potential primarily through epithelial to mesenchymal transition (EMT)-like process during wound healing, tissue regeneration, and organ fibrosis. It has been hypothesized that CSCs may also activate their migration through the process of EMT (Figure-6.2).

During the final stages of cell division, each daughter cell must lose contact with each other to generate independent progeny. The final step in this process occurs within a tube or bridge that is connecting the two daughter cells while a protein structure called the mid-body is essential for the process of separating the two cells. Cancer cells accumulate mid-body derivatives, which enhance the tumorigenicity of cancer cells.[86] Moreover, several microRNAs (miRNAs) also participate in the activation of CSC-like activities.[87]

6.2 DEFINITION AND ROLE OF THE TUMOR MICROENVIRONMENT

Although researchers now have a general understanding of most characteristics of cancer,[88] the characteristics promoting cancer formation remain less well-understood. After the 'ecological therapy' strategy was widely employed,[88] much effort has been devoted to determining how cellular and non-cellular components of the tumoral niche help the tumors to acquire these characters. These cellular and non-cellular components of the tumoral niche comprise the tumor microenvironment (TME).[88–90] It is well accepted that the TME[91,92] comprising of stromal fibroblasts, inflammatory/immune cells, neuronal cells, the vasculature, and the extracellular matrix, etc. influences tumorigenesis, but the potential impact of TME on the origin of cancer cells has only come to light recently (Figure-6.1).[93] Strikingly, inflammation can alter the fate of the cells that are normally refractory to cellular transformation and convert them into stem-like cells capable of tumor initiation. Tumor microenvironment or niche remains a major factor that extrinsically influences the tumor heterogeneity (Figure-6.2). Tumor niche comprises of various cell types, i.e., stromal cells, immune cells, endothelial cells, and cancer cells per se, as well as connective tissue components, growth factors, and cytokines that play an essential role in CSC maintenance/enrichment, preservation of the phenotypic plasticity, immune-surveillance, differentiation/dedifferentiation, angiogenesis activation and invasion/metastasis.[94-96] CSCs reside in the tumor niche which not only provides the much needed physical support for CSCs but also fundamentally influences their functionality. A tumor can locally and metastatically colonize at suitable sites with a central role for CSCs in these processes.

6.2.1 Niche Components that Contribute to the Stemness of CSCs

Using a cell-lineage-tracing approach, Tammela *et al.* [96] found that lung tumors i.e., adenocarcinomas, populate tumor cells that produce a mix of two cell types: tumor cells and (non-tumor) support cells that constitute the tumor niche.[101] The niche cell population derived from the tumor cells expresses the enzyme porcupine that contributes to maturation of Wnt signalling protein in the endoplasmic reticulum which is secreted from the cell. The binding of Wnt protein with its receptor on a tumor cell activates the downstream signalling to drive tumor growth.[97] Lim *et al.*[98] investigated a different type of lung tumor and reported that the niche cells can also support tumor growth through the secretion of a protein that activates the Notch signalling pathway in the tumor cells.[98]

6.2.1.1 *Endothelial cells*

The vascular endothelium is a dynamic cellular "organ" that controls the passage of nutrients into the tissues, maintains the flow of blood, and regulates the trafficking of leukocytes.[99] In tumors, various factors i.e., hypoxia and chronic exposure to growth factor stimulation, result in endothelial dysfunction. Tumor-associated endothelial cells play a key role in the cancer process. On the one hand, they form tumor-associated (angiogenic) vascular structures through sprouting of the locally pre-existing blood vessels or *via* recruitment of bone marrow-derived endothelial progenitor cells, to provide nutritional support to the growing tumor (Figure-6.2).[100]

On the other hand, they are at the interface between circulating blood cells, tumor cells and the extracellular matrix, thereby playing a central role in various functions including controlling leukocyte recruitment, tumor cell behavior, and metastasis. Hypoxia is a critical parameter modulating the tumor microenvironment and endothelial/tumor cell interactions through stimulating tumor cells to produce pro-angiogenic factors and factors supporting the migratory activity of tumor cells, thus promoting metastasis.[100]

It was noticed a long time ago that tumor blood vessels were morphologically deviating from the normal structure.[101] Three-dimensional scanning electron microscopy of vascular plaster casts showed networks of tortuous endothelium that was missing the normal hierarchical arrangement of artery-arteriole capillary.[101] Poor tumor vessel stability is caused by defects in the pericytes, which are in lower abundance and are loosely attached compared to normal vessels, thus effecting the vascular stability and hence the blood flow.[101,102] This is evident from the observations that some tumor vessels remain un-perfused whereas the others are perfused but may have blood flowing in reverse directions. For example, vascular endothelial growth factor (VEGF) and some of the pro-inflammatory chemokines are also immune modulators, which increase angiogenesis and lead to immune suppression.[103] Amongst these pro-angiogenic factors, VEGF, one of the main angiogenic modulators, also plays a critical role in the control of immune tolerance. Albini *et al.* have discussed the regulation of angiogenesis by innate immune cells in the tumor microenvironment, specific features, and roles of major players: macrophages, neutrophils, myeloid-derived suppressor and dendritic cells, mast cells, γδT cells, innate lymphoid cells, and natural killer cells.[103] Anti-VEGF or anti-inflammatory drugs could balance out an immunosuppressive microenvironment into an immune-permissive one. Anti-VEGF, as well as anti-inflammatory drugs, could, therefore, represent partners for combinations with immune checkpoint inhibitors, enhancing the effects of immune therapy.[103]

6.2.1.2 Extracellular matrix (ECM)

The extracellular matrix (ECM) is composed of various proteins including collagen, proteoglycans, laminin, and fibronectin. Even amongst these ECM components, some subtypes that further specify their properties and functions.[104] The function of ECM may be best described in the context of embryonic development. The development of a mammalian embryo to a fully developed organism is a well-orchestrated and meticulously controlled process. It is tightly regulated in terms of the spatiotemporal composition, amount, and characteristics of the ECM. Several studies have shown that mutated ECM components lead to birth defects or even embryonic lethality in some cases, which emphasizes its role in development.[105,106] The geometry, rigidity, and other physical properties of the ECM are sensed by the cells and ultimately direct their adherence, proliferation, migration and differentiation, thus culminating into the complex spatial and structural arrangements they form in the tissues. The ECM influences the migration track and the rate of migrating cells through its topography, composition, and physical properties. The alignment of the underlying ECM directs cell migration and proliferation. The traditional perspective of cancer has shifted to reflect the important role of the ECM in regulating cell proliferation, migration, and apoptosis. As the tumor cells proliferate, the surrounding ECM undergoes significant architectural changes in a dynamic interplay between the microenvironment and resident cells. These changes, including increased secretion of fibronectin and collagens I, III, and IV, show that tumor progression necessitates an uninterrupted and close interaction between the ECM and the tumor cells.[107] Increased deposition of ECM proteins promotes tumor progression by interfering with cell–cell adhesion, cell polarity, and ultimately amplifying growth factor signalling.[108]

6.2.1.3 *Cancer-associated fibroblasts (CAFs)*

Cancer-associated fibroblasts (CAFs) constitute a primary source of the fibrotic ECM. CAFs organize collagen fibrils which undergo biomechanical alterations to provide pathways for the invading tumor cells either under the guidance of CAFs or following their EMT.[109] The increased hyaluronan (HA) metabolism in the tumor microenvironment instructs the cancer cells to initiate and disseminate multiple functions. The key effects of HA reviewed here include its role in activating CAFs during the pre-malignant and malignant stroma and facilitate invasion by promoting motility of both CAFs and tumor cells, thus enabling their invasion to the nearby tissues. The circulating CAFs (cCAFs) also form heterotypic clusters with circulating tumor cells (CTC), which are considered to be precursors of metastatic colonies.[109]

Clinically, CAF-like fibroblast-induced stromal ECM changes precede the process of tumor formation and these early changes in ECM have prognostic significance that permits risk stratification. For example, high mammographic density is a strong risk factor in breast cancer.[110-112] The important clinical features of this condition, which precede the subsequent detectable tumor formation, include adipocyte loss and high ECM production. For example, this condition has been linked to the loss of expression of the mesenchymal differentiation regulator

CD36 in the stromal fibroblasts, which phenocopies the clinical features of high mammographic density breast tissue.[113-115]

A number of studies using mouse models also predict that elevated extracellular matrix component HA production, primarily by fibroblasts, pre-disposes epithelial cells to tumor initiation. Examples include evidence that an HA-rich stroma precedes increased mammary tumor formation in transgenic mice expressing both MMTV-driven HAS2 and a c-neu proto-oncogene.[116] CAFs play a significant role in tumor dissemination by inducing an invasive phenotype in the tumor cells via promoting motile phenotypes and remodeling in the ECM. Invasion is achieved in part by CAF-driven EMT and consequent cell migration that is driven by factors such as TGF-β, HGF-1, and CXCL12/SDF-1α.[116] Paladin-expressing CAF create "tunnels" in the ECM, which cancer cells migrate through.[110] Under CAF guidance, tumor cells also migrate and invade as groups in the absence of apparent EMT. This collective migration and invasion is driven by heterotypic E-cadherin/N-cadherin interactions between tumor cells and CAFs that result in a mechanically active adhesion.[111]

6.2.1.4 *Tumor-associated macrophages*

The tumor microenvironment is a complex assembly of a genetically heterogeneous population of cancer cells supported in the sustenance of their biological activity by different cell types that constitute the local environment.[116] Tumor-associated macrophages (TAMs) are one of the most abundant immune cells in the tumor microenvironment of solid tumors.[117,118] TAMs are one of the most abundant immune cells in the microenvironment of solid tumors and their presence correlates well with reduced survival in most cancers. They are present during all stages of tumor progression and stimulate angiogenesis, tumor cell invasion, and intravasation at the primary site.[116] At the metastatic site, macrophages and monocytes prepare for the arrival of disseminated run-away tumor cells from their primary location and support extravasation and survival by inhibiting their immune-mediated clearance or by directly engaging with tumor cells to activate pro-survival signalling pathways. Moreover, macrophages also promote the growth of the disseminated tumor cells at the metastatic site by organizing and supporting the formation of a supportive metastatic niche. Various researchers have independently reported a strong correlation between the density of macrophages and poor survival in carcinomas of pancreas and breast, lung, cervix, the bladder, and Hodgkin's lymphoma.[119-123] The expression of colony-stimulating factor-1 (CSF-1), the major lineage regulator for macrophages, or its receptor CSF-1R correlates with poor survival in liver, breast and pancreatic cancer,[124,125] respectively. A macrophage transcriptional signature in patients with breast cancer has been reported as a predictor of poor prognosis and reduced survival in the patients.[126,127] They are also involved in the recurrence and mediatization for their several pro-tumorigenic functions that have important roles in cancer development and progression, such as the ability to express and secrete cytokines and induce tumor angiogenesis.[128]

While elucidating the underlying molecular mechanism, it has been reported that the release of CSF by the tumor cells induces EGF expression in TAMs.[120] This autocrine loop leads to the co-migration of tumor cells and TAMs towards the blood vessels where TAMs produce VEGF-A to promote increased vessel permeability. Additionally, TAMs-derived molecules such as secreted protein acidic and rich in cysteine (SPARC; a multifaceted matricellular protein), C-C motif ligand 18 (CCL18), and proteases promote increased tumor cell invasion and migration. At the metastatic site, tumor cell-derived CCL2 recruits inflammatory monocytes to the metastatic site, where they differentiate into metastasis-associated macrophages (MAPs) that produce VEGF-A and cathepsin S to promote cancer cell extravasation. MAPs promote survival at the metastatic site through the expression of integrin $\alpha 4$ that engages VCAM1 on the tumor cells at the metastatic site, which increases tumor cell survival through PI3K/Akt signalling.[120] MAPs also bind with fibrin complexes on the tumor cell-associated platelets, which increase tumor cell survival in the initial phase of metastatic colonization. MAPs promote metastatic niche formation and release granulin that activates HSTC to produce ECM molecules, such as collagen and periostin, which enhances the colony formation abilities of cancer cells in the metastatic niche.[120]

6.3 CSCs AND MICROENVIRONMENT INTERACTION

The modulation of CSCs activities by the tumor microenvironment is still poorly understood. CSCs and tumor microenvironment mutually interact in a unique manner depending on the tumor microenvironment cells (endothelial, epithelial, extracellular matrix (EMT), stromal and macrophage) which respond to signals from the CSC or vice versa.

6.3.1 The Endothelial Compartment

The endothelial cell compartment is considered a key player in supporting CSC-phenotype.[129] It is well established now that endothelial cells maintain stem-like cells and their activities in tumors, exerting their functions by secreting growth factors, such as epidermal growth factor (EGF), that induces EMT and stem cell features in tumors, as previously described for human head and neck squamous cell carcinoma (SCC). Endothelial cells also promote cancer cell conversion towards the endothelial phenotype. Moreover, it has been reported that EGF inhibition in the endothelial cells rendered the *in vivo* xenograft-derived tumors less invasive and contained a lower proportion of ALDH+CD44+ CSCs.[129] Different conditions, i.e., hypoxia and neo-angiogenesis, are the leading causes that confer on cancer cells the ability to behave like endothelial cells as a consequence to promote their adoption of CSC-phenotype. About these conditions:

(a) Hypoxia promotes aerobic glycolysis in the tumor cells in order to survive in the oxygen-free environment, thus contributing to tumor growth and metastasis. Hypoxia-inducible factor (HIF) is the main tissue controller of oxygen homeostasis.[130] Besides hypoxia, changes in pH can also regulate stem cell behavior by modulating their metabolic status and promoting metabolic re-configuration of cancer cells towards glycolysis, induction of the EMT phenotype (including C-X-C-chemokine receptor 4 (CXCR4), Snail and Twist gene expression), increasing in the number and renewal potential of CSCs, as well as induction of pluripotency-associated transcription factors i.e., Oct-3/4, Nanog and Sox-2.[118] This scenario indicates that "stemness" is more a cellular state than a cancer cell characteristic, modulated by the microenvironment.

(b) Neo-angiogenesis ensures the much-needed nutrient and oxygen supply that is essential for cancer cell survival, growth, and dissemination.[131,132] Tumor cells develop their vasculature through different mechanisms, including formation of new vessels from pre-existing ones, simulation of the vasculature through vasculogenic mimicry and recruitment of endothelial progenitor cells. The vasculogenic mimicry (VM) was first reported in melanomas and referred to *de novo* formation of tubular structures that were perfused by plasma and red blood cells.[133] With increasing knowledge of CSCs-phenotypes and functions, there is mounting evidence which supports the notion that CSCs are involved in VM as promoters of tumor vascularization.[134,135] This could be justified by the CSCs' lineage plasticity in generating tumor.

6.3.2 The Extracellular Matrix

The EMT is a potent driving factor in tumor initiation and progression.[136,137] EMT and CSCs have an inherent relation[138] that has been implied in the metastasis of human tumors.[139,140] This interplay contributes to the mechanism through which CSCs reside in a tissue dormant for years, and later primes tumor recurrence or metastasis in cancer patients (Figure-6.3). There is a mechanism which induces

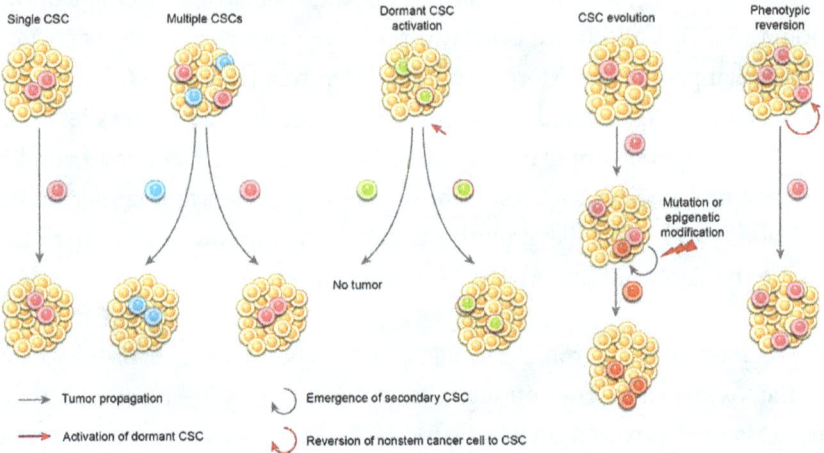

Figure-6.3.   The cancer stem cells interplay: mechanisms of dormancy and of recurrence.

the cancer cells to not only lose their cell–cell adhesions and exhibit elevated motility and invasion but also to gain increased resistance to apoptosis, elevated endurance to chemotherapeutic intervention and develop stem-cell like properties through EMT.[107]

During the progression of tumors, the ECM becomes more disorganized due to the influence of local modulators, thus tipping the ECM as a master regulator in tumor progression through providing sustained proliferative signals, evading growth suppressors, resisting cell death, enabling replicative immortality, inducing angiogenesis and promoting invasion and metastasis. The increased expression and activity of matrix metalloproteinases (MMPs) and collagen cross-linkers are also preponderant for the modulation of ECM within the TME and are generally responsible factors in the poor prognosis.[140,141] Indeed, MMPs are major players in cell invasion, since they are responsible for proteolysis and detachment of tumor cells from the ECM, resulting in CSCs formation and metastasis.[142]

6.3.3 Cancer-Associated Fibroblasts

CAFs in the stroma are influential in reverting differentiated cells towards a de-differentiated phenotype.[143] CAFs support multiple aspects of cancer progression including tumor initiation, invasion, and metastasis. The first evidence in this regard was published by Nai *et al.* who reported that CSCs are one of the key sources of CAFs in the tumor niche.[144] This has been proposed as one of the primary mechanisms in generating CSCs.[15] Recent studies indicate that CAFs have substantial clinical implications in terms of disease staging and cancer recurrence. However, CAFs have not been fully characterized due to several limitations. The first limitation is the uncertainty regarding the origin of CAFs.[144] CAFs have been reported to originate from epithelial cells, mesenchymal stem cells, adipocytes, resident fibroblasts, and bone marrow stem cells.[119] The divergent sources of CAFs account for their broad range of characteristics and molecular markers. Secondly, it is difficult to isolate and maintain CAFs which significantly hampers their characterization. Notably, the microenvironment that supports the growth of CAFs is similar to the microenvironment that supports the viability of CSCs. Recent studies suggest that several types of stromal cells in the niche play a pivotal role in maintenance of the very small population of CSCs which are responsible for cancer recurrence and chemotherapeutic drug resistance.[144,145] However, it remains unclear whether CSCs directly support tumor maintenance and survival by generating CAFs.

First described by Otto Warburg in 1956,[146] metabolic reprogramming in cancer cells involved a shift in energy metabolism away from an oxidative cycle to a glycolytic one — even under aerobic conditions — subsequently termed the "Warburg effect" or "aerobic glycolysis". In this respect, CAFs exert a metabolic reprogramming of cancer cells by inducing a reverse Warburg phenotype.[146] Spreading of tumor from local to distant sites necessitates a supportive and accommodating environment for the disseminating cancer cells. The so-called "metastatic niche" may also be a native stem cell niche of the distant organ, enhancing stem cell properties while repressing differentiation.[146] Overall, the role of CAFs and CSCs regarding the metastatic progression of the tumor has not been fully demarcated. In this regard, CAFs contribute to the metabolic reprogramming of cancer cells by inducing a reverse Warburg phenotype.[144-147]

6.3.4 Tumor-Associated Macrophages (TAMs)

Like normal stem cells, CSCs exist in a cellular niche comprised of numerous cell types including TAMs which provide a unique microenvironment to protect and promote CSC functions.[107] TAMs provide pivotal signals to promote CSCs survival, self-renewal, maintenance, and migratory ability, and in turn, CSCs deliver tumor-promoting cues to TAMs that further enhance tumorigenesis. Studies during the last decade have primarily focused on understanding the molecular mediators of

CSCs and TAMs, and recent advances have begun to elucidate the complex cross-talk that occurs between the two cell types.[107] Another area of intense investigation has been to understand the role of inflammatory cells in the CSCs niche. The tumor microenvironment is characterized by chronic inflammation which favors tumor formation by stimulating cell proliferation, activating CSCs, and promoting metastasis.[147,148] In this regard, TAMs lead the tumor inflammatory response TAMs.[149,150] A correlation between high numbers of TAMs and rapid disease progression has been established with poor patient outcome[39,151]; however, this paradoxical phenotype has been explained only recently. While TAMs in the pre-invasive niche contribute to oncogenic transformation and survival, a growing body of evidence suggests that they are critical for the self-renewal and maintenance of CSCs in the established tumors. STAT3 and NFκB are the key regulators of these processes. Once infiltrated into tumors, TAMs contribute to chronic inflammation by secreting inflammatory cytokines i.e., IL-1β, IL-6, and IL-8 (CXCL8).[119,152] In addition to mediating CSCs' self-renewal and expansion, TAMs are also responsible for the maintenance of the CSCs niche. A recent study by Lu *et al.* demonstrated juxtacrine signaling by TAMs and tumor-associated monocytes with mouse mammary CSCs to support the maintenance of a stem-like state.[153]

While numerous studies have demonstrated that TAMs directly regulate CSCs' self-renewal and maintenance, there is a growing body of research that suggests that CSCs, in turn, recruit macrophages to solid tumors and enhance a pro-tumor phenotype in the TAMs. Zhou *et al.* have reported that the ECM protein periostin is preferentially expressed on CD133$^+$CD15$^+$ glioma stem cells and recruits macrophages through integrin αvβ3 from the peripheral blood to the brain. Deletion of periostin in glioma stem cells decreases M2 TAM density, reduces tumor growth, and consequently increases survival of the glioblastoma xenografts.[154–156]

6.4 FUTURE PERSPECTIVES

Chemotherapy and radiotherapy, either alone or combined with surgery,[157] mainly target the fast-cycling cells. The role of CSCs in tumor formation and progression is being highlighted, however it is still difficult to synthesize CSCs-targeting drugs due to lack of their specific surface markers that could be exploited as targets. Since the discovery of several important mutations that contribute to carcinogenesis, i.e., EGF receptor, p53, and c-Myc, they have been extensively studied as targets for the development of more selective drugs for cancer therapy. Despite the effectiveness of these drugs, multidrug resistance (MDR) is on the rise which often results in tumor relapse.[158] To be therapeutically effective, an anti-cancer agent should be uniformly distributed throughout the tumor circulation, across the vessel wall, and pass through the ECM. On the other hand, tumors create multiple obstacles to drug transport mechanisms, hence, the requirements for effective drug delivery may vary considerably.[158] Expression and proteomic profiling of the individual cell types constituting the cancer microenvironment represent important advances.[158,159] Besides targeting the surviving CSCs, the contemporary oncology research is now mainly focused on innovative therapy targeting both CSCs, TME as well as tumor microenvironment.[158–160]

6.4.1 Targeting CSCs

Genetic variability and genomic instability of the CSCs are the primary hurdles in the development of CSC-specific drugs. Currently, efforts are underway in targeting CSCs' surface markers. One of the most established and commonly used CSC biomarkers is CD44 that has been targeted using anti-CD44 specific antibody to successfully eradicate acute myeloid leukemia.[161,163] Similarly, CD133, a transmembrane glycoprotein well-known in several tumors such as glioblastoma, hepatocellular and colon cancers, has been targeted using anti-CD133 antibody conjugated with a potent cytotoxic drug, monomethyl auristatin. This antibody-drug conjugate was efficiently internalized, co-localized with the lysosome and showed high effectiveness.[163]

Besides antibody-based drug targeting, nanoparticles are being used as an interesting strategy to target CSCs with minimal damage to surrounding normal cells. In this regard, construction of glucose-coated gold nanoparticles (Glu-GNPs) has been shown to facilitate the entry of GNP into leukemia stem cells overexpressing CD44 (TH1-P) with promise.[164] Similarly, mangostin-encapsulated Poly (lactic-co-glycolic acid) nanoparticles (Mang-NPs) have successfully downregulated the known stemness genes c-Myc, Nanog and Oct4, besides abrogation of two CSC-specific markers, i.e., CD24 and CD133, and blocking Shh pathway.[165] Another example of nanoparticle therapy is represented by salinomycin and paclitaxel that are also used to eradicate breast cancer cells including CD44 breast CSCs.[129] Another unraveled aspect is the targeting of CSCs' mitochondrial biogenesis due to their strict dependence on mitochondrial activity. For this reason, specific antibiotics that inhibit mitochondrial biogenesis were studied.[166] An example that supported this new therapeutic approach is doxycycline with positive results in cancer patients,[167] and by metformin which seems to eliminate CSCs. The combination of these two drugs seems to enhance anti- tumor activity. In summary, as the CSCs are more resistant to conventional cancer therapies than non-CSCs their elimination is crucial in treating malignant diseases.[168]

6.4.2 Targeting Tumor Microenvironment

The TMEs are instrumental in mediating the resistance of CSCs to anti-cancer therapies. CSCs in glioblastoma, which are inherently radio-resistant, are "protected" from conventional therapies by factors within the vascular niche, thus enabling CSCs to cause tumor relapse. Hence, treatments that disrupt the aberrant vascular environment may be active against glioblastoma. Various clinical trials of the anti-angiogenic drugs such as bevacizumab[169] and cediranib (AZD2171) have achieved encouraging results in patients with glioblastoma.[170] Moreover, the CSCs associated with the stromal cells are near blood vessels forming a niche characterized by severe hypoxia and increased angiogenesis.[79,83] These aspects of the tumor microenvironment have been explored as possible pharmaceutical targets to eliminate CSCs.[171] The pathways involved in angiogenesis provide a crucial target of cancer therapy. A plethora of anti-angiogenic agents have been developed and tested in preclinical experiments.[163] For example, bevacizumab, a humanized monoclonal antibody that sequesters vascular endothelial growth factor (VEGF) to impair VEGF signaling was approved for the treatment of metastatic colorectal cancer and other metastatic cancers, including non-squamous NSCLC and cervical cancer.[164,165] Another strategy to inhibit tumor angiogenesis is the use of tyrosine kinase inhibitors, such as sorafenib, an inhibitor of VEGFR-1, -2, -3 and PDGFR-b, and sunitinib that blocks VEGFR-2 and PDGFR phosphorylation. The former was used for the treatment of metastatic renal cell carcinoma and un-resectable hepatocellular carcinoma, and the second treatment for gastrointestinal tumor and metastatic renal cell carcinoma.[171] Although anti-angiogenic therapy may potentially have clinical implications, the increase of oral somministration (OS) is insufficient. This is probably due to acquired resistance, the increment of tumor hypoxia and the diminished delivery of chemotherapeutic agents.[172-174]

Concerning the tumor microenvironment ECM, several drugs targeting matrix metallopeptidases (MMPs) have been developed. For example, cyclinide (also known as CMT-3 and COL-3) is an MMP inhibitor that went through several clinical trials for advanced carcinomas (Clinical trials NCT00004147, NCT00003721, NCT00001683, and NCT00020683). The new frontier of cancer treatment is aimed at strengthening the immune system's defense against cancer cells. Targeting T-regulatory lymphocytes (Tregs) directly in the TME has been proposed as another method to re-establish the anti-tumoral immune response. Tregs cells in tumors are immune-enriched for the cell surface markers CTLA-4 and OX-40, to deplete Tregs from the TME. By directly injecting mouse tumors with anti-CTLA-4 and anti-OX-40 antibodies to deplete Tregs, along with the TLR9-activating agonist CpG to trigger the innate immune response, the authors showed the establishment of a systemic antitumor immune response capable of eradicating disseminated disease in mice. Furthermore, this treatment modality was effective against established lymphoma in the central nervous system, which is traditionally considered to be a sanctuary for tumor cells in the face of systemic therapies. This study suggested that antibody therapy could be used to target tumor infiltrating lymphocytes (TILs) locally, thereby inducing an effective systemic immune response. This pioneering therapy has not been tested sufficiently in the clinical setting. More importantly, Tregs produced immunoregulatory factors which might include TGF-β, IL-10, and IL-35. Of these, TGF-β seems a particularly desirable target due to its roles in promoting metastasis and tumor stroma formation, besides its potent inhibitory effects.

Several molecules based on the inhibition of immune checkpoints have been approved by the FDA since 2011. The most promising of these therapies have been represented by the antibodies targeting the cytotoxic T lymphocyte-associated protein 4 (CTLA-4) or the programmed cell death 1 (PD-1) pathway, administered as single or in combination therapy. To induce antitumor responses, T-cells are initially activated in the lymph node in two subsequent steps which are: the engagement of T-cell receptor (TCR) with a tumor antigen MHC complex on antigen-presenting cells (APCs) and the binding of CD28 to the costimulatory molecule B7. Following T-cell activation, CTLA-4 translocates from the intracellular compartment to the cells' surface to compete with the costimulatory molecules, causing the inhibition of T-cell proliferation. The blockade of this essential immune checkpoint with monoclonal antibodies enables T-cells to active, expand and reach the tumor burden, where they can find the cognate antigen presented by cancer cells. These mechanisms are generally implemented to impede the overstimulation of the immune system. Nevertheless, in the context of cancer, they become detrimental to cancer cell elimination. Hence, an immune checkpoint blockade may be exploited to potentiate an anti-tumor immune response.[175–178]

In summary, future therapies need to be optimized for improving their effects inside the tumor microenvironment, efficiently accessing CSCs, with the result to reduce side effects in patients. Gaining a better understanding of the relationship between TME and CSCs at each stage of tumor development and progression, we may discover new approaches to interfere with the TME-CSC cross talk.

REFERENCES

1. Rich JN. Cancer stem cells: understanding tumor hierarchy and heterogeneity. *Medicine*. 2016; 95: S1 (e4764).

2. Huntly BJ, Gilliland DG. Leukaemia stem cells and the evolution of cancer-stem-cell research. *Nat Rev Cancer*. 2005; 5: 311–321.

3. Yu Z, Pestell TG, Lisanti MP, Pestell RG. Cancer stem cells. *Int J Biochem Cell Biol*. 2012; 44(12): 2144–2151.

4. Reya T, Clevers H. Wnt signalling in stem cells and cancer. *Nature*. 2005; 434: 843–850.

5. Rosen JM, Jordan CT. The increasing complexity of the cancer stem cell paradigm. *Science*. 2009; 324: 1670–1673.

6. Shimono Y, Zabala M, Cho RW, Lobo N, Dalerba P, Qian D, Diehn M, *et al*. Downregulation of miRNA-200c links breast cancer stem cells with normal stem cells. *Cell*. 2009; 138: 602–603.

7. Shiozawa Y, Pienta KJ, Taichman RS. Hematopoietic stem cell niche is a potential therapeutic target for bone metastatic tumors. *Clin Cancer Res*. 2011; 17: 5553–5558.

8. Takahashi K, Yamanaka S. Induction of pluripotent stem cells from mouse embryonic and adult fibroblast cultures by defined factors. *Cell*. 2006; 126: 663–676.

9. Sutherland KD, Visvader JE. Cellular mechanisms underlying intertumoral heterogeneity. *Trends Cancer*. 2015; 1(1): 15–23.

10. Burrell RA, McGranahan N, Bartek J, Swanton C. The causes and consequences of genetic heterogeneity in cancer evolution. *Nature*. 2013; 501: 338–345.

11. Kreso A, Dick JE. Evolution of the cancer stem cell model. *Cell Stem Cell*. 2014; 14: 275–291.

12. Gerdes MJ, Sood A, Sevinsky C, Pris AD, Zavodszky MI, Ginty F. Emerging understanding of multiscale tumor heterogeneity. *Front Oncol*. 2014; 4: 366.

13. Michor F, Polyak K. The origins and implications of intratumor heterogeneity. *Cancer Prev Res (Phila)*. 2010; 3: 1361–1364.

14. Stanta G, Bonin S. Overview on clinical relevance of intra-tumor heterogeneity. *Front Med (Lausanne)*. 2018; 5: 85.

15. Ellsworth RE, Blackburn HL, Shriver CD, Soon-Shiong P, Ellsworth DL. Molecular heterogeneity in breast cancer: state of the science and implications for patient care. *Semin Cell Dev Biol*. 2017; 64: 65–72.

16. Biankin AV, Piantadosi S, Hollingsworth SJ. Patient-centric trials for therapeutic development in precision oncology. *Nature*. 2015; 526: 361–370.

17. Davis C, Naci H, Gurpinar E, Poplavska E, Pinto A, Aggarwal A. Availability of evidence of benefits on overall survival and quality of life of cancer drugs approved by European Medicines Agency: retrospective cohort study of drug approvals 2009–2013. *BMJ*. 2017; 359: j4530.

18. Schork NJ. Personalized medicine: time for one-person trials. *Nature*. 2015; 520: 609–611.

19. Bao B, Ahmad A, Azmi AS, *et al*. Overview of cancer stem cells (CSCs) and mechanisms of their regulation: implications for cancer therapy. *Curr Protoc Pharmacol*. 2013; Chapter 14: Unit 14.25.

20. Lathia JD, Hitomi M, Gallagher J, Gadani SP, Adkins J, Vasanji A, Liu L, *et al*. Distribution of CD133 reveals glioma stem cells self-renew through symmetric and asymmetric cell divisions. *Cell Death Dis*. 2011; 2: e200.

21. Easwaran H, Tsai HC, Baylin SB. Cancer epigenetics: tumor heterogeneity, plasticity of stem like states, and drug resistance. *Mol Cell*. 2014; 54: 716–727.

22. Vartanian A, Singh SK, Agnihotri S, Jalali S, Burrel K, Aldape KD, Zadeh G. GBM's multifaceted landscape: highlighting regional and microenvironmental heterogeneity. *Neuro Oncol*. 2014; 16: 1167–1175.

23. Prasetyanti PR, Medema JP. Intra-tumor heterogeneity from a cancer stem cell perspective. *Mol Cancer*. 2017; 16: 41.

24. Driessens G, Beck B, Caauwe A, Simons BD, Blanpain C. Defining the mode of tumor growth by clonal analysis. *Nature [Internet]*. 2012; 488: 527–530.

25. Schepers AG, Snippert HJ, Stange DE, van den Born M, van Es JH, van de Wetering M, *et al*. Lineage tracing reveals Lgr5+ stem cell activity in mouse intestinal adenomas. *Science*. 2012; 337: 730–735.

26. Chen J, Li Y, Yu TS, McKay RM, Burns DK, Kernie SG, *et al*. A restricted cell population propagates glioblastoma growth after chemotherapy. *Nature*. 2012; 488: 522–526.

27. Medema JP. Cancer stem cells: the challenges ahead. *Nat Cell Biol*. 2013; 15: 338–344.

28. Singh SK, Hawkins C, Clarke ID, Squire JA, Bayani J, Hide T, *et al*. Identification of human brain tumor initiating cells. *Nature*. 2004; 432: 396–401.

29. Al-Hajj M, Wicha MS, Benito-Hernandez A, Morrison SJ, Clarke MF. Prospective identification of tumorigenic breast cancer cells. *Proc Natl Acad Sci*. 2003; 100: 3983–3988.

30. Visvader JE, Lindeman GJ. Cancer stem cells. Current status and evolving complexities. *Cell Stem Cell*. 2012; 10: 717–728.

31. Meacham CE, Morrison SJ. Tumor heterogeneity and cancer cell plasticity. *Nature*. 2013; 501: 328–337.

32. Andriani F, Bertolini G, Facchinetti F, Baldoli E, Moro M, Casalini P, Milione M, *et al*. Conversion to stem-cell state in response to microenvironmental cues is regulated by balance between epithelial and mesenchymal features in lung cancer cells. *Mol Oncol*. 2016; 10: 253–271.

33. Chaffer CL, Marjanovic ND, Lee T, Bell G, Kleer CG, Reinhardt F, D'Alesso AC, *et al*. Poised chromatin at the ZEB1 promoter enables breast cancer cell plasticity and enhances tumorigenicity. *Cell*. 2013; 154: 61–74.

34. Vermeulen L, De Sousa EMF, van der Heijden M, Cameron K, deJong JH, Borovski T, Tuynman JB, *et al*. Wnt activity defines colon cancer stem cells and is regulated by the microenvironment. *Nat Cell Biol*. 2010; 12: 468–476.

35. Blagosklonny MV. Why therapeutic response may not prolong the life of a cancer patient: selection for oncogenic resistance. *Cell Cycle*. 2005; 4(12): 1693–1698.

36. G. Frosina. DNA repair in normal and cancer stem cells, with special reference to the central nervous system. *Current Med Chem*. 2009; 16(7): 854–866.

37. Dean M, Fojo T, Bates S. Tumor stem cells and drug resistance. *Nature Rev Cancer*. 2005; 5(4): 275–284.

38. Liu G, Yuan X, Zeng Z, *et al*. Analysis of gene expression and chemoresistance of CD133+ cancer stem cells in glioblastoma. *Mol Cancer*. 2006; 5: 67.

39. Plaks V, Kong N, Werb Z. The cancer stem cell niche: how essential is the niche in regulating stemness of tumor cells? Cell Stem Cell. 2015; 16: 225–238.

40. Bomken S, Fišěr K, Heidenreich O, Vormoor J. Understanding the cancer stem cell. *British J Cancer.* 2010; 103: 439–445.
41. Midde K, Sun N, Rohena C, Josen L, Dhillon H, Ghosh P. Single-cell imaging of metastatic potential of cancer cells. *iScience.* 2018; 10: 53–65.
42. Zobayer N, Kumar Das N, Abir AR. Cancer stem cell: the mastermind of carcinogenesis. *Int J Cancer Res.* 2014; 11(1): 1–18.
43. Nowell PC. The clonal evolution of tumor cell populations. *Science. New Series,* 1976; 194(4260): 23.
44. Espiga de Macedo J, Machado M. Cancer stem cell and its influence in carcinogenesis — an update. *J Neoplasm.* 2017; 2(2): 9.
45. Tomasetti C, Li L, Vogelstein B. Stem cell divisions, somatic mutations, cancer etiology, and cancer prevention. *Science.* 2017; 355(6331): 1330–1334.
46. Frank NY, Schatton T, Frank MH. The therapeutic promise of the cancer stem cell concept. *J Clin Invest.* 2010; 120: 41–50.
47. Matsui WH. Cancer stem cell signalling pathways. *Medicine (Baltimore).* 2016; 95(Suppl 1): S8–S19.
48. Katoh M. Canonical and non-canonical Wnt signalling in cancer stem cells and their niches: cellular heterogeneity, omics reprogramming, targeted therapy and tumor plasticity (Review). *Int J Oncol.* 2017; 51(5): 1357–1369.
49. Katoh M, Katoh M. Wnt signalling pathway and stem cell signalling network. *Clin Cancer Res.* 2007; 13: 4042–4045.
50. Niehrs C. The complex world of Wnt receptor signalling. *Nat Rev Mol Cell Biol.* 2012; 13: 767–779.
51. Holland JD, Klaus A, Garratt AN, Birchmeier W. Wnt signalling in stem and cancer stem cells. *Curr Opin Cell Biol.* 2013; 25: 254–264.
52. Acebron SP, Niehrs C. β-catenin-independent roles of Wnt/LRP6 signalling. *Trends Cell Biol.* 2016; 26: 956–967.
53. Katoh M, Katoh M. Molecular genetics and targeted therapy of Wnt-related human diseases (Review). *Int J Mol Med.* 2017; 40: 587–606.
54. Lui JH, Hansen DV, Kriegstein AR. Development and evolution of the human neocortex. *Cell.* 2011; 146: 18–36.
55. Barker N. Adult intestinal stem cells: Critical drivers of epithelial homeostasis and regeneration. *Nat Rev Mol Cell Biol.* 2014; 15: 19–33.
56. Van Camp JK, Beckers S, Zegers D, Van Hul W. Wnt signalling and the control of human stem cell fate. *Stem Cell Rev.* 2014; 10: 207–229.
57. Yang K, Wang X, Zhang H, Wang Z, Nan G, Li Y, Zhang F, et al. The evolving roles of canonical Wnt signalling in stem cells and tumorigenesis: implications in targeted cancer therapies. *Lab Invest.* 2016; 96: 116–136.
58. Qin L, Yin YT, Zheng FJ, Peng LX, Yang CF, Bao YN, Liang YY, et al. Wnt5A promotes stemness characteristics in nasopharyngeal carcinoma cells leading to metastasis and tumorigenesis. *Oncotarget.* 2015; 6: 10239–10252.
59. Webster MR, Kugel CH, Weeraratna AT. The Wnts of change: how Wnts regulate phenotype switching in melanoma. *Biochim Biophys Acta.* 2015; 1856: 244–251.

60. Kumawat K, Gosens R. Wnt-5A: signaling and functions in health and disease. *Cell Mol Life Sci.* 2016; 73: 567–587.

61. Wang MT, Holderfield M, Galeas J, Delrosario R, To MD, Balmain A, McCormick F. K-Ras promotes tumorigenicity through suppression of non-canonical Wnt signalling. *Cell.* 2015; 163: 1237–1251.

62. Milla LA, González-Ramírez CN, Palma V. Sonic hedgehog in cancer stem cells: a novel link with autophagy. *Biol Res.* 2012; 45(3): 223–230.

63. Petrova R, Joyner AL. Role for Hedgehog signalling in adult organ homeostasis and repair. *Development.* 2014; 141: 3445–3457.

64. Li Y, Laterra J. Cancer stem cells: distinct entities or dynamically regulated phenotypes? *Cancer Res.* 2012; 72(3): 576–580.

65. Venkatesh V, Nataraj R, Thangaraj GS, Karthikeyan M, Gnanasekaran A, Kahinelli SB, Kuppana G, *et al.* Targeting Notch signalling pathway of cancer stem cells. *Stem Cell Investig.* 2018; 5: 5.

66. Patel NS, Li JL, Generaloi D, Poulsom R, Cranston DW, Harris AL. Up-regulation of Delta-like 4 ligand in human tumor vasculature and the role of basal expression in endothelial cell function. *Cancer Res.* 2005; 65: 8690–8697.

67. Al-Hajj M, Wicha SM, Benito-Hernandez A, Morrison SJ, Clarke MF. Prospective identification of tumorigenic breast cancer cells. *Proc Natl Acad Sci.* 2003; 100: 3983–3988.

68. Galoczova M, Coates P, Vojtesek B. STAT3, stem cells, cancer stem cells and p63. *Cell Mol Biol Lett.* 2018; 23: 12.

69. Carpenter RL, Lo HW. STAT3 target genes relevant to human cancers. *Cancers (Basel).* 2014; 6: 897–925.

70. Yu H, Pardoll D, Jove R. STATs in cancer inflammation and immunity: a leading role for STAT3. *Nat Rev Cancer.* 2009; 9: 798–809.

71. Nekulova M, Holcakova J, Coates P, Vojtesek B. The role of p63 in cancer, stem cells and cancer stem cells. *Cell Mol Biol Lett.* 2011; 16: 296–327.

72. Orzol P, Holcakova J, Nekulova M, Nenutil R, Vojtesek B, Coates PJ. The diverse oncogenic and tumor suppressor roles of p63 and p73 in cancer: a review by cancer site. *Histol Histopathol.* 2015; 30: 503–521.

73. Pawlus MR, Wang L, Hu CJ. STAT3 and HIF1α cooperatively activate HIF1 target genes in MDA-MB-231 and RCC4 cells. *Oncogene.* 2014; 33: 1670–1679.

74. Corcoran RB, Contino G, Deshpande V, Tzatsos A, Conrad C, Benes CH, Levy DE, *et al.* STAT3 plays a critical role in KRAS-induced pancreatic tumor genesis. *Cancer Res.* 2011; 71: 5020–5029.

75. Fofaria NM, Srivastava SK. STAT3 induces anoikis resistance, promotes cell invasion and metastatic potential in pancreatic cancer cells. *Carcinogenesis.* 2014; 36: 142–150.

76. Chang JC. Cancer stem cells: role in tumor growth, recurrence, metastasis, and treatment resistance. *Medicine (Baltimore).* 2016; 95(1 Suppl 1): S20–5.

77. Shiozawa Y, Nie B, Pienta KJ, *et al.* Cancer stem cells and their role in metastasis. *Pharmacol Ther.* 2013. May; 138(2): 285–293.

78. Pallini R, Ricci-Vitiani L, Banna GL, Signore M, Lombardi D, Todaro M, Stassi G, et al. Cancer stem cell analysis and clinical outcome in patients with glioblastoma multiforme. *Clin Cancer Res.* 2008; 14: 8205–8212.

79. Woo T, Okudela K, Mitsui H, Yazawa T, Ogawa N, Tajiri M, Yamamoto T, et al. Prognostic value of CD133 expression in stage I lung adenocarcinomas. *Int J Clin Exp Pathol.* 2010; 4: 32–42.

80. Beier D, Wischhusen J, Dietmaier W, Hau P, Proescholdt M, Brawanski A, Bogdahn U, et al. CD133 expression and cancer stem cells predict prognosis in high-grade oligodendroglial tumors. *Brain Pathol.* 2008; 18: 370–377.

81. Wang Q, Chen ZG, Du CZ, Wang HW, Yan L, Gu J. Cancer stem cell marker CD133+ tumor cells and clinical outcome in rectal cancer. *Histopathology.* 2009; 55: 284–293.

82. Zhao P, Li Y, Lu Y. Aberrant expression of CD133 protein correlates with Ki-67 expression and is a prognostic marker in gastric adenocarcinoma. *BMC Cancer.* 2010; 10: 218.

83. Shien K, Toyooka S, Ichimura K, Soh J, Furukawa M, Maki Y, Muraoka T, et al. Prognostic impact of cancer stem cell-related markers in non-small cell lung cancer patients treated with induction chemoradiotherapy. *Lung Cancer.* 2012; 77: 162–167.

84. Horst D, Kriegl L, Engel J, Kirchner T, Jung A. Prognostic significance of the cancer stem cell markers CD133, CD44, and CD166 in colorectal cancer. *Cancer Invest.* 2009; 27: 844–850.

85. De Sousa EMF, Colak S, Buikhuisen J, Koster J, Cameron K, de Jong JH, et al. Methylation of cancer-stem-cell-associated Wnt target genes predicts poor prognosis in colorectal cancer patients. *Cell Stem Cell.* 2011; 9: 476–485.

86. Kuo TC, Chen CT, Baron D, Onder TT, Loewer S, Almeida S, Weismann CM, et al. Midbody accumulation through evasion of autophagy contributes to cellular reprogramming and tumorigenicity 3. *Nat Cell Biol.* 2011; 13: 1214–1223.

87. Cheung JY, Miller BA. Molecular mechanisms of erythropoietin signalling. *Nephron.* 2001; 87: 215–222.

88. Wang M, Zhao J, Zhang L, Wei F, Lian Y, Wu Y, Gong Z, et al. Role of tumor microenvironment in tumorigenesis. *J Cancer.* 2017; 8(5): 761–773.

89. Wang Y, Mo Y, Gong Z, Yang X, Yang M, Zhang S, Liao Q, et al. Circular RNAs in human cancer. *M Cancer.* 2017; 16: 25.

90. Pienta KJ, McGregor N, Axelrod R. Axelrod DE. Ecological therapy for cancer: defining tumors using an ecosystem paradigm suggests new opportunities for novel cancer treatments. *Transl Oncol.* 2008; 1: 158–164.

91. Berdiel-Acer M, Sanz-Pamplona R, Calon A. Cuadras D, Berenguer A, Sanjuan X, Paules MJ, et al. Differences between CAFs and their paired NCF from adjacent colonic mucosa reveal functional heterogeneity of CAFs, providing prognostic information. *Mol Oncol.* 2014; 8: 1290–1305.

92. Wendling O, Bornert JM, Chambon P, Metzger D. Efficient temporally-controlled targeted mutagenesis in smooth muscle cells of the adult mouse. *Genesis.* 2009; 47: 14–8.

93. Saunders NA, Simpson F, Thompson EW, Hill MM, Endo-Munoz L, Leggatt G, Minchin RF, et al. Role of intratumoral heterogeneity in cancer drug resistance: molecular and clinical perspectives. *EMBO Mol Med.* 2012; 4: 675–684.

94. Kise K, Kinugasa-Katayama Y, Takakura N. Tumor microenvironment for cancer stem cells. *Adv Drug Del Rev.* 2016; 99: 197–205.

95. Nguyen LV, Vanner R, Dirks P, Eaves CJ. Cancer stem cells: an evolving concept. *Nature Reviews Cancer.* 2012; 12: 133–143.

96. Tammela T, Sanchez-Rivera FJ, Cetinbas NM, Wu K, Joshi NS, Helenius K, Park Y, et al. A Wnt-producing niche drives proliferative potential and progression in lung adenocarcinoma. *Nature.* 2017; 545(7654): 355–359.

97. Huch M, Rawlins EL. Tumors build their niche. *Nature.* 2017; 545: 292–293.

98. Lim JS, Ibaseta A, Fischer MM, Cancilla B, O'Young G, Cristea S, Luca VC, et al. Intratumoral heterogeneity generated by Notch signalling promotes small-cell lung cancer. *Nature.* 2017; 545(7654): 360–364.

99. Dudley AC. Tumor endothelial cells. *Cold Spring Harb Perspect Med.* 2012; 2(3): a006536.

100. Chouaib S, Kieda C, Benlalam H, Noman MZ, Mami-Chouaib F, Ruegg C. Endothelial cells as key determinants of the tumor microenvironment: interaction with tumor cells, extracellular matrix and immune killer cells. *Crit Rev Immunol.* 2010; 30(6): 529–545.

101. Nagy JA, Chang S-H, Dvorak AM, and Dvorak HF. Why are tumor blood vessels abnormal and why is it important to know? *Br J Cancer.* 2009; 100(6): 865–869.

102. Warren BA. The vascular morphology of tumors. In Tumor Blood Circulation: Angiogenesis, Vascular Morphology and Blood Flow of Experimental and Human Tumors (ed.) Peterson H. Boca Raton, FL: CRC Press, pp. 1–47.

103. Albini A, Bruno A, Noonan DM, Mortara L. Contribution to tumor angiogenesis from innate immune cells within the tumor microenvironment: implications for immunotherapy. *Front Immunol.* 2018; 9: 527.

104. Cameron W, Elijah M, del Río HA. Role of extracellular matrix in development and cancer progression. *Int J Mol Sci.* 2018; 19(10): 3028.

105. Bonnans C, Chou J, Werb Z. Remodelling the extracellular matrix in development and disease. *Nat Rev Mol Cell Biol.* 2014; 15: 786–801.

106. Rozario T, DeSimone DW. The extracellular matrix in development and morphogenesis: a dynamic view. *Dev Biol.* 2010; 341: 126–140.

107. Malik R, Lelkes PI, Cukierman E. Biomechanical and biochemical remodeling of stromal extracellular matrix in cancer. *Trends Biotechnol.* 2015; 33: 230–236.

108. Paszek MJ, Zahir N, Johnson KR, Lakins JN, Rozenberg GI, Gefen A, Reinhart-King CA, et al. Tensional homeostasis and the malignant phenotype. *Cancer Cell.* 2005; 8: 241–254.

109. McCarthy JB, El-Ashry D, and Turley EA. Hyaluronan, cancer-associated fibroblasts and the tumor microenvironment in malignant progression. *Front Cell Dev Biol.* 2018; 6: 48.

110. Brentnall TA. Arousal of cancer-associated stromal fibroblasts: palladin-activated fibroblasts promote tumor invasion. *Cell Adh Migr.* 2012; 6: 488–494.

111. Labernadie A, Kato T, Brugues A, Serra-Picamal X, Derzsi S, Arwert E, *et al.* A mechanically active heterotypic E-cadherin/N-cadherin adhesion enables fibroblasts to drive cancer cell invasion. *Nat Cell Biol.* 2017; 19: 224–237.

112. Nielsen SR, Schmid MC. Macrophages as key drivers of cancer progression and metastasis. *Mediators of Inflammation.* 2017; Article ID 9624760.

113. de Filippis RA, Chang H, Dumont N, Rabban JT, Chen YY, Tlsty TD, *et al.* CD36 repression activates a multicellular stromal program shared by high mammographic density and tumor tissues. *Cancer Discov.* 2012; 2: 826–839.

114. Ghosh K, Vierkant RA, Frank RD, Winham S, Visscher DW, Vachon CM, *et al.* Association between mammographic breast density and histologic features of benign breast disease. *Breast Cancer Res.* 2017; 19: 134.

115. Vinnicombe SJ. Breast density: why all the fuss? *Clin Radiol.* 2017; 73: 334–357.

116. Kalluri R. The biology and function of fibroblasts in cancer. *Nat Rev Cancer.* 2017; 16: 582–598.

117. Qian BZ, Pollard JW. Macrophage diversity enhances tumor progression and metastasis. *Cell.* 2010; 141(1): 39–51.

118. Hanahan D, Weinberg RA. Hallmarks of cancer: the next generation. *Cell.* 2011; 144(5): 646–674.

119. Bingle L, Brown NJ, Lewis CE. The role of tumorassociated macrophages in tumor progression: implications for new anticancer therapies. *The J Pathol.* 2002; 196(3): 254–265.

120. Chen JJ, Lin YC, Yao PL, *et al.* Tumor-associated macrophages: the double-edged sword in cancer progression. *J Clin Oncol.* 2004; 23(5): 953–964.

121. DeNardo DG, Brennan DJ, Rexhepaj E, *et al.* Leukocyte complexity predicts breast cancer survival and functionally regulates response to chemotherapy. *Cancer Discov.* 2011; 1(1): 54–67.

122. Di Caro G, Cortese N, Castino GF, Grizzi F, Gavazzi F, Ridolfi C, Capretti G, *et al.* Dual prognostic significance of tumor-associated macrophages in human pancreatic adenocarcinoma treated or untreated with chemotherapy. *Gut.* 2016; 65; 1710–1720.

123. Steidl C, Lee T, Shah SP, Farinha P, Han G, Nayar T, Delaney A, *et al.* Tumor-associated macrophages and survival in classic Hodgkin's lymphoma. *The New Engl J Med.* 2010; 362: 875–885.

124. Kluger HM, Dolled-Filhart M, Rodov S, Kacinski BM, Camp RL, Rimm DL. Macrophage colony-stimulating factor-1 receptor expression is associated with poor outcome in breast cancer by large cohort tissue microarray analysis. *Clin Cancer Res.* 2004; 10(1): 173–177.

125. Zhu Y, Knolhoff BL, Meyer MA, Nywening TM, West BL, Lou J, Wang-Gillam A, *et al.* CSF1/CSF1R blockade reprograms tumor-infiltrating macrophages and improves response to T-cell checkpoint immunotherapy in pancreatic cancer models. *Cancer Res.* 2014; 74(18): 5057–5069.

126. Ojalvo LS, King W, Cox D, Pollard JW. High-density gene expression analysis of tumor-associated macrophages from mouse mammary tumors. *Am J Pathol.* 2009; 174(3): 1048–1064.

127. Zabuawala T, Taffany DA, Sharma SM, Merchant A, Adair B, Sirnivasan R, *et al.* An Ets2-driven transcriptional program in tumor-associated macrophages promotes tumor metastasis. *Cancer Res.* 2010; 70(4): 1323–1333.

128. Grivennikov SI, Greten FR, Karin M. Immunity, inflammation, and cancer. *Cell.* 2010; 140(6): 883–899.

129. Zhang Z, Dong Z, Lauxen IS, Filho MS, Nor JE. Endothelial cell secreted EGF induces epithelial to mesenchymal transition and endows head and neck cancer cells with stem-like phenotype. *Cancer Res.* 2014; 74: 2869–2881.

130. Parks SK, Mazure NM, Counillon L, Pouyssegur J. Hypoxia promotes tumor cell survival in acidic conditions by preserving ATP levels. *J Cell Physiol.* 2013; 228: 1854–1862.

131. Chung AS, Lee J, Ferrara N. Targeting the tumor vasculature: insights from physiological angiogenesis. *Nat Rev Cancer.* 2010; 10: 505–514.

132. Kerbel RS. Tumor angiogenesis. *N Engl J Med.* 2008; 358: 2039–2049.

133. Maniotis AJ, Folberg R, Hess A, Seftor EA, Gardner LM, Pe'er J, Trent JM, *et al.* Vascular channel formation by human melanoma cells in vivo and in vitro: vasculogenic mimicry. *Am J Pathol.* 1999; 155: 739–752.

134. Fan Y-L, Zheng M, Tang Y-L and Liang X-H. A new perspective of vasculogenic mimicry: EMT and cancer stem cells. *Oncol Lett.* 2013; 6: 1174–1180.

135. Wang R, Chadalavada K, Wilshire J, Kowalik U, Hovinga KE, Geber A, Fligelman B, *et al.* Glioblastoma stem-like cells give rise to tumor endothelium. *Nature.* 2010; 468: 829–833.

136. Sloane R, Priest L, Lancashire L, Hou JM, Greystoke A, Ward TH, Blackhall FH. Evaluation and prognostic significance of circulating tumor cells in patients with non-small-cell lung cancer. *J Clin Oncol.* 2011; 29: 1556–1563.

137. Mani SA, Guo W, Liao MJ, Elinor Ng, Eaton EN, Ayyanan A, Alicia Y, *et al.* The epithelial-mesenchymal transition generates cells with properties of stem cells. *Cell.* 2008; 133(4): 704–715.

138. Heerboth S, Housman G, Leary, Longacre M, Byler S, Lapinska K, Willbanks A, *et al.* EMT and tumor metastasis. *Clin Transl Med.* 2015; 4: 6.

139. Wang Y, Zhong Y, Hou T, Liao J, Zhang C, Sun C, *et al.* PM2.5 induces EMT and promotes CSC properties by activating Notch pathway in vivo and vitro. *Ecotoxicol Env Safety.* 2019; 178: 159–167.

140. Pickup MW, Mouw JK, Weaver VM. The extracellular matrix modulates the hallmarks of cancer. *EMBO Rep.* 2014; 15: 1243–1253.

141. PujadaA, Walter L, Patel A, Bui TA, Zhang Z, Zhang Y, Denning TL, *et al.* Matrix metalloproteinase MMP9 maintains epithelial barrier function and preserves mucosal lining in colitis associated cancer. *Oncotarget.* 2017, 8, 94650–94665.

142. Turunen SP, Tatti-Bugaeva O, Lehti K. Membrane-type matrix metalloproteases as diverse effectors of cancer progression. *Biochim Biophys Acta.* 2017; 1864: 1974–1988.

143. Shiga K, Hara M, Nagasaki T, Sato T, Takahashi H, Takeyama H. Cancer associated fibroblasts: their characteristics and their roles in tumor growth. *Cancers (Basel).* 2015; 7(4): 2443–2458.

144. Nair N, Calle AS, Zahra MH, Prieto-Vila M, Oo AKK, Hurley L, Vaidyanath A, *et al.* A cancer stem cell model as the point of origin of cancer-associated fibroblasts in tumor microenvironment. *Sci Rep.* 2017; 7(1): 6838.

145. Buchsbaum, RJ, Oh SY. Breast cancer-associated fibroblasts: where we are and where we need to go. *Cancers (Basel).* 2016; 8:19.

146. Warburg O. On the origin of cancer cells. *Science.* 1956; 123: 309–314.

147. Avagliano A, Granato G, Ruocco MR, Romano V, Belviso I, Carfora A, Montagnani S, *et al.* Metabolic reprogramming of cancer associated fibroblasts: the slavery of stromal fibroblasts. *Biomed Res Int.* 2018; 2018: 6075403.

148. Sainz B Jr., Carron E, Vallespinós M, Machado HL, *et al.* Cancer stem cells and macrophages: implications in tumor biology and therapeutic strategies. *Mediators Inflamm.* 2016; 2016: 9012369.

149. Cabarcas SM, Mathews LA, Farrar WL. The cancer stem cell niche — there goes the neighborhood? *International Journal of Cancer.* 2011; 129(10): 2315–2327.

150. Noy R, Pollard JW. Tumor-associated macrophages: from mechanisms to therapy. *Immunity.* 2014; 41(1): 49–61.

151. DeNardo DG, Brennan DJ, Rexhepaj E, Shiao SL, Madden SF, Gallagher WM, Wadhwani N, *et al.* Leukocyte complexity predicts breast cancer survival and functionally regulates response to chemotherapy. *Cancer Discov.* 2011; 1(1): 54–67.

152. Qian B-Z, Pollard JW. Macrophage diversity enhances tumor progression and metastasis. *Cell.* 2010; 141(1): 39–51.

153. Fang W, Ye L, Shen L, Cai J, Huang F, Wei Q, Fei X, *et al.* Tumor-associated macrophages promote the metastatic potential of thyroid papillary cancer by releasing CXCL8. *Carcinogenesis.* 2014; 35(8): 1780–1787.

154. Balkwill F, Mantovani A. Inflammation and cancer: back to Virchow? *The Lancet.* 2001; 357(9255): 539–545.

155. Lu H, Clauser KR, Tam WL, Frose J, Ye X, Eaton EN, Reinhardt F, *et al.* A breast cancer stem cell niche supported by juxtacrine signalling from monocytes and macrophages. *Nat Cell Biol.* 2014; 16(11): 1105–1117.

156. Zhou W, Ke SQ, Huang Z, Flavahan W, Fang X, Paul J, Sloan AE, *et al.* Periostin secreted by glioblastoma stem cells recruits M2 tumor-associated macrophages and promotes malignant growth. *Nat Cell Biol.* 2015; 17(2): 170–182.

157. Bjornmalm M, Thurecht KJ, Michael M, Scott AM, Caruso F. Bridging bio-nano science and cancer nanomedicine. *ACS Nano.* 2017; 11: 9594–9613.

158. Tsai MJ, Chang WA, Huang MS, Kuo PL. Tumor microenvironment: A new treatment target for cancer. *ISRN Biochem.* 2014; 2014: 351959.

159. Barnes TA, Amir E. Hype or hope: The prognostic value of infiltrating immune cells in cancer. *Br J Cancer.* 2017; 117: 451–460.

160. Adorno-Cruz V, Kibria G, Liu X, Doherty M, Junk DJ, Guan D, Hubert C, *et al.* Cancer stem cells: targeting the roots of cancer, seeds of metastasis, and sources of therapy resistance. *Cancer Res.* 2015; 75(6).

161. Papetti M. Mechanisms of normal and tumor-derived angiogenesis. *Am J Physiol Cell Physiol.* 2002; 282: C947–C970.

162. Ferrara N, Hillan KJ, Gerber HP, Novotny W. Discovery and development of bevacizumab, an anti-VEGF antibody for treating cancer. *Nat Rev Drug Discov*. 2004; 3: 391–400.

163. Hurwitz H, Fehrenbacher L, Novotny W, Cartwright T, Hainsworth J, Heim W, Berlin J, *et al*. Bevacizumab plus irinotecan, fluorouracil, and leucovorin for metastatic colorectal cancer. *N Engl J Med*. 2004; 350, 2335–2342.

164. Tewari KS, Sill MW, Long HJ 3rd, Penson RT, Huang H, Ramondetta LM, Landrum LM, *et al*. Improved survival with bevacizumab in advanced cervical cancer. *N Engl J Med*. 2014; 370(8): 734–743.

165. Kvinlaug BT, Huntly BJP. Targeting cancer stem cells. *Expert Opin Ther Tar*. 2007; 11(7): 915–927.

166. Rich JN, Bao S. Chemotherapy and cancer stem cells. *Cell Stem Cell*. 2007; 1(4): 353–355.

167. Deonarain MP, Kousparou CA, Epenetos AA. Antibodies targeting stem cells: a new paradigm in immunotherapy? MAbs. 2009; 1: 12–25.

168. Chen K, Huang YH, Chen JL. Understanding and targeting cancer stem cells: therapeutic implications and challenges. *Acta Pharmacol*. 2013; 34(6): 732–740.

169. Kim YJ, Siegler EL, Siriwon N, Wang P. Therapeutic strategies for targeting cancer stem cells. *J Cancer Metastasis Treatment*. 2016; 2: 233–242.

170. Vredenburgh JJ, Desjardins A, Herndon JE, Dowell JM, Reardon DA, Quinn JA, Rich JN, *et al*. Phase II trial of bevacizumab and irinotecan in recurrent malignant glioma. *Clin Cancer Res*. 2007; 13(4): 1253–1259.

171. Batchelor TT, Sorensen AG, di Tomaso E, Zhang WT, Duda DG, Cohen KS, Kozak KR, *et al*. AZD2171, a pan-VEGF receptor tyrosine kinase inhibitor, normalizes tumor vasculature and alleviates edema in glioblastoma patients. *Cancer Cell*. 2007; 11(1): 83–95.

172. Ghotra EPS, Jordi C. Puigvert JC, Danen EHJ. The cancer stem cell microenvironment and anti-cancer therapy. *Int J Radiat Biol*. 2009; 85(11): 955–962.

173. Wilhelm S, Carter C, Lynch M, Lowinger T, Dumas J, Smith RA, Schwartz B, *et al*. Discovery and development of sorafenib: a multikinase inhibitor for treating cancer. *Nat Rev Drug Discov*. 2006; 5(10): 835–844.

174. Escudier B, Pluzanska A, Koralewski P, Ravaud A, Bracarda S, Szczylik C, Chevreau C, *et al*. Bevacizumab plus interferon alfa-2a for treatment of metastatic renal cell carcinoma: a randomised, double-blind phase III trial. *Lancet*. 2017; 370(9605): 2103–2111.

175. Lu P, Weaver VM, Werb Z. The extracellular matrix: a dynamic niche in cancer progression. *JCB*. 2012; 196 (4): 395.

176. Albini A, Bruno A, Gallo C, Pajardi G, Noonan DM, Dallaglio K. Cancer stem cells and the tumor microenvironment: interplay in tumor heterogeneity. *Connect Tissue Res*. 2015; 56(5): 414–425.

177. Ribas A, Wolchok JD. Cancer immunotherapy using checkpoint blockade. *Science*. 2018; 359(6382): 1350–1355.

178. Okazaki, Chikuma S, Iwai Y, Fagarasan S, Honjo T. A rheostat for immune responses: the unique properties of PD-1 and their advantages for clinical application. *Nat Immunol. 2013*; 14(12): 1212–1218.

CHAPTER 7

Mesenchymal stem cells for the treatment of immune-mediated diseases

Ling Ling Liau*, Qi Hao Looi‡, Sue Ping Eng‡, Muhammad Dain Yazid†,
Nadiah Sulaiman†, Mohd Fauzi Mh Busra†, Min Hwei Ng†, Jia Xian Law†,§

*Department of Physiology
†Tissue Engineering Centre
Faculty of Medicine, Universiti Kebangsaan Malaysia Medical Centre,
Jalan Yaacob Latif, 56000 Kuala Lumpur, Malaysia
‡Ming Medical Sdn. Bhd., Pusat Perdagangan Dana 1, 47301 Petaling Jaya, Selangor, Malaysia

ABSTRACT

Mesenchymal stem cells (MSCs) can be isolated from many different tissue sources and have several superior characteristics that render them more ideal for therapeutic application compared to other types of stem cells. MSCs have excellent immune-modulatory properties for reducing local and systemic inflammation. MSCs modulate the immune system by secreting paracrine factors and direct cell-to-cell contact with immune cells. MSCs interact with multiple immune cells, including T-cells, B-cells, natural killer cells, macrophages and dendritic cells to modulate the host immune response. Due to their ability to suppress inflammation, preclinical and clinical studies have investigated MSCs extensively for treating autoimmune and inflammatory diseases such as multiple sclerosis, systemic lupus erythematosus, diabetes mellitus, graft-versus-host disease, and graft rejection. Thus far, most studies have reported that MSC therapy yields positive results in ameliorating immune-mediated diseases. Both autologous and allogeneic MSCs can be transplanted into humans safely without eliciting the host immune response and with minimal risk of tumor formation. However, a few limitations such as dosage, route, number, and timing of transplantation require more extensive study to derive the maximum therapeutic effect.

KEYWORDS

Mesenchymal stem cells; Immunology; Treatment; Autoimmune disease; Anti-inflammation.

---

§ Corresponding author. Email: danieljx08@gmail.com; lawjx@ppukm.ukm.edu.my

LIST OF ABBREVIATIONS

ADSCs	=	Adipose-derived MSCs
APCs	=	Antigen-presenting cells
BCR	=	B cell receptor
BILAG	=	British Isles Lupus Assessment Group
Blipm1	=	B lymphocyte-induced maturation protein 1
BMSCs	=	Bone marrow-derived MSCs
C1P	=	Ceramide-1-phosphate
CCR5	=	C-C chemokine receptor type 5
COX2	=	Cyclooxygenase 2
CSF	=	Colony-stimulating factor
CXCR3	=	CXC chemokine receptor 3
DM	=	Diabetes mellitus
EDSS	=	Expanded Disability Status Scale
GM-CSF	=	Granulocyte-macrophage-colony-stimulating factor
GvHD	=	Graft-versus-host disease
HGF	=	Hepatocyte growth factor
HSCs	=	Hematopoietic stem cells
ICAM-1	=	Intercellular adhesion molecule 1
IDO	=	Indoleamine 2,3-dioxygenase
IFNγ	=	Interferon-gamma
IL-1α	=	Interleukin-1alpha
iNOS	=	Inducible nitric oxide synthase
ITP	=	Immune thrombocytopenic purpura
LPS	=	Lipopolysaccharide
MCP-1	=	Monocyte chemoattractant protein-1
MHC	=	Major histocompatibility complex
MIP-1α	=	Macrophage inflammatory protein-1α
MMEP	=	Multimodal evoked potential
MS	=	Multiple sclerosis
MSCs	=	Mesenchymal stem cells
NK cells	=	Natural killer cells
NO	=	Nitric oxide
PAMPs	=	Pathogen-associated molecular patterns
PDSCs	=	Placenta-derived decidual stromal cells
PGE2	=	Prostaglandin E2
S1P	=	Sphingosine-1-phosphate

SCF   = Stem cell factor
SDF-1   = Stromal cell-derived factor-1
SLE   = Systemic lupus erythematosus
SLEDAI   = SLE Disease Activity Index
STAT3   = Signal transducer and activator of transcription 3
TGF-β   = Transforming growth factor-β
Th1   = T helper-1
Th2   = T helper 2
TNF-α   = Tumor necrosis factor-α
TNFR1   = Tumor necrosis factor receptor 1
Treg   = Regulatory T cells
VCAM-1   = Vascular cell adhesion molecule 1
VEGF   = Vascular endothelial growth factor
WJ-MSCs = Wharton's jelly-derived MSCs

7.1 INTRODUCTION

Mesenchymal stem cells (MSCs) harvested from both adult (e.g. bone marrow and adipose tissue) and neonatal tissue (e.g. umbilical cord and cord blood) have been extensively studied to understand their therapeutic potential in treating immune-mediated diseases.[1,2] MSCs have immune-modulatory properties, which can reduce local and systemic inflammation. MSCs are immune gate-keepers that can function as both antigen-presenting cells (APCs) and immune suppressors. MSCs affect both the innate [dendritic cells, macrophages, neutrophils, mast cells, natural killer (NK) cells] and adaptive immune response (T-cells, B-cells) by interacting with immune cells, either directly through cell-to-cell contact or indirectly via the secretion of paracrine factors.[3] The number of publications studying the interaction between MSC-derived microvesicles (e.g. ectosomes and exosomes) and immune cells has increased recently. Generally, microvesicles have immune-modulatory properties just like the parent cells.

MSCs can evade the host immune response as they do not express major histocompatibility complex (MHC) class II and have low expression of the co-stimulatory molecules CD80, CD86 and CD40.[4] Thus, both autologous and allogeneic MSCs can be used clinically. Nonetheless, MSCs must meet 3 basic consensus criteria before clinical application: plastic adherence ability; phenotypically, >95% of isolated MSCs should express CD73 (SH3), CD90 and CD105 (HS2) and >98% of isolated MSCs should not express CD14, D19, CD34, CD45, CD11b, CD79a, and human leukocyte antigen (HLA)-DR; and the capacity to differentiate into osteoblastic, chondrogenic and adipogenic lineages.[5,6]

Multiple clinical trials/studies have evaluated MSCs for treating immune-mediated diseases such as multiple sclerosis (MS), systemic lupus erythematosus (SLE), graft rejection, diabetes mellitus (DM), immune thrombocytopenic purpura (ITP) and graft-versus-host disease (GvHD). MSCs are commonly administered intravenously and occasionally administered directly to the target site. MSCs administered intravenously can respond to chemoattractants release and migrate to the target tissue. However, only a low percentage of cells will reach the injured tissue, as the majority of the cells will be trapped in the lung. More cells will accumulate at the target site if the cells are administered locally.

7.2 MSC MECHANISMS OF IMMUNOSUPPRESSION

Preclinical and clinical studies have shown that MSCs have potent anti-inflammatory and immune evasion properties.[7,8] MSCs exert immunomodulatory effects via cell-to-cell contact and by secreting biologically active substances such as growth factors, cytokines, and chemokines.[9] MSCs can suppress the activation and functionality of several types of immune cells such as T-cells,[10] B-cells,[11] NK cells[12] and dendritic cells.[13] Furthermore, MSCs also induce the conversion of T-cells to regulatory T-cells (Treg).[14] Treg play a modulatory role in the immune system, maintain tolerance to self-antigens, and prevent autoimmune diseases.[15] Understanding of the mechanisms of MSC-based immunomodulation remains incomplete to date.

MSCs adopt several mechanisms to evade the host immune response. First, MSCs lack MHC class II and the co-stimulatory molecules CD40, CD80 and CD86 that are required for T-cell activation.[16,17] Second, MSCs directly and indirectly (by modulating dendritic cells) disrupt the functionality of T-cells and NK cells.[18] Third, MSCs secrete prostaglandin E2 (PGE2), interleukin-10 (IL-10) as well as indoleamine 2, 3-dioxygenase (IDO), which depletes local tryptophan, creating an immunosuppressive local environment.[19] Even though these mechanisms enable MSCs to evade the host immune response, the cells will eventually be rejected as well, but at a slower rate compared to other types of cells. Typically, CD4$^+$ T-cells will be activated upon being presented with antigens by the MHC class II molecules expressed on the surface of APCs. Activated T-cells will secrete interferon-γ (IFN-γ), tumor necrosis factor-α (TNF-α), IL-1α and IL-1β, which induce MSC secretion of various immunomodulatory soluble factors that can suppress the host immune response.[20] Figure-7.1 summarises the effects of MSCs on immune cells.

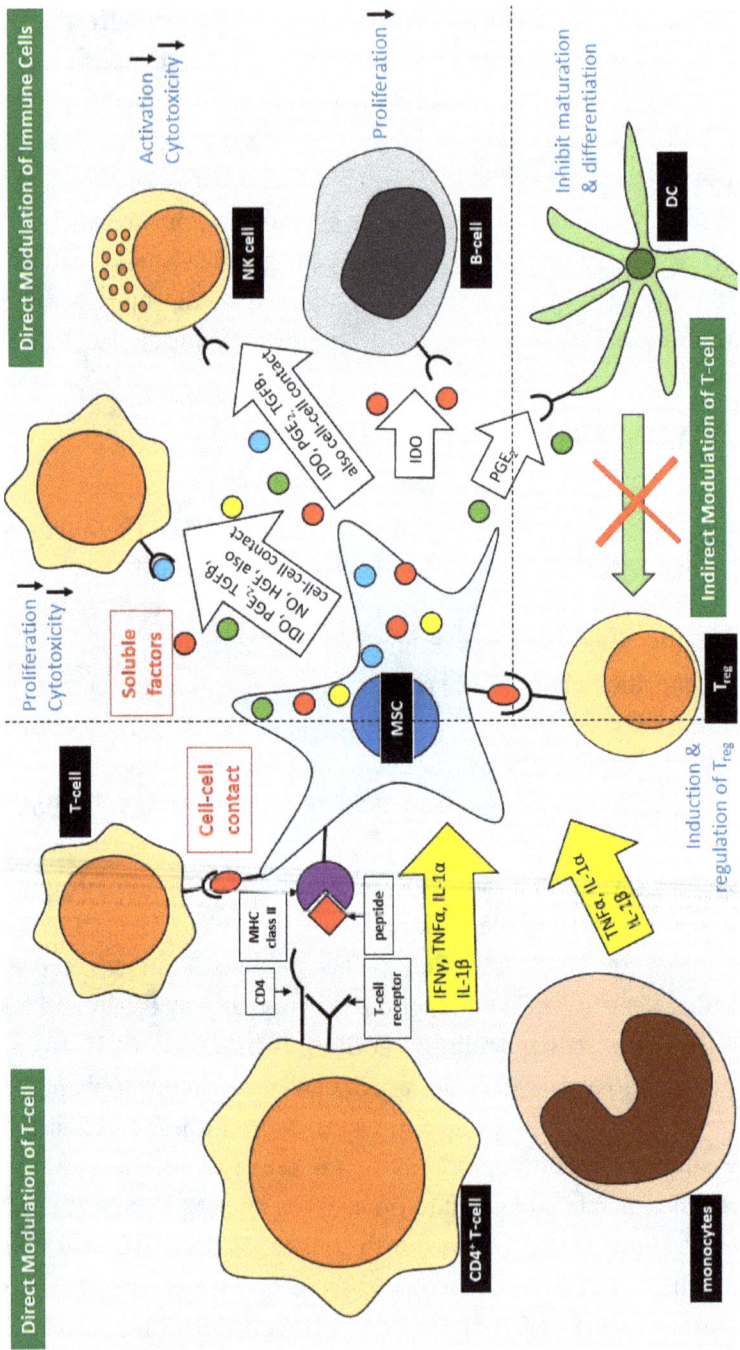

Figure-7.1.   Summary of the immunomodulatory effects of MSCs on various immune cells.

Table-7.1.   Immune-modulatory effects of MSC-derived soluble factors.

Soluble factors	Types of immune cells				
	T-cell	NK cell	B-cell	DC	Treg
Indoleamine 2,3-dioxygenase (IDO)	√	√	√		
Prostaglandin E2 (PGE2)	√	√		√	
Transforming growth factor-β (TGF-β)	√	√			
Nitric oxide (NO)	√				
Hepatocyte growth factor (HGF)	√				
Cell-cell contact	√	√			√

Abbreviations: NK: natural killer; DC: dendritic cell; Treg: regulatory T-cell.

Table-7.1 shows the major immunomodulatory soluble factors [IDO, PGE2, nitric oxide (NO), transforming growth factor-β (TGF-β), hepatocyte growth factor (HGF)] secreted by activated MSCs and their target immune cells. These factors suppress T-cell proliferation through cell cycle arrest at the G0/G1 phase.[21,22] Also, MSCs suppress T-cell proliferation by upregulating inducible NO synthase (iNOS), which produces NO.[23] Moreover, the MSC-derived cytokines increase the proportion of Treg, which in turn suppresses T-cell proliferation. T-cell conversion into Treg plays an important role in halting the T-cell–mediated immune reaction and in suppressing autoreactive T-cells.[15] MSCs also suppress T-cell proliferation indirectly by modulating monocytes and dendritic cells.[24]

Among the key factors secreted by MSCs, IDO and PGE2 have consistently been reported to target various immune cells, including B-cells and T-cells, by downregulating their proliferation as well as suppressing NK cell functionality. PGE2 inhibits dendritic cell maturation, activation and antigen presentation to prevent MSCs from being recognized as foreign bodies.[25] IDO inhibits B-cell proliferation by catalyzing the conversion of tryptophan to kynurenine.[22]

MSCs can be primed by the activated immune cells to boost the anti-inflammatory properties. This can be done by exposing MSCs to the pro-inflammatory cytokines secreted by the immune cells, such as IFN-γ, TNF-α, IL-1α and IL-1β.[3] This priming is beneficial for clinical applications, as it generates functionally improved MSCs to modulate the host immune system. The large amount of data collected thus far clearly shows that MSCs have excellent anti-inflammatory properties for therapeutic use. However, further investigation of their behaviour, especially in the complex cross-talk between MSCs and immune cells, is important for better mechanistic understanding.

7.3 IMMUNOMODULATORY EFFECT OF MSCs ON THE INNATE AND ADAPTIVE IMMUNE SYSTEMS

MSCs profoundly affect the immune response through their interactions with the cellular components of both the innate immune system (phagocytic cells, dendritic cells, NK cells) and the adaptive immune system (T-lymphocytes and B-lymphocytes). MSC immune regulation can take place through contact-dependent methods via receptor–ligand recognition and/or contact-independent mechanisms via the secretion of diverse soluble factors. MSCs express pattern recognition receptors known as Toll-like receptors (TLR) and tumor necrosis factor receptor 1 (TNFR1) on their surface membranes.[26,27] The presence of the receptors suggests that MSCs can be activated by pathogen-associated molecular patterns (PAMPs) to initiate the innate immune response. In addition, MSCs prevent inappropriate autoimmune response activation during organ transplantation and generate a tolerogenic environment during wound repair, thus contributing to the maintenance of immune homeostasis. The following sections describe the interaction of both innate and adaptive immune cells with MSCs during initial activation of the immune response.

7.3.1 Innate Immune System

7.3.1.1 *Phagocytic cells*

Three types of phagocytic cells respond to and destroy foreign particles and cellular debris through a process known as phagocytosis: macrophages, granulocytes and dendritic cells. Phagocytic cells arise from hematopoietic stem cells (HSCs) and develop under the stimulation of colony-stimulating factors (CSF) and various transcription factors. Bone marrow–derived MSCs (BMSCs) mediate the differentiation of the surrounding HSCs by secreting a wide range of CSFs such as macrophage CSF, granulocyte CSF (G-CSF) and granulocyte–macrophage CSF (GM-CSF). In addition, BMSCs also secrete various HSC-promoting factors, including stem cell factor (SCF), CXCL12, vascular cell adhesion molecule 1 (VCAM1), N-cadherin and angiopoietin 1 to ensure HSC differentiation (haematopoiesis).[28]

7.3.1.2 *Macrophages*

Macrophages are mature monocytes that serve as scavengers and constantly circulate in the bloodstream to remove foreign particles, dead cell debris as well as to function as APCs involved in the adaptive immune system.[29] Macrophages can be polarised to either pro-inflammatory M1 macrophages or anti-inflammatory

M2 macrophages that promote tissue remodelling.[31,32] During inflammation, both TLR and TNFR1 on the MSCs will respond to IFN or microbial stimuli such as lipopolysaccharide (LPS), thus initiating the secretion of various C-C motif chemokine ligands such as CCL2, CCL3, and CCL12 to recruit monocytes from the bone marrow to the tissue inflammation site and promote wound healing.[32] Subsequently, the monocytes either differentiate into pro-inflammatory cytokine–producing M1 macrophages or into anti-inflammatory M2 macrophages depending on local chemical factors such as TGF-β or IL-10. Co-culture of MSCs with activated macrophages stimulates the shift towards M2 macrophages that secrete fewer inflammatory cytokines such as IFN-γ, TNF-α, IL-1β and IL-12, consequently accelerating tissue regeneration.[33,34] Similar results have been observed when activated macrophages were co-cultured with MSC-derived exosomes.[35,36] MSCs induce the conversion of M1 macrophages to M2 macrophages by activating the nuclear factor-κB (NF-κB) and signal transducer and activator of transcription 3 (STAT3) pathways as well as altering the macrophage metabolic status via a PGE2-dependent mechanism.[34,37] The STAT2 pathway has also been linked with the polarization towards M2 macrophages induced by MSC-derived microvesicles.[38] Surprisingly, monocytes or primitive macrophages co-cultured with MSCs are also prone to differentiation into M1 macrophages, which are beneficial for fighting infection.[34] An experiment using an indirect co-culture model suggested the involvement of MSC-derived factors that skewed macrophage differentiation and polarization towards the immunosuppressive M2 lineage.[39] M2 macrophages release high levels of IL-10 and IL-4 and low levels of IL-6 and IL-1β.

7.3.1.3 Neutrophils

Granulocytes are a group of cells whose cytoplasm contains granules and include neutrophils, eosinophils, basophils and mast cells. Neutrophils are the most abundant type of non-tissue resident granulocyte that circulate continuously in the bloodstream; they are a vital component of the innate immune system. They migrate to the site of inflammation and engulf the foreign particle. MSCs can recruit neutrophils by secreting chemotactic cytokines such as IL-6, IL-8, and GM-CSF.[18] IL-6 acts as pro-inflammatory stimuli that enhance the effect of the acute-phase response during inflammation by increasing the body temperature and the secretion of acute-phase proteins such as opsonins, and promoting neutrophil production in the bone marrow.[40] IL-8 is a neutrophil chemotactic factor, given its main function of trafficking neutrophils to the injury site, promoting neutrophil survival as well as enhancing their phagocytic activity.[41] Also, GM-CSF secreted by LPS-activated MSCs can promote neutrophil proliferation and maturation to eliminate foreign particles. Lastly, Maqbool *et al.* have suggested that MSCs can inhibit neutrophil apoptosis to preserve the naive neutrophil pool in the bone marrow sinusoids.[42] In summary, microbial-activated MSCs produce various factors required for maintaining neutrophil survivability and activity.

7.3.1.4 *Mast cells*

Mast cells are granulocytes that greatly mediate the allergic response. Instead of circulating in the bloodstream, mast cells localize in the connective tissues close to the external environment. Mast cells originate from HSCs and express a marker known as CD34+. Once exposed to an allergen, mast cells migrate to the allergen site. Mature mast cells produce large amounts of T helper 2 (Th2) cytokines that both induce parasite killing by eosinophils and facilitate the transformation of B-cell isotopes in the adaptive immune response. Furthermore, mast cells also secrete histamine-containing granules to increase blood vessel permeability for intruding white blood cell to the site of infection. *In vitro* study has shown that co-culture of MSCs with mast cells reduces mast cell–derived TNF-α and SCF through a cyclooxygenase-2 (COX-2)-dependent pathway.[43] TNF-α and SCF are cytokines that stimulate the acute-phase reaction of immunity and promote mast cell survival, proliferation, adhesion, migration as well as histamine secretion. Thus, MSCs can ameliorate the allergic response by inhibiting mast cell migration and degranulation by upregulating COX-2 in MSCs and the production of MSC-derived PGE2, which suppresses mast cell activation.[44]

7.3.1.5 *Dendritic cells*

Similar to macrophages, dendritic cells are derived from monocytes. Dendritic cells are potent APCs and a vital component of the body immune system, as they bridge the innate and adaptive immune systems. Immature dendritic cells constantly perform surveillance around the body through the bloodstream. During maturation and activation, dendritic cells increase their expression of co-stimulatory molecules such as CD40, CD80, CD83, and CD86, and migrate to the lymph nodes as APCs, which then present the pathogen-derived antigens on their surface to activate T-lymphocytes. Furthermore, during T-cell priming, dendritic cells produce cytokines that affect the downstream development of T-cells.[45] MSCs affect most of these processes. Co-culture of MSCs with monocytes inhibits monocyte differentiation into dendritic cells and impairs their antigen-presenting functionality by blocking the monocytes from entering the G1 phase of the cell cycle and by downregulating the production of IL-12, thus removing their T-cell–activating capability.[46,47] Furthermore, MSCs downregulate the expression of MHC class II and co-stimulatory molecules (CD80, CD86) by mature dendritic cells.[48,49] MSCs also downregulate dendritic cell maturation and function by interfering with the dendritic cell endocytosis process and decreasing the dendritic cell capacity for releasing IL-12 and galectin-1, resulting in poorer T-cell activation.[13,50,51] MSCs can downregulate IFN-γ and TNF-α expression on dendritic cells, consequently causing the loss of their ability to activate lymphocytes.[52]

Apart from cell-to-cell contact and paracrine factor secretion, MSC-derived microvesicles can also inhibit dendritic cell maturation and function.[53] In that study, the authors found that 4 of the top 10 most highly expressed microRNAs (miRNAs) found in MSC-derived microvesicles, i.e. miR-21-5p, miR-142-3p, miR-223-3p and miR-126-3p, regulate dendritic cell maturation and function. The dendritic cells treated with MSC-derived microvesicles demonstrated poor antigen uptake and maturation, reduced pro-inflammatory cytokine secretion, higher anti-inflammatory cytokine production and impaired ability to migrate towards the C-C chemokine receptor type 7 (CCR7) ligand CCL21.

7.3.1.6 Natural killer cells

NK cells are innate immune cells that release pro-inflammatory cytokines that directly act on virally infected cells and tumor cells. NK cells also secrete a large number of cytokines and chemokines such as IL-3, IL-10, TNF-α, GM-CSF, G-CSF, CCL2, CCL3, CCL4, CCL5, and CXCL8. These cytokines and chemokines allow NK cells to co-localize with other immune cells at the site of inflammation and to stimulate the activation of native T cells residing within the lymph nodes.[54,55] Three factors regulate the cytotoxic activity of NK cells: surface receptors (activating and inhibitory receptors), the MHC expression levels of the targeted cells and stimulation via antibody-dependent cytotoxicity.

The immunosuppressive effects of MSCs on NK cells depend greatly on the activation state of the NK cells and the duration of co-culture of the two cell types.[45] Spaggiari et al. reported that freshly isolated NK cells were not cytotoxic towards allogeneic MSCs, but acquired cytotoxicity after 20-hour activation with IL-2.[56] MSCs exert an inhibitory effect on NK cell proliferation and cytokine secretion.[12,56,57] MSCs suppress the expression of NK cell activating receptors, i.e. 2B4, NKp44, NKG2D and NKp30, thus reducing NK cell cytotoxic activity. MSC-derived PGE2, TGF-β and IDO also contribute to the immunosuppression of NK cells. Fan et al. reported that MSC-derived exosomes contained TGF-β that inhibited NK cell proliferation, differentiation and cytotoxicity.[58] Neutralization of exosomal TGF-β attenuated the suppression of NK cell proliferation, differentiation and cytotoxicity.

7.3.2 Adaptive Immune System

7.3.2.1 *T-cells*

T-cells emerge from the thymus and are later distributed throughout the lymphoid tissue. In the thymus, HSC-derived progenitors develop into T-cells through a series of distinct differentiation steps.[59] The activation of progenitor T-cells requires two signals, namely T-cell receptor signalling and co-stimulatory signalling.[60,61] Upon activation, CD4+ T-cells can differentiate into IFN-γ–producing Th1, IL-4- and IL-13-producing Th2, IL-19–producing Th9, IL-17–producing Th17 or IL-10–producing Treg subsets, depending on the intensity of stimulation and the cytokine milieu.[62–65] Additionally, CD8+ T-cells further develop into cytotoxic T-lymphocytes that secrete granzymes, perforins, and cytokines to eliminate cancerous cells and infected cells.[66] T-cell–mediated immunity is a critical component of the adaptive immune system, providing lifelong protection against infections at diverse sites of infection, controlling malignancies as well as mediating several autoimmune diseases.[67] Over the last decade, the interplay between MSCs and T-cells has been thoroughly studied. The results indicate that MSCs potently inhibit T-cell proliferation and activation.[68,69]

The proliferation-inhibiting effect of MSCs on T-cells appears to involve both contact-independent and contact-dependent components.[70] MSCs release various factors, including TGF-β, which decrease both the mRNA and protein levels of cyclin D2 and HGF, which induces the redistribution of p27kip1 expression in T-cells, resulting in the arrest of proliferation in the G0/G1 phase.[71] Further study has suggested that this inhibition is also MHC-independent, as both autologous and allogeneic MSCs exert similar anti-proliferative effects when co-cultured with T-cells.[45] Besides, MSCs can suppress T-cell populations by inducing the apoptosis of activated T-cells, a process associated with the Fas/Fas ligand–dependent pathway and tryptophan conversion into kynurenine.[72] Besides affecting T-cell proliferation and inducing apoptosis, MSCs can also alter T-cell activation and differentiation. Luz-Crawford *et al.* studied the role of MSCs during T-cell differentiation and found that MSCs suppressed the differentiation of CD4+ T-cells into Th1 and Th17 and promoted the development of Treg.[73] Recent studies have reported that MSC-derived exosomes increase Treg production and attenuate Th1 and Th17 development via an APC-mediated pathway.[74,75] Treg are a subpopulation of T-cells that are vital to maintaining homeostasis and regulating tolerance to self-antigens. Furthermore, Treg suppress the effector T-cells.[76] Consentius *et al.* suggested that MSCs suppress effector T-cell priming indirectly by regulating the body's innate immune system, mainly through dendritic cells and NK cells, as mentioned above.[77] MSC-derived exosomes also attenuate T-cell proliferation, differentiation, and activation.[78,79]

Despite showing promising results in the immune regulation of T-cell populations, MSCs can only suppress T-cell proliferation and activation when pre-stimulated by certain inflammatory cytokines, such as IFN-γ, TNF-α, IL-1α, or IL-1β.[23,80] Pre-activation of these inflammatory cytokines increases iNOS and COX-2 expression levels in MSCs, which are associated with the secretion of immunosuppressive molecules such as intercellular adhesion molecule 1 (ICAM-1), CXC chemokine receptor 3 (CXCR3) ligands, VCAM1, CCR5 ligands, NO and PGE2.[81-83] However, the immunosuppressive ability of MSCs is not always observed, as several contradictory findings have shown that MSCs were unable to suppress T-cell responses under several conditions. Li *et al.* studied MSC plasticity in immunomodulation by investigating how different concentrations of cytokines affected MSC functions in immune regulation.[84] Their results suggested that the immunomodulatory potential of MSCs is generally dependent on the inflammatory signal type and strength. For example, high pro-inflammatory cytokine levels significantly increase NO production by MSCs, resulting in a higher degree of inhibitory effects on T-cells. In contrast, low levels of pro-inflammatory cytokines led to inadequate production of NO and subsequently enhanced T-lymphocyte proliferation. A similar result was observed in IDO knockdown MSCs, suggesting that both NO and IDO serve as a switch to regulate the immune activity of MSCs.[84]

7.3.2.2 B-cells

B-cells constitute an important part of adaptive/humoral immunity. B-cells originate from HSCs through a series of coordinated steps.[85] Following the recognition of specific antigens through B-cell receptor (BCR) ligation, CD40/CD40L (CD40 ligand) binding and TLR, progenitor B-cells will proliferate and differentiate into antibody-producing cells, memory cells or IL-10–producing regulatory B-cells (Breg), which play multiple roles in the body's immune response, including opsonization, neutralization, phagocytic uptake and the activation of other immune cells.[86,87] Despite the divergent results arising from the interaction between MSCs and B-cells being mainly due to the inconsistent experimental conditions,[71,88] there is plausible evidence that MSCs can suppress B-cell proliferation, differentiation and activation. B-cells co-cultured with MSCs undergo cell cycle arrest and have impaired plasma cell generation, compromised immunoglobulin-secreting ability, and reduced chemotactic properties.[89,90] Moreover, B-cell proliferation can be attenuated by MSC-derived exosomes.[79] An initial study in a mouse model suggested that MSCs exhibit a dampening effect on B-cell proliferation,[88] which is concordant with most published data.[90-92] Similar to T-cell suppression, both contact-independent and contact-dependent mechanisms are involved in MSC immune regulation of B-cells.

The primary mechanism of B-cell suppression is through the production of soluble factors, as indicated by indirect co-culture experiments whereby MSCs significantly reduced B-cell secretion of immunoglobulin (Ig)M, IgG and IgA.[93,94] CCL2 mediates these actions, as MSC-derived CCL2 was able to inhibit STAT3 in plasma cells, leading to the suppression of immunoglobulin synthesis.[90] In addition to soluble factors, cell–cell contact is critical in this regard. In recent years, several signalling pathways have been identified as being involved in B-cell suppression. These signalling pathways include the PD-1/PD-L1 (programmed cell death 1/PD-1 ligand 1) pathway, Akt pathway, extracellular response kinase 1/2 pathway, p38 pathway and B-lymphocyte–induced maturation protein 1 (Blimp1) signalling.[89,94,95] Furthermore, MSCs modify the chemotactic properties of B-lymphocytes by altering the expression of their chemokine receptors, including CXCR4, CXCR5 and CCR7.[88] However, several published studies have reported conflicting results whereby MSCs enhance the proliferation, activation, differentiation and antibody production potential of B-cells.[96-98] These disparities are mainly attributable to inconsistent B-cell purity, stimulus variation and strength, MSC source and the MSC/B-cell ratio. Similar to T-cells, inflammatory stimulation of MSCs enhances their inhibitory effects on B-cells. Potent IFN-γ signalling is crucial for stimulating the suppressive function of MSCs, while insufficient inflammation will downregulate MSC functionality, leading to increased B-cell proliferation and differentiation or a higher number of antibody-secreting B-cells.[80]

7.4 ADMINISTRATION OF MSCs

MSCs expanded *in vitro* can be administered to patients via different routes to modulate their immune response. Most studies prefer the intravenous route of delivering the cells when they are only required in the circulatory system to modulate the host immune system. However, MSCs can also be administered locally to the injured tissue as a non-systemic homing approach to reduce the localized inflammation and to prevent further damage. For systemic homing, exogenous MSCs are administered systemically, and then migrate to the injured tissue in response to the secreted paracrine factors that provide chemical cues for the homing of the delivered cells. The essential steps for systemic homing include cell administration into the circulation, vascular rolling and adhesion and endothelial transmigration, followed by extracellular space movement towards the injured tissue.[99]

MSC mobilization and homing is tightly regulated by the chemokines released by the injured tissue and the expression of specific adhesion receptors on the cell surface.[100] One of the main chemokines that play a major role in MSC mobilization and homing to injured tissue is stromal cell–derived factor-1 (SDF-1), which binds to the CXCR4 receptors on the MSC surface.[101] Additionally, monocyte chemoattractant protein-1 (MCP-1) and macrophage inflammatory protein-1α (MIP-1α) also contribute significantly to cell recruitment from the circulatory system.[101,102] Furthermore, growth factors such as vascular endothelial growth factor (VEGF) also contribute to stem cell mobilization and homing.[103] Apart from the chemotactic peptides, ceramide 1-phosphate (C1P) and sphingosine-1-phosphate (S1P), a bioactive lipid released by injured cells and a bioactive sphingolipid metabolite, respectively, also promote MSC migration.[104] MSCs are also activated by the complement 3a (C3a) and C5a proteins to migrate to the injured tissue.[105]

MSCs migrating from the peripheral circulation are required to adhere and cross the endothelial layer and migrate through the extracellular matrix to reach the injured tissue. MSCs migrate via selectin-mediated rolling, chemokine-derived retention and integrin-dependent arrest on the endothelium.[106,107] The contributory factors aiding MSC rolling on the endothelial surface include the selectins, i.e. P-selectin and E-selectin.[106] P-selectin is highly expressed on the activated endothelium, which controls MSC rolling. Platelets affect MSC extravasation, where massive depletion of platelets reduced MSC migration and trafficking to the injury site in an *in vivo* model.[108] The inflammatory molecules secreted by injured cells also stimulate the protein/integrin interactions (VCAM1–VLA-4 [integrin subunit alpha 4], ICAM-1–β2 integrin) for cell adhesion[106,107] and the stimulate chemokines (CCL20, CCL25, CXCL9, CXCL16) for transendothelial migration.[109] The final step in MSC migration to the injured tissue is facilitated by the basal membrane and by enzymatic matrix degradation, i.e. by matrix metalloproteinase-2 (MMP-2) and MMP-9, as well as local chemokines that favor cell locomotion and tissue reconstitution.[106,107]

7.5 CLINICAL APPLICATION OF MSCs FOR AMELIORATING IMMUNE-MEDIATED DISEASES

After the encouraging data from *in vitro* and pre-clinical studies, MSCs have been evaluated clinically for treating immune-mediated diseases. MSCs are typically administered intravenously when treating immune-mediated diseases. Thus far, data from clinical studies have shown that MSC-based therapy is safe and effective for ameliorating immune-mediated diseases (Table-7.2).

Table-7.2.　Clinical trials/studies applying MSCs to treat immune-mediated diseases.

Diseases	Cell sources	No. of patients	Key findings	References
aGvHD	WJ-MSCs	2	CR: 2	[111]
aGvHD	WJ-MSCs	19	CR: 11, PR: 4	[112]
aGvHD & cGvHD	WJ-MSCs	10	CR: 2, PR: 4	[113]
cGvHD	ADSCs	14	CR: 8, PR: 2	[114]
aGvHD	PDSCs	38	CR: 16, PR: 15	[115]
aGvHD	BMSCs	33	CR: 18, PR: 7	[155]
aGvHD	BMSCs	69	CR: 42, PR: 17	[156]
aGvHD & cGvHD	BMSCs	40	CR: 11, PR: 16	[157]
aGvHD	BMSCs	25	CR: 11, PR: 6	[158]
aGvHD	BMSCs	37	CR: 24, PR: 8	[117]
aGvHD	BMSCs	50	CR: 17, PR: 16	[159]
aGvHD	BMSCs	55	CR: 30, PR: 9	[160]
aGvHD	BMSCs	9	CR: 5	[161]
aGvHD	BMSCs	13	CR: 1, PR: 1	[162]
aGvHD & cGvHD	BMSCs	11	CR: 3, PR: 5	[163]
aGvHD & cGvHD	BMSCs	12	CR: 7, PR: 4	[164]
aGvHD	BMSCs	12	CR: 7, PR: 2	[165]
aGvHD	BMSCs	14	CR: 8, PR: 5	[166]
aGvHD	BMSCs	75	CR + PR: 46	[167]
aGvHD	BMSCs	28	CR: 17, PR: 4	[168]
		19 (control)	CR: 5, PR: 3	
aGvHD	BMSCs	25	CR: 6, PR: 9	[169]
aGvHD & cGvHD	BMSCs	18	CR: 2, PR: 9	[170]
aGvHD	BMSCs	46	CR: 3, PR: 20	[171]
aGvHD	BMSCs	58	CR: 5, PR: 22	[172]
Graft rejection	BMSCs	2	Safe and have higher Treg	[123]
Graft rejection	BMSCs	2	Safe and higher Treg/memory T-cell ratio	[124]
Graft rejection	WJ-MSCs	1	Rejection of transplanted liver ameliorated	[125]
Graft rejection	BMSCs	10	No difference in rate of graft survival	[118, 119]
		10 (control)		
Graft rejection	BMSCs	6	Prevented and ameliorated graft rejection	[121]

Table-7.2.   (*Continued*)

Diseases	Cell sources	No. of patients	Key findings	References
Graft rejection	BMSCs	6 6 (control)	No rejection 1 acute rejection	[122]
Graft rejection	BMSCs	7	3 acute rejections	[120]
Graft rejection	BMSCs	105 51 (control)	8 acute rejections 11 acute rejections	[126]
Graft rejection	BMSCs	16 16 (control)	BMSC therapy reduced dosage of immunosuppressive drug needed	[127]
T1DM	WJ-MSCs	15 14 (control)	Better disease control compared to control group	[128]
T1DM	WJ-MSCs	17	8 patients responded to the therapy	[129]
T1DM	ADSCs and ADSCs-derived IPC	1	Decrease in daily insulin requirement	[131]
T1DM	ADSCs-derived IPC	10	Decrease in daily insulin requirement	[132]
T1DM	ADSCs-derived IPC	20	Autologous cells gave better disease control	[133]
T1DM	BMSCs	3	Decrease in daily insulin requirement	[134]
T1DM	BMSCs	10 10 (control)	Better disease control compared to control group	[130]
SLE	WJ-MSCs	16	All patients showed improvement	[135]
	WJ-MSCs	40	24 patients showed improvement	[136]
SLE	BMSCs	15	All patients showed improvement	[139]
SLE	BMSCs	2	Treg increased but no effect on disease activity	[140]
SLE	WJ-MSCs and BMSCs	87	43 patients had CR 20 patients had disease relapse	[137]
SLE	WJ-MSCs and BMSCs	81	37 patients had CR and PR 9 patients had disease relapse	[138]

(*Continued*)

Table-7.2.   (*Continued*)

Diseases	Cell sources	No. of patients	Key findings	References
Lupus nephritis	WJ-MSCs	12	9 patients showed improvement	[141]
		6 (control)	5 patients showed improvement	
SLE with refractory cytopenia	BMSCs	35	Most patients with leukopenia, anaemia and thrombocytopenia showed improvement	[142]
MS	WJ-MSCs	23	Slower disease progression compared to untreated patients	[143]
MS	WJ-MSCs	1	Improvement in EDSS score	[144]
MS	BMSCs	10	EDSS score: improved in 1 patient, no change in 4 patients, deteriorated in 5 patients MRI lesions: decreased in one patient, no change in 7 patients, increased in 2 patients	[145]
MS	BMSCs	25	EDSS score: improved in 4 patients, no change in 12 patients, deteriorated in 6 patients MRI lesions: no change in 15 patients, increased in 6 patients	[146]
MS	BMSCs	26	No significant change in EDSS score and MRI lesions	[147]
MS	BMSCs	10	Improvement in visual acuity and visual evoked response latency, increase in optic nerve area	[150]
MS	BMSCs	15	Improvement in EDSS score	[151]
MS	BMSCs	5	Fewer MRI lesions	[152]
		4 (control)		
MS	BMSCs	6	Improvement in MMEP, no change in EDSS and MRI lesions	[153]

Table-7.2.   (*Continued*)

Diseases	Cell sources	No. of patients	Key findings	References
MS	BMSCs	10	EDSS score: improved in 5 patients, no change in 1 patient, deteriorated in 1 patient MRI lesions: increased in 5 patients	[149]
MS	PMSCs	12 4 (control)	No significant change in EDSS score and MRI lesions	[148]
ITP	WJ-MSCs	4	All patients achieved CR	[154]

Abbreviations: aGvHD: acute graft-versus-host disease; cGvHD: chronic graft-versus-host disease; WJ-MSCs: Wharton's jelly-derived mesenchymal stem cells; BMSCs: bone marrow-derived mesenchymal stem cells; CB-MSCs: cord blood-derived mesenchymal stem cells; CR: complete response; PR: partial response; NR: no response; T1DM: type 1 diabetes mellitus; SLE- systemic lupus erythematosus, MS-multiple sclerosis, ITP-immune thrombocytopenic purpura, PDSCs-placenta-derived decidual stromal cells; IPC: insulin producing cells; T-reg: regulatory T cells; EDSS: Expanded Disability Status Scale; PMSCs: placenta-derived mesenchymal stem cells; MMEP: multimodal evoked potential.

It is generally believed that in the near future, clinical trials/studies using MSC secretory products such as conditioned medium and microvesicles will emerge as therapeutic modalities for a cell-free therapy approach.[75,110] Furthermore, the production and application of the MSC secretome is less technically demanding and is associated with fewer regulatory issues and ethical concerns compared to cell-based therapy.

7.5.1 Graft-Versus-Host Disease

The majority of studies use BMSCs to treat GvHD, while only a few studies have used Wharton's jelly–derived MSCs (WJ-MSCs) and adipose-derived MSCs (ADSCs). In 3 separate studies, a total of 28 patients with acute GvHD (aGvHD) and 3 patients with chronic GvHD (cGvIID) were treated with WJ-MSCs.[111-113] A total of 15 and 6 patients with aGvHD achieved complete response (CR) and partial response (PR), respectively. Two of the 3 patients with cGvHD achieved PR. Jurado *et al.* treated 14 patient with cGvHD with ADSCs, and reported that 8 and 2 patients achieved CR and PR, respectively.[114] The use of placenta-derived decidual stromal cells (PDSCs) has also been reported: among 38 patients, 16 and 15 patients achieved CR and PR, respectively.[115] Most of the studies that used BMSCs have reported that the therapy is safe and might be effective for treating GvHD, as shown in Table-7.2.

Thus far, 2 phase III clinical trials (NCT00366145, sponsored by Osiris Therapeutics; and NCT02336230, sponsored by Mesoblast) using allogeneic BMSCs for treating GvHD have been completed and the outcomes have been disclosed in a public forum.[116] Osiris Therapeutics used prochymal (2×10^6 cells/kg b.w. [body weight] twice a week for 4 weeks) to treat aGvHD and found no difference in the ratio of patients who achieved CR compared to the control group. Mesoblast used Remestemcel-L (2×10^6 cells/kg b.w. twice a week for 4 weeks) to treat paediatric patients with aGvHD and reported that 69% of the MSC-treated patients achieved CR and PR, which was significantly higher compared to 45% in the control group.

Overall, multiple infusions appear to be useful for some patients when single administration is not effective. In addition, the timing of MSC therapy might be crucial for treating GvHD, as early administration has been linked with better prognosis.[117]

7.5.2 Graft Rejection

In a study that administered BMSCs together with standard immunosuppressive treatment to 10 patients who had undergone liver transplantation, the authors found that the rate of graft acceptance was similar to that of the 10 patients in the control group, who had received standard immunosuppressive treatment only.[118,119] Lee et al. reported that 3 of 7 kidney transplant recipients who received BMSC therapy showed acute rejection within a year.[120] These results indicate that MSCs might not be able to provide additional benefit for preventing graft rejection after solid organ transplantation. However, Reinders et al. reported positive results in 6 patients who had undergone kidney transplantation, where BMSC therapy prevented and ameliorated graft rejection.[121] Similarly, Peng et al. found that BMSC therapy was safe and promoted kidney graft acceptance, as no rejection was reported for the BMSC group, while there was one case of acute rejection in the control group.[122] Perico et al. also reported that BMSC therapy is safe and may help in preventing kidney graft rejection when the cells are administered pre- and post-transplantation.[123,124] Zhang et al. used WJ-MSCs to treat a patient with acute graft rejection after liver transplantation, and reported that the graft rejection was ameliorated.[125] A larger study by Tan et al. found that MSC therapy reduced the ratio of kidney transplant recipients with graft rejection (8/105, 7.6%) as compared to the control group (11/51, 21.6%).[126] BMSC therapy has also been linked with lower immunosuppressive drug dosage needed upon kidney transplantation.[127]

7.5.3 Type 1 Diabetes Mellitus (T1DM)

Hu *et al.* treated 15 patients with T1DM using WJ-MSCs. The WJ-MSC–treated patients had better disease control and lower daily insulin requirement compared to the 14 patients in the control group, who received normal saline.[128] Li *et al.* reported that 8 of 15 patients with T1DM who received WJ-MSCs responded to the therapy, as indicated by the ≥30% reduction in daily insulin requirement, ≥30% increase in fasting C-peptide or ≥50% increase in postprandial C-peptide for at least 3 months.[129] Carlson *et al.* used BMSCs to treat T1DM[130] and found that BMSC-treated patients had better disease control compared to those who received insulin only.

MSCs have also been differentiated into insulin-producing cells for treating T1DM. Thadani *et al.* treated a patient with T1DM using ADSCs and ADSC-derived insulin-producing cells,[131] and the patient's daily insulin requirement decreased markedly after 3 months. Dave *et al.* reported similar results after treating 10 patients with T1DM using ADSC-derived insulin-producing cells and bone marrow–derived HSCs.[132] The same group also reported that autologous ADSC-derived insulin-producing cells and bone marrow–derived HSCs are better than their allogeneic counterparts for treating T1DM.[133] Wang *et al.* co-transplanted BMSCs with pancreatic islets into 3 patients with T1DM and found that the diabetes was controlled, as the patients had lower daily insulin intake as well as lower fasting blood glucose and smaller decrease in C-peptide.[134]

7.5.4 Systemic Lupus Erythematosus

In two studies, a total of 56 patients with SLE were treated with WJ-MSCs, and 40 patients showed improvement while 16 patients were not responsive to the WJ-MSC therapy.[135,136] In another study, 87 patients with SLE were treated with either BM-MSCs or WJ-MSCs, and the patients had improved renal function and SLE disease activity index (SLEDAI) score for up to 4 years.[137] A total of 43 patients achieved CR and PR, with 20 patients subsequently relapsing. The same group treated 81 patients with SLE with either BM-MSCs or WJ-MSCs and found that 37 patients had disease remission; subsequently, 9 patients had disease relapse.[138] Similarly, Liang *et al.* reported that all patients showed improved SLEDAI score and proteinuria upon being treated with BMSCs.[139] However, not all studies have reported positive results: Carrion *et al.* found that BMSC therapy increased Treg but failed to suppress disease activity, as no changes were detected for the British Isles Lupus Assessment Group (BILAG) index and SLEDAI score.[140]

Deng *et al.* treated 18 patients with lupus nephritis, where 12 patients received WJ-MSCs plus standard immunosuppressive treatment, and the remaining patients received standard immunosuppressive treatment only.[141] The authors found that WJ-MSCs did not confer additional benefits when applied together with standard immunosuppression, as the same percentages of patients in both groups achieved CR. Nonetheless, BMSC therapy can improve the blood cell count in patients with SLE who have leukopenia, anaemia, and thrombocytopenia.[142]

7.5.5 Multiple Sclerosis

Li *et al.* treated 13 patients with MS using WJ-MSCs together with anti-inflammatory and immunosuppressive agents, and treated 10 patients with anti-inflammatory and immunosuppressive agents only.[143] They found that the WJ-MSC–treated patients had slower disease progression, as indicated by the Expanded Disability Status Scale (EDSS) score and fewer cases of disease relapse. Liang *et al.* transplanted WJ-MSCs into one patient with MS and observed that the EDSS score improved from 8.5 to 5.5.[144]

Bonad *et al.* reported 2 separate studies on the management of MS with BMSCs. The results varied, as the majority of the patients either showed no improvement or showed deterioration in the EDSS score and the number of brain lesions.[145,146] Similarly, Cohen *et al.* and Lublin *et al.* observed no improvement after BMSC therapy.[147,148] Yamout *et al.* reported improved EDSS scores but not magnetic resonance imaging (MRI) lesions.[149] Connick *et al.* reported improvement in some visual endpoints after BMSC therapy.[150] Several other studies have reported more promising results whereby there was improvement in the EDSS score,[151] gadolinium-enhancing lesions [152] and multimodal evoked potential (MMEP).[153]

7.5.6 Immune Thrombocytopenic Purpura

WJ-MSC therapy has been effective for treating ITP, whereby all 4 WJ-MSC–treated patients achieved CR within 24 months.[154]

7.6 CONCLUSION

Studies on the interaction between MSCs and immune cells, either individually or collectively, present a new opportunity to better understand the overall immune reaction of the body. Currently, the available knowledge of MSC–immune cell interaction is being applied to understand and develop new treatments for autoimmune-mediated diseases. Furthermore, it also provides important information for understanding the mechanism and treatment of GvHD and allograft rejection, and for laying the foundation for the application of MSCs in tissue engineering and regenerative medicine to enhance tissue repair and regeneration. MSCs have been tested clinically for treating many immune-mediated diseases that are yet to have satisfactory therapy. Nonetheless, most of the reported studies have been unable to provide conclusive results in terms of safety and efficacy due to the study limitations, especially small sample size. Furthermore, more work is needed to optimize MSC functionality *in vitro* (cell expansion procedure, cell pre-conditioning) before transplantation and to develop the ideal treatment protocol (e.g. route, number and timing of cell administration) for each disease. Hopefully, in the near future, more effective MSC therapy can be developed to benefit humankind, supported by new findings.

REFERENCES

1. Liau LL, Makpol S, Azurah AGN, Chua KH. Human adipose-derived mesenchymal stem cells promote recovery of injured HepG2 cell line and show sign of early hepatogenic differentiation. *Cytotechnology*. 2018; 70: 1221–1233.

2. Hafez P, Chowdhury SR, Jose S, Law JX, Ruszymah BHI, Ramzisham ARM, *et al*. Development of an in vitro cardiac ischemic model using primary human cardiomyocytes. *Cardiovasc Eng Technol*. 2018; 9: 529–538.

3. Wang M, Yuan Q, Xie L. Mesenchymal stem cell-based immunomodulation: Properties and clinical application. *Stem Cells Int*. 2018; 3057624.

4. Jacobs SA, Roobrouck VD, Verfaillie CM, Van Gool SW. Immunological characteristics of human mesenchymal stem cells and multipotent adult progenitor cells. *Immunol Cell Biol*. 2013; 91: 32–39.

5. Wang S, Qu X, Zhao RC. Clinical applications of mesenchymal stem cells. *J Hematol Oncol*. 2012; 5: 19.

6. Hocking AM, Gibran NS. Mesenchymal stem cells: Paracrine signaling and differentiation during cutaneous wound repair. *Exp Cell Res*. 2010; 316: 2213–2219.

7. Lim J, Razi ZRM, Law JX, Nawi AM, Idrus RBH, TG Chin TG, *et al*. Mesenchymal stromal cells from the maternal segment of human umbilical cord is ideal for bone regeneration in allogenic setting. *Tissue Eng Regen Med*. 2018; 15: 75–87.

8. Cunningham CJ, Redondo-Castro E, Allan SM. The therapeutic potential of the mesenchymal stem cell secretome in ischaemic stroke. *J Cereb Blood Flow Metab*. 2018; 38: 1276–1292.

9. Kyurkchiev D, Ivan B, Ivanova-Todorova E, Mourdjeva M, Oreshkova T, Belemezova K, *et al*. Secretion of immunoregulatory cytokines by mesenchymal stem cells. *World J Stem Cells*. 2014; 6: 552–570.

10. Krampera M, Glennie S, Dyson J, Scott D, Laylor R, Simpson E, *et al*. Bone marrow mesenchymal stem cells inhibit the response of naive and memory antigen-specific T cells to their cognate peptide. *Blood*. 2003; 101: 3722–3729.

11. Bochev I, Elmadjian G, Kyurkchiev D, Tzvetanov L, Altankova I, Tivchev P, *et al*. Mesenchymal stem cells from human bone marrow or adipose tissue differently

modulate mitogen-stimulated B-cell immunoglobulin production in vitro. *Cell Biol Int*. 2008; 32: 384–393.

12. Spaggiari GM, Capobianco A, Abdelrazik H, Becchetti F, Mingari MC, Moretta L. Mesenchymal stem cells inhibit natural killer–cell proliferation, cytotoxicity, and cytokine production: role of indoleamine 2, 3-dioxygenase and prostaglandin E2. *Blood*. 2008; 111: 1327–1333.

13. Zhang Y, Ge X, Guo XJ, Guan S, Li X, Gu W, *et al*. Bone marrow mesenchymal stem cells inhibit the function of dendritic cells by secreting galectin-1. *Biomed Res Int*. 2017; 2017: 3248605.

14. Khosravi M, Bidmeshkipour A, Moravej A, Hojjat-Assari S, Naserian S, Karimi MH. Induction of CD4+CD25+Foxp3+ regulatory T cells by mesenchymal stem cells is associated with RUNX complex factors. *Immunol Res*. 2018; 66: 207–218.

15. Sakaguchi S, Miyara M, Costantino CM, Hafler DA. FOXP3 + regulatory T cells in the human immune system. *Nat Rev Immunol*. 2010; 10: 490–500.

16. Lim J, Razi ZRM, Law J, Nawi AM, Idrus RBH, Ng MH. MSCs can be differentially isolated from maternal, middle and fetal segments of the human umbilical cord. *Cytotherapy*. 2016; 18: 1493–1502.

17. Ankrum JA, Ong JF, Karp JM. Mesenchymal stem cells: Immune evasive, not immune privileged. *Nat Biotechnol*. 2014; 32: 252–260.

18. Blanc K, Le Davies LC. Mesenchymal stromal cells and the innate immune response. *Immunol Lett*. 2015; 168: 140–146.

19. Ryan JM, Barry FP, Murphy JM, Mahon BP. Mesenchymal stem cells avoid allogeneic rejection. *J Inflamm (Lond)*. 2005; 2: 8.

20. William TT, Pendleton JD, Beyer WM, Egalka MC, Guinan EC. Suppression of allogeneic T-cell proliferation by human marrow stromal cells: implications in transplantation. *Transplantation*. 2003; 75: 389–397.

21. Kim JH, Lee YT, Hong JM, Hwang Y. Suppression of in vitro murine T cell proliferation by human adipose tissue-derived mesenchymal stem cells is dependent mainly on cyclooxygenase-2 expression. *Anat Cell Biol*. 2013; 46: 262.

22. Jarvinen L, Badri L, Wettlaufer S, Ohtsuka T, Standiford TJ, Toews GB, *et al*. Lung resident mesenchymal stem cells isolated from human lung allografts inhibit T cell proliferation via a soluble mediator. *J Immunol*. 2014; 181: 4389–4396.

23. Ren G, Zhang L, Zhao X, Xu G, Zhang Y, Roberts AI, *et al*. Mesenchymal stem cell-mediated immunosuppression occurs via concerted action of chemokines and nitric oxide. *Cell Stem Cell*. 2008; 2: 141–150.

24. Duffy MM, Ritter T, Ceredig R, Griffin MD. Mesenchymal stem cell effects on T-cell effector pathways. *Stem Cell Res Ther*. 2011; 2: 34.

25. Krause P, Singer E, Darley PI, Klebensberger J, Groettrup M, Legler DF. Prostaglandin E2 is a key factor for monocyte-derived dendritic cell maturation: enhanced T cell stimulatory capacity despite IDO. *J Leukoc Biol*. 2007; 82: 1106–1114.

26. Tomchuck SL, Zwezdaryk KJ, Coffelt SB, Waterman RS, Danka ES, Scandurro AB. Toll-like receptors on human mesenchymal stem cells drive their migration and immunomodulating responses. *Stem Cells*. 2008; 26: 99–107.

27. Berk LCJ, Jansen BJH, Siebers-Vermeulen KGS, Roelofs H, Figdor CG, Adema GJ, et al. Mesenchymal stem cells respond to TNF but do not produce TNF. *J Leukoc Biol.* 2010; 87: 283–289.

28. Le Blanc K, Mougiakakos D. Multipotent mesenchymal stromal cells and the innate immune system. *Nat Rev Immunol.* 2012; 12: 383.

29. Hirayama D, Iida T, Nakase H. The phagocytic function of macrophage-enforcing innate immunity and tissue homeostasis. *Int J Mol Sci.* 2017; 19: 92.

30. Italiani P, Boraschi D. From monocytes to M1/M2 macrophages: Phenotypical vs. functional differentiation. *Front Immunol.* 2014; 5: 514.

31. Hettinger J, Richards DM, Hansson J, Barra MM, Joschko AC, Krijgsveld J, et al. Origin of monocytes and macrophages in a committed progenitor. *Nat Immunol.* 2013; 14: 821–830.

32. Du Rocher B, Mencalha AL, Gomes BE, Abdelhay E. Mesenchymal stromal cells impair the differentiation of CD14++ CD16– CD64+ classical monocytes into CD14++ CD16+ CD64++ activate monocytes. *Cytotherapy.* 2012; 14: 12–25.

33. Selleri S, Bifsha P, Civini S, Pacelli C, Dieng MM, Lemieux W, et al. Human mesenchymal stromal cell-secreted lactate induces M2-macrophage differentiation by metabolic reprogramming. *Oncotarget.* 2016; 7: 30193–30210.

34. Vasandan AB, Jahnavi S, Shashank C, Prasad P, Kumar A, Prasanna SJ. Human mesenchymal stem cells program macrophage plasticity by altering their metabolic status via a PGE 2-dependent mechanism. *Sci Rep.* 2016; 6: 38308.

35. Jiang L, Zhang S, Hu H, Yang J, Wang X, Ma Y, et al. Exosomes derived from human umbilical cord mesenchymal stem cells alleviate acute liver failure by reducing the activity of the NLRP3 inflammasome in macrophages. *Biochem Biophys Res Commun.* 2019; 508: 735–741.

36. Lo Sicco C, Reverberi D, Balbi C, Ulivi V, Principi E, Pascucci L, et al. Mesenchymal stem cell-derived extracellular vesicles as mediators of anti-inflammatory effects: Endorsement of macrophage polarization. *Stem Cells Transl Med.* 2017; 6: 1018–1028.

37. Gas S, Mao F, Zhang B, Zhang L, Zhang X, Wang M, et al. Mouse bone marrow-derived mesenchymal stem cells induce macrophage M2 polarization through the nuclear factor-κB and signal transducer and activator of transcription 3 pathways. *Exp Biol Med.* 2014; 239: 366–375.

38. Zhao H, Shang Q, Pan Z, Bai Y, Li Z, Zhang H, et al. Exosomes from adipose-derived stem cells attenuate adipose inflammation and obesity through polarizing M2 macrophages and beiging in white adipose tissue. *Diabetes.* 2018; 67: 235–247.

39. Cho DI, Kim MR, Jeong H, Jeong HC, Jeong MH, Yoon SH, et al. Mesenchymal stem cells reciprocally regulate the M1/M2 balance in mouse bone marrow-derived macrophages. *Exp Mol Med.* 2014; 46: e70.

40. Biffl WL, Moore EE, Moore FA, Barnett CC, Silliman CC, Peterson VM. Interleukin-6 stimulates neutrophil production of platelet-activating factor. *J Leukoc Biol.* 1996; 59: 569–574.

41. Cassatella MA, Mosna F, Micheletti A, Lisi V, Tamassia N, Cont C, et al. Toll-like receptor-3-activated human mesenchymal stromal cells significantly prolong the survival and function of neutrophils. *Stem Cells.* 2011; 29: 1001–1011.

42. Maqbool M, Vidyadaran S, George E, Ramasamy R. Human mesenchymal stem cells protect neutrophils from serum-deprived cell death. *Cell Biol Int*. 2011; 35: 1247–1251.

43. Brown JM, Nemeth K, Kushnir-Sukhov NM, Metcalfe DD, Mezey E. Bone marrow stromal cells inhibit mast cell function via a COX2-dependent mechanism. *Clin Exp Allergy*. 2011; 41: 526–534.

44. H Kim, J Yun, T Shin, S Lee, B Lee, K Yu, *et al.* Human umbilical cord blood mesenchymal stem cell-derived PGE 2 and TGF-β1 alleviate atopic dermatitis by reducing mast cell degranulation. *Stem Cells*. 2015; 33: 1254–1266.

45. Glenn JD, Whartenby KA. Mesenchymal stem cells: Emerging mechanisms of immunomodulation and therapy. *World J Stem Cells*. 2014; 6: 526.

46. Ramasamy R, Fazekasova H, Lam EW-F, Soeiro I, Lombardi G, Dazzi F. Mesenchymal stem cells inhibit dendritic cell differentiation and function by preventing entry into the cell cycle. *Transplantation*. 2007; 83: 71–76.

47. Spaggiari GM, Abdelrazik H, Becchetti F, Moretta L. MSCs inhibit monocyte-derived DC maturation and function by selectively interfering with the generation of immature DCs: Central role of MSC-derived prostaglandin E2. *Blood*. 2009; 113: 6576–6583.

48. Djouad F, Charbonnier LM, Bouffi C, Louis-Plence P, Bony C, Apparailly F, *et al.* Mesenchymal stem cells inhibit the differentiation of dendritic cells through an interleukin-6-dependent mechanism. *Stem Cells*. 2007; 25: 2025–2032.

49. Jiang XX, Zhang Y, Liu B, Zhang SX, Wu Y, Yu XD, *et al.* Human mesenchymal stem cells inhibit differentiation and function of monocyte-derived dendritic cells. *Blood*. 2005; 105: 4120–4126.

50. Zhang W, Ge W, Li C, You S, Liao L, Han Q, *et al.* Effects of mesenchymal stem cells on differentiation, maturation, and function of human monocyte-derived dendritic cells. *Stem Cells Dev*. 2004; 13: 263–271.

51. Nauta AJ, Kruisselbrink AB, Lurvink E, Willemze R, Fibbe WE. Mesenchymal stem cells inhibit generation and function of both CD34+-derived and monocyte-derived dendritic cells. *J Immunol*. 2006; 177: 2080–2087.

52. Gao WX, Sun YQ, Shi J, Li CL. Fang SB, Wang D, *et al.* Effects of mesenchymal stem cells from human induced pluripotent stem cells on differentiation, maturation, and function of dendritic cells. *Stem Cell Res Ther*. 2017; 8: 48.

53. Reis M, Mavin E, Nicholson L, Green K, Dickinson AM, Wang X. Mesenchymal stromal cell-derived extracellular vesicles attenuate dendritic cell maturation and function. *Front Immunol*. 2018; 9: 2538.

54. Moretta A, Marcenaro E, Sivori S, Chiesa MD, Vitale M, Moretta L. Early liaisons between cells of the innate immune system in inflamed peripheral tissues. *Trends Immunol*. 2005; 26: 668–675.

55. Martín-Fontecha A, Thomsen LL, Brett S, Gerard C, Lipp M, Lanzavecchia A, *et al.* Induced recruitment of NK cells to lymph nodes provides IFN-γ for T(H)1 priming. *Nat Immunol*. 2004; 5: 1260–1265.

56. Spaggiari GM, Capobianco A, Becchetti S, Mingari MC, Moretta L. Mesenchymal stem cell-natural killer cell interactions: Evidence that activated NK cells are capable

of killing MSCs, whereas MSCs can inhibit IL-2-induced NK-cell proliferation. *Blood*. 2006; 107: 1484–1490.

57. Sotiropoulou PA, Perez SA, Gritzapis AD, Baxevanis CN, Papamichail M. Interactions between human mesenchymal stem cells and natural killer cells. *Stem Cells*. 2006; 24: 74–85.

58. Fan Y, Herr F, Vernochet A, Mennesson B, Oberlin E, Durrbach A. Human fetal liver mesenchymal stem cell-derived exosomes impair natural killer cell function. *Stem Cells Dev*. 2018; 28: 44–55.

59. Koch U, Radtke F. Mechanisms of T cell development and transformation. *Annu Rev Cell Dev Biol*. 2011; 27: 539–562.

60. June CH, Ledbetter JA, Gillespie MM, Lindsten T, Thompson CB. T-cell proliferation involving the CD28 pathway is associated with cyclosporine-resistant interleukin 2 gene expression. *Mol Cell Biol*. 1987; 7: 4472–4481.

61. Mueller DL, Jenkins MK, Schwartz RH. Clonal expansion versus functional clonal inactivation: a costimulatory signalling pathway determines the outcome of T cell antigen receptor occupancy. *Annu Rev Immunol*. 1989; 7: 445–480.

62. O'Garra A. Cytokines induce the development of functionally heterogeneous T helper cell subsets. *Immunity*. 1998; 8: 275–283.

63. Hori S, Nomura T, Sakaguchi S. Control of regulatory T cell development by the transcription factor Foxp3. *Science*. 2003; 299: 1057–1061.

64. Langrish CL, Chen Y, Blumenschein WM, Mattson J, Basham B, Sedgwick JD, *et al*. IL-23 drives a pathogenic T cell population that induces autoimmune inflammation. *J Exp Med*. 2005; 201: 233–240.

65. Soroosh P, Doherty TA. Th9 and allergic disease. *Immunology*. 2009; 127: 450–458.

66. Kaech SM, Cui W. Transcriptional control of effector and memory CD8+ T cell differentiation. *Nat Rev Immunol*. 2012; 12: 749.

67. Dimeloe S, Burgener A, Grählert J, Hess C. T-cell metabolism governing activation, proliferation and differentiation; a modular view. *Immunology*. 2017; 150: 35–44.

68. Niu J, Yue W, Le-Le Z, Bin L, Hu X. Mesenchymal stem cells inhibit T cell activation by releasing TGF-β1 from TGF-β1/GARP complex. *Oncotarget*. 2017; 8: 99784–99800.

69. Cuerquis J, Romieu-Mourez R, François M, Routy JP, Young YK, Zhao J, *et al*. Human mesenchymal stromal cells transiently increase cytokine production by activated T cells before suppressing T-cell proliferation: Effect of interferon-γ; and tumor necrosis factor-α stimulation. *Cytotherapy*. 2014; 16: 191–202.

70. Groh ME, Maitra B, Szekely E, Koç ON. Human mesenchymal stem cells require monocyte-mediated activation to suppress alloreactive T cells. *Exp Hematol*. 2005; 33: 928–934.

71. Glennie S, Soeiro I, Dyson PJ, Lam EW-F, Dazzi F. Bone marrow mesenchymal stem cells induce division arrest anergy of activated T cells. *Blood*. 2005; 105: 2821–2827.

72. Akiyama K, Chen C, Wang D, Xu X, Qu C, Yamaza T, *et al*. Mesenchymal-stem-cell-induced immunoregulation involves FAS-ligand-/FAS-mediated T cell apoptosis. *Cell Stem Cell*. 2012; 10: 544–555.

73. Luz-Crawford P, Kurte M, Bravo-Alegría J, Contreras R, Nova-Lamperti E, Tejedor G, *et al*. Mesenchymal stem cells generate a CD4+ CD25+ Foxp3+ regulatory T cell population during the differentiation process of Th1 and Th17 cells. *Stem Cell Res Ther*. 2013; 4: 65.

74. Zhang B, Yeo RWY, Lai RC, Sim EWK, Chin KC, Lim SK. Mesenchymal stromal cell exosome–enhanced regulatory T-cell production through an antigen-presenting cell–mediated pathway. *Cytotherapy*. 2018; 20: 687–696.

75. Shigemoto-Kuroda T, Oh JY, Kim D, Jeong HJ, Park SY, Lee HJ, *et al*. MSC-derived extracellular vesicles attenuate immune responses in two autoimmune murine models: Type 1 diabetes and uveoretinitis. *Stem Cell Rep*. 2017; 8: 1214–1225.

76. Schmidt A, Oberle N, Krammer PH. Molecular mechanisms of treg-mediated T cell suppression. *Front Immunol*. 2012; 3: 51.

77. Consentius C, Akyüz L, Schmidt-Lucke JA, Tschöpe C, Pinzur L, Ofir R, *et al*. Mesenchymal stromal cells prevent allostimulation in vivo and control checkpoints of Th1 priming: migration of human DC to lymph nodes and NK cell activation. *Stem Cells*. 2015; 33: 3087–3099.

78. Blazquez R, Sanchez-Margallo FM, de la Rosa O, Dalemans W, Alvarez V, Tarazona R, *et al*. Immunomodulatory potential of human adipose mesenchymal stem cells derived exosomes on in vitro stimulated T cells. *Front Immunol*. 2014; 5: 556.

79. Khare D, Or R, Resnick I, Barkatz C, Almogi-Hazan O, Avni B. Mesenchymal stromal cell-derived exosomes affect mRNA expression and function of B-lymphocytes. *Front Immunol*. 2018; 9: 3053.

80. Li N, Hua J. Interactions between mesenchymal stem cells and the immune system. *Cell Mol Life Sci*. 2017; 74: 2345–2360.

81. Crop MJ, Baan CC, Korevaar SS, Ijzermans JNM, Pescatori M, Stubbs AP, *et al*. Inflammatory conditions affect gene expression and function of human adipose tissue-derived mesenchymal stem cells. *Clin Exp Immunol*. 2010; 162: 474–486.

82. Ma S, Xie N, Li W, Yuan B, Shi Y, Wang Y. Immunobiology of mesenchymal stem cells. *Cell Death Differ*. 2014; 21: 216.

83. Wang Y, Chen X, Cao W, Shi Y. Plasticity of mesenchymal stem cells in immunomodulation: pathological and therapeutic implications. *Nat Immunol*. 2014; 15: 1009.

84. Li W, Ren G, Huang Y, Su J, Han Y, Li J, *et al*. Mesenchymal stem cells: A double-edged sword in regulating immune responses. *Cell Death Differ*. 2012; 19: 1505.

85. Pieper K, Grimbacher B, Eibel H. B-cell biology and development. *J Allergy Clin Immunol*. 2013; 131: 959–971.

86. Hayakawa K, Hardy RR, Herzenberg LA. Progenitors for Ly-1 B cells are distinct from progenitors for other B cells. *J Exp Med*. 1985; 161: 1554–1568.

87. De Silva NS, Klein U. Dynamics of B cells in germinal centres. *Nat Rev Immunol*. 2015; 15: 137.

88. Corcione A, Benvenuto F, Ferretti E, Giunti D, Cappiello V, Cazzanti F, *et al*. Human mesenchymal stem cells modulate B-cell functions. *Blood*. 2006; 107: 367–372.

89. Tabera S, Pérez-Simón JA, Díez-Campelo M, Sánchez-Abarca LI, Blanco B, López A, *et al*. The effect of mesenchymal stem cells on the viability, proliferation and differentiation of B-lymphocytes. *Haematologica*. 2008; 93: 1301–1309.

90. Che N, Li X, Zhou S, Liu R, Shi D, Lu L, *et al*. Umbilical cord mesenchymal stem cells suppress B-cell proliferation and differentiation. *Cell Immunol*. 2012; 274: 46–53.

91. Luk F, Carreras-Planella L, Korevaar SS, de Witte SFH, Borràs FE, Betjes MGH, *et al*. Inflammatory conditions dictate the effect of mesenchymal stem or stromal cells on B cell function. *Front Immunol*. 2017; 8: 1042.

92. Budoni M, Fierabracci A, Luciano R, Petrini S, Di Ciommo V, Muraca M. The immunosuppressive effect of mesenchymal stromal cells on B lymphocytes is mediated by membrane vesicles. *Cell Transplant*. 2013; 22: 369–379.

93. Asari S, Itakura S, Ferreri K, Liu CP, Kuroda Y, Kandeel F, *et al*. Mesenchymal stem cells suppress B-cell terminal differentiation. *Exp Hematol*. 2009; 37: 604–615.

94. Luz-Crawford P, Djouad F, Toupet K, Bony C, Franquesa M, Hoogduijn MJ, *et al*. Mesenchymal stem cell-derived interleukin 1 receptor antagonist promotes macrophage polarization and inhibits B cell differentiation. *Stem Cells*. 2016; 34: 483–492.

95. Schena F, Gambini C, Gregorio A, Mosconi M, Reverberi D, Gattorno M, *et al*. Interferon-γ-dependent inhibition of B cell activation by bone marrow–derived mesenchymal stem cells in a murine model of systemic lupus erythematosus. *Arthritis Rheum*. 2010; 62: 2776–2786.

96. Bonnaure G, Gervais-St-Amour C, Néron S. Bone marrow mesenchymal stem cells enhance the differentiation of human switched memory B lymphocytes into plasma cells in serum-free medium. *J. Immunol. Res*. 2016; 2016: 7801781.

97. Traggiai E, Volpi S, Schena F, Gattorno M, Ferlito F, Moretta L, *et al*. Bone marrow-derived mesenchymal stem cells induce both polyclonal expansion and differentiation of B cells isolated from healthy donors and systemic lupus erythematosus patients. *Stem Cells*. 2008; 26: 562–569.

98. Healy ME, Bergin R, Mahon BP, English K. Mesenchymal stromal cells protect against caspase 3-mediated apoptosis of CD19+ peripheral B cells through contact-dependent upregulation of VEGF. *Stem Cells Dev*. 2015; 24: 2391–2402.

99. Mauney J, Olsen BR, Volloch V. Matrix remodeling as stem cell recruitment event: A novel in vitro model for homing of human bone marrow stromal cells to the site of injury shows crucial role of extracellular collagen matrix. *Matrix Biol*. 2010; 29: 657–663.

100. Hocking AM. The role of chemokines in mesenchymal stem cell homing to wounds. *Adv Wound Care*. 2015; 4: 623–630.

101. Marquez-Curtis LA, Janowska-Wieczorek A. Enhancing the migration ability of mesenchymal stromal cells by targeting the SDF-1/CXCR4 axis. *Biomed Res Int*. 2013; 2013: 561098.

102. Lin W, Xu L, Zwingenberger S, Gibon E, Goodman SB. Mesenchymal stem cells homing to improve bone healing. *J Orthop Transl*. 2017; 9: 19–27.

103. Li Q, Xia S, Fang H, Pan J, Jia Y, Deng G. VEGF treatment promotes bone marrow-derived CXCR4+ mesenchymal stromal stem cell differentiation into vessel endothelial cells. *Exp Ther Med*. 2017; 13: 449–454.

104. Kim C, Schneider G, Abdel-Latif A, Mierzejewska K, Sunkara M, Borkowska S, *et al*. Ceramide-1-phosphate regulates migration of multipotent stromal cells and

endothelial progenitor cells — implications for tissue regeneration. *Stem Cells.* 2013; 31: 500–510.

105. Schraufstatter IU, DiScipio RG, Zhao M, Khaldoyanidi SK. C3a and C5a are chemotactic factors for human mesenchymal stem cells, which cause prolonged ERK1/2 phosphorylation. *J Immunol.* 2009; 182: 3827–3836.

106. Atkinson K (ed.). The Biology and Therapeutic Application of Mesenchymal Cells. John Wiley & Sons, New Jersey, 2016.

107. Nitzsche F, Müller C, Lukomska B, Jolkkonen J, Deten A, Boltze J. Concise review: MSC adhesion cascade-insights into homing and transendothelial migration. *Stem Cells.* 2017; 35: 1446–1460.

108. Sheriff L, Alanazi A, Ward LSC, Ward C, Munir H, Rayes J, *et al.* Origin-specific adhesive interactions of mesenchymal stem cells with platelets influence their behavior after infusion. *Stem Cells.* 2018; 36: 1062–1074.

109. Rennert RC, Sorkin M, RK Garg, Gurtner GC. Stem cell recruitment after injury: Lessons for regenerative medicine. *Regen Med.* 2012; 7: 833–850.

110. Cosenza S, Toupet K, Maumus M, Luz-Crawford P, Blanc-Brude O, Jorgensen C, *et al.* Mesenchymal stem cells-derived exosomes are more immunosuppressive than microparticles in inflammatory arthritis. *Theranostics.* 2018; 8: 1399–1410.

111. Wu KH, Chan CK, Tsai C, Chang YH, Sieber M, Chiu TH, *et al.* Effective treatment of severe steroid-resistant acute graft-versus-host disease with umbilical cord-derived mesenchymal stem cells. *Transplantation.* 2011; 91: 1412–1416.

112. Chen GH, Yang T, Tian H, Qiao M, Liu HW, Fu CC, *et al.* Clinical study of umbilical cord-derived mesenchymal stem cells for treatment of nineteen patients with steroid-resistant severe acute graft-versus-host disease. Zhonghua Xue Ye Xue Za Zhi. 2012; 33: 303–306.

113. Boruczkowski D, Gladysz D, Ruminski S, Czaplicka-Szmaus I, Murzyn M, Olkowicz A, *et al.* Third-party Wharton's jelly mesenchymal stem cells for treatment of steroid-resistant acute and chronic graft-versus-host disease: A report of 10 cases. *Turkish J Biol.* 2016; 40: 493–500.

114. Jurado M, De La Mata C, Ruiz-García A, López-Fernández E, Espinosa O, Remigia MJ, *et al.* Adipose tissue-derived mesenchymal stromal cells as part of therapy for chronic graft-versus-host disease: A phase I/II study. *Cytotherapy.* 2017; 19: 927–936.

115. Ringden O, Baygan A, Remberger M, Gustafsson B, Winiarski J, Khoein B, *et al.* Placenta-derived decidua stromal cells for treatment of severe acute graft-versus-host disease. *Stem Cells Transl Med.* 2018; 7: 325–331.

116. Galipeau J, Sensébé L. Mesenchymal stromal cells: Clinical challenges and therapeutic opportunities. *Cell Stem Cell.* 2018; 22: 824–833.

117. Ball LM, Bernardo ME, Roelofs H, van Tol MJD, Contoli B, Zwaginga JJ, *et al.* Multiple infusions of mesenchymal stromal cells induce sustained remission in children with steroid-refractory, grade III-IV acute graft-versus-host disease. *Br J Haematol.* 2013; 163: 501–509.

118. O Detry, M Vandermeulen, M-H Delbouille, J Somja, N Bletard, A Briquet, *et al.* Infusion of mesenchymal stromal cells after deceased liver transplantation: A phase I–II, open-label, clinical study. *J Hepatol.* 2017; 67: 47–55.

119. Erpicum P, Weekers L, Detry O, Bonvoisin C, Delbouille MH, Grégoire C, *et al.* Infusion of third-party mesenchymal stromal cells after kidney transplantation: A phase I-II, open-label, clinical study. *Kidney Int.* 2019; 95: 693–707.

120. Lee H, Park JB, Lee S, Baek S, Kim H, Kim SJ. Intra-osseous injection of donor mesenchymal stem cell (MSC) into the bone marrow in living donor kidney transplantation; a pilot study. *J Transl Med.* 2013; 11: 96.

121. Reinders MEJ, de Fijter JW, Roelofs H, Bajema IM, de Vries DK, Schaapherder AF, *et al.* Autologous bone marrow-derived mesenchymal stromal cells for the treatment of allograft rejection after renal transplantation: Results of a phase I study. *Stem Cells Transl Med.* 2013; 2: 107–111.

122. Peng Y, Ke M, Xu L, Liu L, Chen X, Xia W, *et al.* Donor-derived mesenchymal stem cells combined with low-dose tacrolimus prevent acute rejection after renal transplantation: A clinical pilot study. *Transplant.* 2013; 95: 161–168.

123. Perico N, Casiraghi F, Introna M, Gotti E, Todeschini M, Cavinato RA, *et al.* Autologous mesenchymal stromal cells and kidney transplantation: A pilot study of safety and clinical feasibility. *Clin J Am Soc Nephrol.* 2011; 6: 412–422.

124. N Perico, F Casiraghi, E Gotti, M Introna, M Todeschini, RA Cavinato, *et al.* Mesenchymal stromal cells and kidney transplantation: Pretransplant infusion protects from graft dysfunction while fostering immunoregulation. *Transpl Int.* 2013; 26: 867–878.

125. Zhang Y, Li S, Wang G, Peng Y, Zhang Q, Li H, *et al.* Mesenchymal stem cells for treatment of steroid-resistant acute rejection after liver transplantation. *Liver Res.* 2017; 1: 140–145.

126. Tan J, Wu W, Xu X, Liao L, Zheng F, Messinger S, *et al.* Induction therapy with autologous mesenchymal stem cells in living-related kidney transplants: A randomized controlled trial. *JAMA.* 2012; 307: 1169–1177.

127. Pan GH, Chen Z, Xu L, Zhu JH, Xiang P, Ma JJ, *et al.* Low-dose tacrolimus combined with donor-derived mesenchymal stem cells after renal transplantation: A prospective, non-randomized study. *Oncotarget.* 2016; 7: 12089–12101.

128. Hu J, Yu X, Wang Z, Wang F, Wang L. Gao H. Long term effects of the implantation of Wharton's jelly-derived mesenchymal stem cells from the umbilical cord for newly-onset type 1 diabetes mellitus. *Endocr J.* 2013; 60: 347–357.

129. Li L, Lu J, Shen S, Jia X, Zhu D. Wharton's jelly-derived mesenchymal stem cell therapy to improve β-cell function in patients with type 1 diabetes and ketoacidosis: A single-centre, single-group, open-label, phase 2 trial. *Lancet Diabetes Endocrinol.* 2016; 4: S17.

130. Carlsson PO, Schwarcz E, Korsgren O, Le Blanc K. Preserved β-cell function in type 1 diabetes by mesenchymal stromal cells. *Diabetes.* 2015; 64: 587–592.

131. Thadani JM, Marathe A, Vakodikar S, Kshatriya P, Modi D, Vyas R, *et al.* Treatment of type I diabetes using autologous adipose derived mesenchymal stem cells translated to insulin secreting Islet like cell aggregates. *J Case Rep.* 2017; 7: 235–238.

132. Dave SD, Vanikar AV, Trivedi HL, Thakkar UG, Gopal SC, Chandra T. Novel therapy for insulin-dependent diabetes mellitus: Infusion of in vitro-generated insulin-secreting cells. *Clin Exp Med.* 2015; 15: 41–45.

133. Thakkar UG, Trivedi HL, Vanikar AV, Dave SD. Insulin-secreting adipose-derived mesenchymal stromal cells with bone marrow–derived hematopoietic stem cells from autologous and allogenic sources for type 1 diabetes mellitus. *Cytotherapy*. 2015; 17: 940–947.

134. Wang H, Strange C, Nietert PJ, Wang J, Turnbull TL, Cloud C, et al. Autologous mesenchymal stem cell and islet cotransplantation: safety and efficacy. *Stem Cells Transl Med*. 2018; 7: 11–19.

135. Sun L, Wang D, Liang J, Zhang H, Feng X, Wang H, et al. Umbilical cord mesenchymal stem cell transplantation in severe and refractory systemic lupus erythematosus. *Arthritis Rheum*. 2010; 62: 2467–2475.

136. Wang D, Li J, Zhang Y, Zhang M, Chen J, Li X, et al. Umbilical cord mesenchymal stem cell transplantation in active and refractory systemic lupus erythematosus: A multicenter clinical study. *Arthritis Res Ther*. 2014; 16: R79.

137. D Wang, H Zhang, J Liang, X Li, X Feng, H Wang, et al. Allogeneic mesenchymal stem cell transplantation in severe and refractory systemic lupus erythematosus: 4 years of experience. *Cell Transplant*. 2013; 22: 2267–2277.

138. Wang D, Zhang H, Liang J, Wang H, Hua B, Feng X, et al. A long-term follow-up study of allogeneic mesenchymal stem/stromal cell transplantation in patients with drug-resistant systemic lupus erythematosus. *Stem Cell Rep*. 2018; 10: 933–941.

139. Liang J, Zhang H, Hua B, Wang H, Lu L, Shi S, et al. Allogenic mesenchymal stem cells transplantation in refractory systemic lupus erythematosus: A pilot clinical study. *Ann Rheum Dis*. 2010; 69: 1423–1429.

140. Carrion F, Nova E, Ruiz C, Diaz F, Inostroza C, Rojo D, et al. Autologous mesenchymal stem cell treatment increased T regulatory cells with no effect on disease activity in two systemic lupus erythematosus patients. *Lupus*. 2010; 19: 317–322.

141. Deng D, Zhang P, Guo Y, Lim TO. A randomised double-blind, placebo-controlled trial of allogeneic umbilical cord-derived mesenchymal stem cell for lupus nephritis. *Ann Rheum Dis*. 2017; 76: 1436–1439.

142. Li X, Wang D, Liang J, Zhang H, Sun L. Mesenchymal SCT ameliorates refractory cytopenia in patients with systemic lupus erythematosus. *Bone Marrow Transplant*. 2013; 48: 544.

143. Li JF, Zhang DJ, Geng T, Chen L, Huang H, Yin HL, et al. The potential of human umbilical cord-derived mesenchymal stem cells as a novel cellular therapy for multiple sclerosis. *Cell Transplant*. 2014; 23: S113–S122.

144. Liang J, Zhang H, Hua B, Wang H, Wang J, Han Z, et al. Allogeneic mesenchymal stem cells transplantation in treatment of multiple sclerosis. *Mult Scler J*. 2009; 15: 644–646.

145. Bonab MM, Yazdanbakhsh S, Lotfi J, Alimoghaddom K, Talebian F, Hooshmand F, et al. Does mesenchymal stem cell therapy help multiple sclerosis patients? Report of a pilot study. *Iran J Immunol*. 2007; 4: 50–57.

146. Bonab MM, Sahraian MA, Aghsaie A, Karvigh SA, Hosseinian SM, Nikbin B, et al. Autologous mesenchymal stem cell therapy in progressive multiple sclerosis: An open label study. *Curr Stem Cell Res Ther*. 2012; 7: 407–414.

147. JA Cohen, PB Imrey, SM Planchon, RA Bermel, E Fisher, RJ Fox, et al. Pilot trial of intravenous autologous culture-expanded mesenchymal stem cell transplantation in multiple sclerosis. *Mult Scler J*. 2018; 24: 501–511.

148. Lublin FD, Bowen JD, Huddlestone J, Kremenchutzky M, Carpenter A, Corboy JR, *et al.* Human placenta-derived cells (PDA-001) for the treatment of adults with multiple sclerosis: A randomized, placebo-controlled, multiple-dose study. *Mult Scler Relat Disord.* 2014; 3: 696–704.

149. Yamout B, Hourani R, Salti H, Barada W, El-Hajj T, Al-Kutoubi A, *et al.* Bone marrow mesenchymal stem cell transplantation in patients with multiple sclerosis: A pilot study. *J Neuroimmunol.* 2010; 227: 185–189.

150. Connick P, Kolappan M, Crawley C, Webber DJ, Patani R, Michell AW, *et al.* Autologous mesenchymal stem cells for the treatment of secondary progressive multiple sclerosis: An open-label phase 2a proof-of-concept study. *Lancet Neurol.* 2012; 11: 150–156.

151. Karussis D, Karageorgiou C, Vaknin-Dembinsky A, Gowda-Kurkalli B, Gomori JM, Kassis I, *et al.* Safety and immunological effects of mesenchymal stem cell transplantation in patients with multiple sclerosis and amyotrophic lateral sclerosis. *Arch Neurol.* 2010; 67: 1187–1194.

152. S Llufriu, M Sepúlveda, Y Blanco, P Marín, B Moreno, J Berenguer, *et al.* Randomized placebo-controlled phase II trial of autologous mesenchymal stem cells in multiple sclerosis. *PLoS One* 9. 2014; e113936.

153. Rice CM, Mallam EA, Whone AL, Walsh P, Brooks DJ, Kane N, *et al.* Safety and feasibility of autologous bone marrow cellular therapy in relapsing-progressive multiple sclerosis. *Clin Pharmacol Ther.* 2010; 87: 679–685.

154. Wang X, Yin X, Sun W, Bai J, Shen Y, Ao Q, *et al.* Intravenous infusion umbilical cord-derived mesenchymal stem cell in primary immune thrombocytopenia: A two-year follow-up. *Exp Ther Med.* 2017; 13: 2255–2258.

155. Erbey F, Atay D, Akcay A, Ovali E, Ozturk G. Mesenchymal stem cell treatment for steroid refractory graft-versus-host disease in children: A pilot and first study from Turkey. *Stem Cells Int.* 2016; 2016: 1641402.

156. Bader P, Kuçi Z, Bakhtiar S, Basu O, Bug G, Dennis M, *et al.* Effective treatment of steroid and therapy-refractory acute graft-versus-host disease with a novel mesenchymal stromal cell product (MSC-FFM). *Bone Marrow Transplant.* 2018; 53: 852–862.

157. Introna M, Lucchini G, Dander E, Galimberti S, Rovelli A, Balduzzi A, *et al.* Treatment of graft versus host disease with mesenchymal dtromal vells: A phase I Study on 40 adult and pediatric patients. *Biol Blood Marrow Transplant.* 2014; 20: 375–381.

158. Sánchez-Guijo F, Caballero-Velázquez T, López-Villar O, Redondo A, Parody R, Martínez C, *et al.* Sequential third-party mesenchymal stromal cell therapy for refractory acute graft-versus-host disease. *Biol Blood Marrow Transplant.* 2014; 20: 1580–1585.

159. Resnick IB, Barkats C, Shapira MY, Stepensky P, Bloom AI, Shimoni A, *et al.* Treatment of severe steroid resistant acute GVHD with mesenchymal stromal cells (MSC). *Am J Blood Res.* 2013; 3: 225–238.

160. Le Blanc K, Frassoni F, Ball L, Locatelli F, Roelofs H, Lewis I, *et al.* Mesenchymal stem cells for treatment of steroid-resistant, severe, acute graft-versus-host disease: A phase II study. *Lancet.* 2008; 371: 1579–1586.

161. Ringdén O, Uzunel M, Rasmusson I, Remberger M, Sundberg B, Lönnies H, *et al.* Mesenchymal stem cells for treatment of therapy-resistant graft-versus-host disease. *Transplantation.* 2006; 81: 1390–1397.

162. Von Bonin M, Stölzel F, Goedecke A, Richter K, Wuschek N, Hölig K, *et al.* Treatment of refractory acute GVHD with third-party MSC expanded in platelet lysate-containing medium. *Bone Marrow Transplant.* 2009; 43: 245.

163. Lucchini G, Introna M, Dander E, Rovelli A, Balduzzi A, Bonanomi S, *et al.* Platelet-lysate-expanded mesenchymal stromal cells as a salvage therapy for severe resistant graft-versus-host disease in a pediatric population. *Biol Blood Marrow Transplant.* 2010; 16: 1293–1301.

164. Herrmann R, Sturm M, Shaw K, Purtill D, Cooney J, Wright M, *et al.* Mesenchymal stromal cell therapy for steroid-refractory acute and chronic graft versus host disease: A phase 1 study. *Int J Hematol.* 2012; 95: 182–188.

165. Prasad VK, Lucas KG, Kleiner GI, Talano JAM, Jacobsohn D, Broadwater G, *et al.* Efficacy and safety of ex vivo cultured adult human mesenchymal stem cells (Prochymal™) in pediatric patients with severe refractory acute graft-versus-host disease in a compassionate use study. *Biol Blood Marrow Transplant.* 2011; 17: 534–541.

166. Muroi K, Miyamura K, Ohashi K, Murata M, Eto T, Kobayashi N, *et al.* Unrelated allogeneic bone marrow-derived mesenchymal stem cells for steroid-refractory acute graft-versus-host disease: a phase I/II study. *Int J Hematol.* 2013; 98: 206–213.

167. Kurtzberg J, Prockop S, Teira P, Bittencourt H, Lewis V, Chan KW, *et al.* Allogeneic human mesenchymal stem cell therapy (remestemcel-L, Prochymal) as a rescue agent for severe refractory acute graft-versus-host disease in pediatric patients. *Biol Blood Marrow Transplant.* 2014; 20: 229–235.

168. Zhao K, Lou R, Huang F, Peng Y, Jiang Z, Huang K, *et al.* Immunomodulation effects of mesenchymal stromal cells on acute graft-versus-host disease after hematopoietic stem cell transplantation. *Biol Blood Marrow Transplant.* 2015; 21: 97–104.

169. Muroi K, Miyamura K, Okada M, Yamashita T, Murata M, Ishikawa T, *et al.* Bone marrow-derived mesenchymal stem cells (JR-031) for steroid-refractory grade III or IV acute graft-versus-host disease: A phase II/III study. *Int J Hematol.* 2016; 103: 243–250.

170. Pérez-Simon JA, López-Villar O, Andreu EJ, Rifón J, Muntion S, Diez Campelo M, *et al.* Mesenchymal stem cells expanded in vitro with human serum for the treatment of acute and chronic graft-versus-host disease: Results of a phase I/II clinical trial. *Haematologica.* 2011; 96: 1072–1076.

171. Dotoli GM, De Santis GC, Orellana MD, de Lima Prata K, Caruso SR, Fernandes TR, *et al.* Mesenchymal stromal cell infusion to treat steroid-refractory acute GvHD III/IV after hematopoietic stem cell transplantation. *Bone Marrow Transplant.* 2017; 52: 859.

172. von Dalowski F, Kramer M, Wermke M, Wehner R, Röllig C, Alakel N, *et al.* Mesenchymal stromal cells for treatment of acute steroid-refractory graft versus host disease: clinical responses and long-term outcome. *Stem Cells.* 2016; 34: 357–366.

CHAPTER 8

Chimeric Antigen Receptor (CAR) T-cells as a therapeutic modality

Mehmet Özen*, Mehmet Gündüz†, Khawaja Husnain Haider‡

*Bayindir Sogutozu Hospital, Department of Hematology and Bone Marrow Transplantation, Kizilirmak M, 1443. S. No:17, Ankara, Turkey
Email: kanbilimci@gmail.com
†Ankara Atatürk Education and Research Hospital, Clinics of Hematology, Universiteler M Bilkent C. No. 1 Cankaya, Ankara, Turkey
Email: drmgunduz02@gmail.com
‡Cellular and Molecular Pharmacology, Department of Basic Sciences, Sulaiman AlRajhi Medical School, PO Box 777, Al Bukairiyah 51941, Kingdom of Saudi Arabia

ABSTRACT

In the past decade, several major breakthroughs and groundbreaking discoveries have been reported from medical science researchers involved in the areas of hematology and oncology, due to improved understanding of the human immune system. Novel amongst these discoveries is the biopharmaceuticals such as monoclonal antibodies and immune checkpoint regulators that have revolutionized our options of therapeutic interventions for patients with diseases once considered untreatable. Autologous/allogeneic/mesenchymal stem cells are collected by Apheresis machines and are used in regenerative medicine by physicians to treat previously considered incurable diseases. Eventually, it has all progressed to the reprogramming of somatic cells by genetic engineering techniques. Chimeric Antigen Receptor (CAR) T-cells are reprogrammed cells with the capacity to kill malignant cells via binding of their new, synthetic and specific receptors to their targets on tumor cells. Some of the CAR T-cell preparations are currently approved by the Food and Drug Agency for the treatment of refractory acute lymphocytic leukemia and diffuse large B cell lymphoma. Their potentials are under investigation in treating chronic lymphocytic leukemia, acute myeloid leukemia, multiple myeloma, and other lymphomas. After constructing a CAR gene and transferring it into patients' own T-cells' DNA via a vector, genetically modified CAR T-cells are created. Currently, CAR T-cells are more popular and effective than dendritic cell-based tumor vaccines. Development and usage of CAR T-cells in clinical studies have also been reviewed.

---

* Corresponding author. *kanbilimci@gmail.com*

KEYWORDS

Antigen; B-cells; CAR, Chimeric; FcR; Leukemia; Receptors; T-cells.

LIST OF ABBREVIATIONS

ALL   = Acute lymphoblastic leukemia
BCMA = B-Cell maturation antigen
CAR   = Chimeric Antigen Receptor
CES   = CAR T-cell related encephalopathy syndrome
CLL   = Chronic lymphocytic leukemia
CRS   = Cytokine release syndrome
FcR   = Fc receptor
HBSS   = Hank's balanced salt solution
HER 2 = Human epidermal growth factor receptor 2
HLA   = Human leukocyte antigen
HSA   = Human serum albumin
HV   = Variable regions of heavy chain
LeY   = Lewis antigen
MHC   = Major histocompatibility complex
PBMC = Peripheral blood mononuclear cells
PBS   = Phosphate buffered saline
PIDD   = Primary immune deficiency diseases
RSCs   = Reed Steinberg Cells
SCFv   = Single chain variable fragment
TLS   = Tumor Lysis Syndrome
TNC   = Total nucleated cells
VL   = Variable regions of light chain

8.1 INTRODUCTION

8.1.1 Cancer and the Immune System

The human immune system is programmed to attack and destroy harmful cells, i.e., microorganisms and tumor cells, from our body as part of the body's defense mechanism. Cancer cells are normally destroyed by concomitant activation of both cellular as well as humoral immune responses.[1] With thorough research and understanding of the working principle of the immune system in performing this role, researchers have attempted to mimic the intrinsic defense mechanisms for the treatment of cancer.

The humoral immune system includes antibodies that bind to the specific antigen that culminate in the activation of the complement system *via* Fc part of the antibody. The Fc part of the antibody is also recognized by the Fc receptors expressed on macrophages and lymphocytes. Hence, the antigen-expressing cells are destroyed by the immune system *via* antibodies.[2] Commercially produced monoclonal antibodies using hybridoma technology, akin to the naturally produced antibodies, bind to the antigens, stimulate the immune system and destroy the cells expressing the antigen/s. Thus, identification of the antigens expressed by the cancer cells will primarily allow us to use this knowledge and produce specific monoclonal antibodies against these cancer cells. For example, most of the lymphoma cells express CD20 antigens on their surface and hence, anti-CD20 monoclonal antibodies (e.g. rituximab, ofatumumab, obinutuzumab) are commercially available to eliminate lymphoma cells in the clinical setting. The clinical use of anti-CD20 monoclonal antibodies has significantly improved the survival rate of patients with lymphoma. The successful clinical application of monoclonal antibodies and advances in the monoclonal antibody technology has prompted the search for new targets on cancer cells. Trastuzumab, marketed under the brand name of Herceptin, is being successfully used as an adjunct to chemotherapy against HER2 (human epidermal growth factor receptor 2) positive breast cancer cells to treat breast cancer patients. Currently, the production of humanized monoclonal antibodies against cancer cells is an area of intense research interest with great promise and hope for cancer patients.[3]

8.1.2 Cellular Immune System

The cellular immune system plays a significant role in eliminating cancer cells. In this regard, the most important cells for the cellular immune system are T-lymphocytes; B-lymphocytes produce antibodies while the T-lymphocytes manage the B-lymphocytes. Thus, the cellular immune system has control over the humoral immune system. Besides, the cellular immune system can also work against the tumor cells. The cytotoxic T-lymphocytes and the natural killer cells can destroy cancer cells alone.[1] In the light of the effective and well-defined role of lymphocytes in the elimination of cancer cells, a novel hypothesis has pronounced cancer as an immune system disorder that results from immune deficiency against cancer cells.[4] Some of the published data supporting this novel hypothesis has originated from the primary immune deficiency diseases (PIDD) studies. According to a recently published report on the cancer incidence in the USA Primary Immune Deficiency Network Registry involving 3658 subjects, the relative risk of cancer increases by 1.42-fold in the patients with

PIDD.[5] This increased relative risk was pronounced in both men and women for the development of lymphoma. However, there was no increased risk in PIDD subjects for most common solid tumor malignancies. These data also pointed to limited participation of the immune system in protection against specific cancers. Probably, the first immune therapy against cancer was allogeneic stem cell transplantation which involved transplantation of donor bone marrow stem cells that lead to immune reconstitution in the recipient after a heavy dose of chemotherapy and radiotherapy. The approach now has made immense progress in the treatment of both malignant as well as non-malignant pathologies. Understanding how reduced intensity conditioning regimens work and the mechanism of graft versus leukemia supports this hypothesis.[6]

To involve the cellular immune system to destroy cancer cells, an alternative approach based on the use of tumor vaccines has been developed. The tumor vaccines are anticipated to induce protective immunity against tumor-specific antigens expressed on the tumor cells. In this respect, various avenues are being explored. Although there are no preventive vaccines to alleviate the development of tumors, some of the available tumor vaccines have shown selective therapeutic promise and therefore provide a viable option for active immunotherapy of cancers that actively engages a patient's immune system. In the majority of the cases, the tumor vaccines are dendritic cells with enhanced capacity to recognize tumor associated antigens and improve their presentation to the naive recipient T-lymphocytes. However, most tumor vaccines have limited success due to the possibility of faulty antigen presentation on tumor cells, thus leading to the development of tolerance towards the antigens present on tumor cells. Some tumor cells also secrete molecules that inhibit T-lymphocytes to enhance their own survival. Immune checkpoint regulators are commercially produced to prevent T-cell inhibition originating from tumor cells.[7] The most popular immune checkpoint inhibitors available in the market include programmed death-1 (nivolumab, pembrolizumab), programmed death ligand-1 (atezolizumab, durvalumab) and CTLA-4 inhibitors (Ipilimumab, tremelimumab). These immune check-point inhibitors are currently being effectively used for solid tumors, albeit with limited success.[8]

The strategy of stimulation and/or prevention of inhibition of T-cells to treat cancer is deficient in many aspects. To overcome these deficiencies, Chimeric Antigen Receptor (CAR) T-cells-based therapy is a novel emerging strategy for cancer therapy that has been developed. This strategy involves reprogramming T-lymphocytes to express artificial T-cell receptors *via* genetic engineering techniques. The genetically modified T-lymphocytes are called chimeric antigen receptor (CAR) T-cells. The chimeric antigen receptors expressed on the genetically reprogrammed T-cells impart new potential to the T-lymphocytes carrying these receptors to target a new antigen for which they have been engineered.[9] This chapter discusses the present scope of CAR T-cell therapy with a special focus on the advancements in the field, future applications, and limitations.

8.2 STRUCTURE OF CAR T-CELLS

As described earlier, CAR T-cells possess new, genetically engineered receptors specific for the tumor-associated antigens. To produce CAR T-cells, firstly a CAR gene is constructed which encodes for a new effective receptor specific for the tumor-associated antigen of interest. A typical First-Generation CAR, encoded by a transgene, has one each of extracellular, transmembrane and intracellular domain wherein the intracellular domain consists of CD3-zeta.[10] Of these three domains, the extracellular domain is an antibody binding region and classically includes the Fab region of a monoclonal antibody, which has increased binding capacity to the tumor-associated antigen. The single chain variable fragment (scFv target domain) of variable regions of heavy (VH) and light (VL) chains of a monoclonal antibody determines the CAR specificity. Subsequent to the fusion of the extracellular domain with specific antigen on the tumor cell surface, it undergoes a conformational change that initiates a signal transduction downstream to the intracellular domain *via* the transmembrane domain (Figure-8.1).

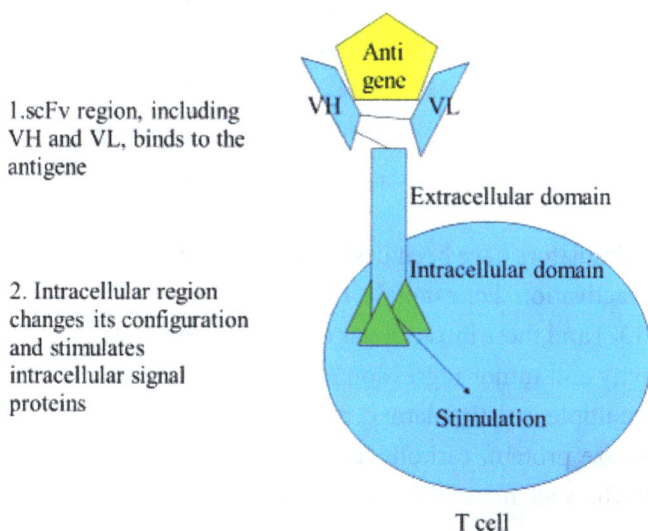

1.scFv region, including VH and VL, binds to the antigene

2. Intracellular region changes its configuration and stimulates intracellular signal proteins

Anti gene

VH   VL

Extracellular domain

Intracellular domain

Stimulation

T cell

Figure-1:   CAR-T cells.

Once engaged with the target cell, CAR T-cells develop a non-classical synapse which is essential for their normal effector function and destroy the cancer cells having the specific antigen. Their anti-tumor activity is facilitated by the involvement of perforin and granzyme axis, Fas-Fas ligand axis and through the release of cytokines which sensitize the tumor stroma.[11]

Given the limited anti-tumor activity of the 1st generation CAR as a result of low-level activation due to lack of co-stimulators, 2nd and 3rd generation of CAR have been developed which show more efficient activation as compared to their 1st generation counterparts. For all the 3 generations developed, CAR T-cells have the same extracellular and transmembrane domains, including scFv, VH, and VL. The difference lies in the constitution of the intracellular domain.[7,10] While intracellular signal domain in all the three generations have CD3ζ, 2nd and 3rd generation CAR T-cells also have CD28 in their intracellular domains and the 3rd generation CAR T-cells additionally have other costimulatory proteins including CD3zeta-CD28-OX40 (tumor necrosis factor receptor)(Figure-8.2).[10]

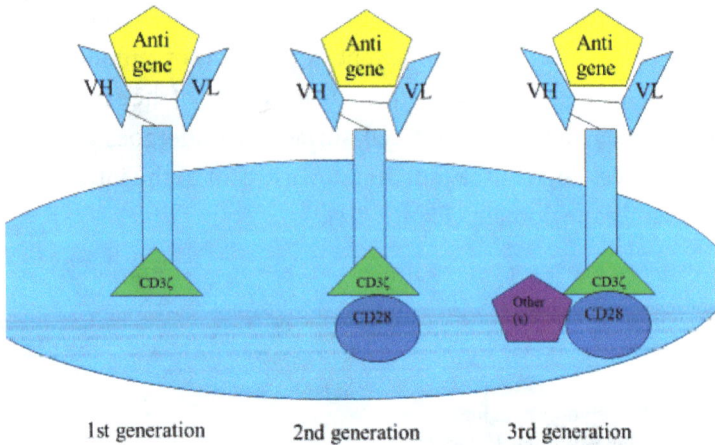

Figure-2:   First, second and third generation CAR-T cells.

These co-stimulators have been designed to provide T-cells with more efficient and prolonged activation. For example, the 2nd and 3rd generation CAR T-cells have CD28 (or CD137) and these intracellular co-stimulatory proteins increase the CAR T-cells' effectivity and tumor regression effects.[7] Similarly, the 3rd generation CAR T-cells have multiple co-stimulatory molecules. Third-generation CAR T-cells can also recognize protein, carbohydrate or glycolipid structures of the various antigens.[7,12] As the specificity of CAR-mediated T-cell recognition is determined by the specificity of the antibody domain, antigen presentation is not necessary for CAR T-cells and they are independent of MHC and/or HLA.[13] Hence, their structure also allows commercially produced CAR T-cells to be used in all patients who have the same tumor-associated antigen.

8.3 PRODUCTION OF CAR T-CELLS

The development and manufacturing of superior quality, safe and clinical-grade CAR-T-cells and logistic requirements for product characterization and optimization remain fundamental to its use in the clinical settings for optimal prognosis in cancer patients. All CAR T-cells recognize a specific tumor-associated antigen target and hence, the CAR T-cells are produced according to their target specificity which is determined by the scFv part of the receptor. To this end, CAR T-cells are genetically manipulated using recombinant viral vectors carrying the CAR T transgene of interest. More recently, sleeping beauty, TALEN and CRISPR methods are being employed for placing the CAR T gene in T-cells.[14]

The development of a recombinant viral vector carrying CAR T transgene of interest (scFv, CD3ζ etc.) is the first step towards the production of CAR T-cells with the required specificity. For example, if the target is CD19, CD19 specific CAR T-cells MSVG-FMC63-28z (mouse stem cell virus-based splice-gag vector) recombinant retroviral vector has been developed (Surgery Branch, NCI, and NIH Vector Production Facility SBVPF).[15] The vector encodes for the CAR T gene consisting of both light and heavy chains of anti-CD-19 monoclonal antibody FMC63 coupled with CD28 as well as zeta chain of the T-cell receptor (FMC63-28z). The retroviral vector is then used for transduction of T-lymphocytes (allogeneic or autologous) which then replaces the genetic potential to recognize CD19 expressing tumor cells. It is pertinent to mention that CD19 is B-lymphocyte specific antigen which is involved in all B-lymphocyte lineage malignancies.

As part of the protocol for scalable manufacture of CAR T-cells, T-cells initiation and ex-vivo expansion is imperative to achieve maximum yield of cells without compromising their phenotype and functionality. It is pertinent to mention that minimally differentiated T-lymphocytes show high anti-tumor activity as compared to the differentiated T-lymphocyte subsets. The media requirements for both steps are similar and use AIM V medium (Gibco, Grand Island, NY), supplemented with 5% heat-inactivated human AB-Serum (Valley Biomedical, Winchester, VA), and 1% Gluta-Max (Gibco, Grand Island, NY). The only difference between the medium requirement for the two steps is that initiation medium has 40 IU/mL IL-2 (Novartis Vaccines and Diagnostics, Inc. Emeryville, CA) while the expansion medium has 300 IU/mL IL-2 (Novartis Vaccines and Diagnostics, Inc. Emeryville, CA).[16] Peripheral blood mononuclear cells (PBMC) are collected from a donor (for allogeneic cells) or from the patient himself/herself (for autologous cells) by apheresis and used for T-cells enrichment. Enriched T-cells are then transfected with retroviral vector. After tranduced T-cells becomes CAR T-cells, they are subjected to extensive quality control before use in the patient (Figure-8.3).

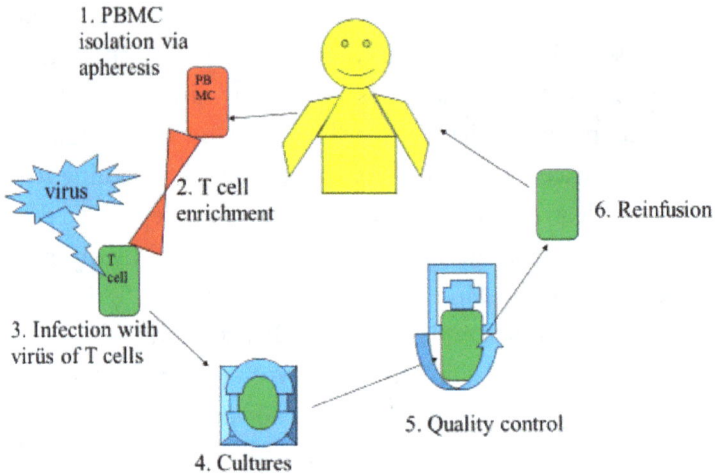

Figure-3:   The production and usage of CAR-T cells.

On day 0, fresh PBMC products should be enriched for CD3+ and CD28+ cells using anti-CD3 and anti-CD28 antibodies paramagnetic beads (Dynabeads ClinExVivo CD3/CD28, Invitrogen, Camarillo, CA) at a ratio of 3:1 (beads:cells). Cells are diluted to 20-30 × 10^6/ml nucleated cell (TNC) concentration using PermaLife bags made of FEP (OriGen Biomedical, Austin, TX). Then the cells are incubated with anti-CD3 or anti-CD28 tagged beads for 2 hours at room temperature and cell enrichment is performed with Dynal ClinExVIVO MPC magnet (Invitrogen, Camarillo, CA). Enriched cell are then re-suspended in initiation media at a concentration of 1 × 10^6 cells/ml in PermaLife bags (OriGen Biomedical, Austin, TX).[16] On day 1, clinical-grade RetroNectin (Takara Bio Division, Shiga, Japan) is reconstituted in sterile water at 1 mg/ml which is then used for coating of PermaLife bags (OriGen Biomedical, Austin, TX) to reach a concentration of 2 µg/cm^2 in a solution of 10 µg/ml in PBS (Lonza, Walkerville, MD). These coated bags are incubated overnight at 4°C.[16] On day 2, the RetroNectin solution is aspirated from bags and the same volume of 2.5% human serum albumin (HSA) (Baxter, Westlake Village, CA) in PBS is added to each bag. After 30 minute incubation, the HSA solution is aspirated and the bag is washed with 2.5% HEPES (Lonza, Walkerville, MD) diluted in Hank's balanced salt solution (HBSS) (Lonza, Walkerville, MD). For transduction, the retroviral supernatant is rapidly thawed and added to each bag while a RetroNectin coated bag is also filled with the expansion medium without retroviral supernatant for use as un-transduced control (UT CNTR bag). After incubation for 2 hours at 37°C with 5% CO_2, CD3+ cells are mixed 1:1 with the clinical-grade vector containing RetroNectin coated bags having IL-2 concentration raised to 300 IU/ml. The cultures are left for 24 hours in incubator. On day 3, transduction is re-performed with new coated bags which is ended on day 4 and the transduced cells are counted, transferred to a new culture bag and re-suspended in fresh T cell expansion medium at a concentration of 0.4×10^6 cells/ml.[16] The cultures are continued until day 11/12 and fed every other day with fresh expansion media to maintain cell concentration at 0.4×10^6 cells/ml. On day 11/12, cells are harvested, the anti-CD3/anti-CD28 paramagnetic beads are removed using the Dynal ClinExVIVO MPC magnet (Invitrogen, Camarillo, CA), washed, concentrated and subjected to quality control. The cells are packaged for infusion at a concentration of 2.5×10^6 cells/mL in the infusion media (plasmalyte A (Baxter, Deerfield, IL) containing 4% HSA (Baxter, Westlake Village, CA). In future, automation and robotic culture technologies may improve the effectivity of CAR T-cells' production.[10,17]

8.4 CLINICAL USE OF CAR T-CELLS FOR ANTI-TUMOR ACTIVITY

Since the publication of the first results of a human clinical study using CAR T-cells in 2006, more than 200 CAR T-cell based clinical trials have been reported, most of which are focused on the treatment of lymphoma and leukemia patients using CD19 specific CAR T-cells.[18,19] Additionally, there are also some clinical trials which target non-CD19 antigens for hematological and solid malignancies which represent an effort to expand the horizon of using this approach beyond B-cell malignancies.[19] CAR T-cell therapy is a typical example of the fast-emerging group of biopharmaceutical drugs also known as "living drug".[20] Although early studies on CAR T-cell therapy was on solid tumor era, hematologic malignancies remains a major area for their application due to most exciting data emanating from hematologic malignancy patient studies.[11,21] CAR T-cells therapy in patients is generally combined with lymphodepleting chemotherapy that includes 25 mg/m² fludarabine from day -5 to day -3 and 60 mg/kg cyclophosphamide from day -5 to day -3 using central venous line and the patients should be monitored for hydration status.[2] However, the method of lymphodepleting remains an area of debate. Additionally, Paracetamol, antihistamine and levetiracetam prophylaxis should be done before infusion. Cytokine Release Syndrome (CRS) and neurotoxicity are the most important and signature side effects of CAR T-cell treatment. Therefore tocilizumab should also be available when it is needed for CRS and neurotoxicity.[23]

8.4.1   CAR T-cell Therapy for B-cell Malignancies

The first-generation CAR T-cells (alfaCD20-CD3zeta CARs-modified T-cells) have shown effectiveness against B-cell lymphomas.[24] The CAR T-cells are also efficient in chronic lymphocytic leukemia (CLL) patients. However, they can present tumor lysis syndrome due to "cytokine release" after CAR T-cell infusion.[25] B-Cell aplasia occurs nearly in all cases that were treated with CAR-T cell. Therefore, the effectiveness of CAR T-cells on B-cell Acute Lymphoblastic Leukemia (ALL) made them an important research area.[11] Despite the expression of antigens by the malignant T-lymphocytes on their surface membrane akin to their normal counterparts, most of the malignant B-cells express CD19 thus rendering CD19 as a logical target for antitumor therapy.[27] An alternative target may be CD22, which is also commonly expressed by the malignant B-Cells.[28] However, CD22 is expressed on mature lymphocytes as well. The other B-cell antigens used in pre-clinical experimental studies include CD23, ROR1, and the light chain. CD23 is expressed on CLL cells but not on the normal B-cells. ROR1 is detected on malignant B-cells in CLL and MCL while light chains may also be logical in mature B-cell malignancies.[19] T-cell malignancies are less common and more aggressive than B cell malignancies. Currently, therapy for T-cell malignancy patients is limited to chemotherapy and allogeneic hematopoietic stem cell transplantation. Although many patients require more effective treatment for T-cell malignancies, there is no clear target for T-cell neoplasia which can be used for engineering CAR T-cells. Targeting CD5 and CD7 for CAR T-cells seem to result in fratricide and ineffectivity.[26] As some T-cell lymphomas express CD30, theoretically an anti-CD30 CAR T-cell could be used for this indication. One possible approach would be to use natural killer cells to treat T-cell malignancies targeting non-shared antigens, but the persistence of the NK cells requires close monitoring to prevent prolonged T-cell lymphopenia. There is little pre-clinical data evidence that target CAR T-cells against T-cell malignancies thus rendering this as a challenging but significant avenue of investigation.

The first CAR T-cell therapy reported in a CLL patient was performed with low dose 1.46×10^5 CAR-T-19 cells/kg.[29] The infused cells persisted for 6 months, remission continued for 10 months after treatment and CRS was reported 14 days after infusion.[29] A subsequent study by Turtle *et al.* showed that CAR-T cells targeting CD19 was also effective in relapsed CLL patients after prior ibrutinib treatment. The study included 24 CLL patients who received lymphodepletion chemotherapy and anti-CD19 CAR T-cells at either of the three incremental doses, i.e., 2×10^5, 2×10^6 or 2×10^7 cells/kg. Four weeks follow-up after the treatment showed 71% overall response rate and that was stable until 3 years. Twenty of the 24 patients (83%) developed CRS and 8 patients developed neurotoxicity which was reversible in all patients except one patient with fatal outcome.[30]

The addition of CD28 part to CD19 to produce 2nd generation CAR T-cells seemed a better option for ALL cases.[11] Their use has produced promising results including remission in relapsed and multiple chemotherapy-resistant B-cell ALL after hematopoietic stem cell transplantation. In a study involving 30 children and adult patients (ClinicalTrials.gov identifier: NCT01626495 and NCT01029366) with refractory or relapsed ALL, they received 0.76×10^6 to 20.6×10^6 cells/kg autologous genetically modified CAR T-cells having for CD19 specific CAR gene (CTL019) using a lentiviral vector.[31] Complete remission was observed in 90% of patients, 2 of whom had blinatumomab-refractory disease and 15 patients had undergone stem cell transplantation. Sustained remission was observed until 24 months. Given the encouraging results in human patients, the U.S. Food and Drug Agency approved this for the treatment of ALL. The same group of researchers later reported longer-term outcome from a study involving 53 children and adults with relapsed/refractory ALL. A median of 43×10^6 CTL019 cells/kg was infused into the patients over 1-2 days. The patients did not show any signs of infusion toxicity. Long-term disease control and persistence of CAR T-cells was achieved up to 39 months after cell infusion.[32,33] Maude et al. designed a phase-II study (ClinicalTrials.gov identifier: NCT02435849) with CD19 targeted CAR T-cells in children and young adults (3–21 years) with ALL.[32] A total of 75 patients were evaluated to determine the efficacy of CD19 targeted CAR T-cells (tisagenlecleucel). CAR T infusion dose was a median weight-adjusted dose of 3.1×10^6 viable CAR T-cells/kg. Patients had a median of 3 previous therapies and a median bone marrow blast percentage of 74%. Also, 61% of patients had undergone prior allogeneic hematopoietic stem cell transplantation. Lymphodepleting chemotherapy was given to 96% of patients prior to CAR T-cell infusion. The overall remission rate was 81% at 3 months after infusion. All patients giving a response to the treatment also had negative minimal residual disease. Event-free survival rate was 73% and the overall survival rate was 90% at 6 months and 50% and 76% at 12 months, respectively while CAR T-cells were observed in the blood for up to 20 months of follow-up.[34] The authors concluded that CAR-T-cell therapy was safe and effective even after a single infusion of tisagenleceucel which provided a stable remission and long-term persistence of the infused cells in both children and young patients with relapsed/refractory B-cell ALL.

In another phase I/II study (ZUMA-3), 10 adult (≥18 years) relapse/refractory ALL patients received 1 or 2 × 10^6 CAR T-cells/kg after chemotherapy with cyclophosphamide and fludarabine. Patients had relapse/ refractory disease with prior 1-4 lines chemotherapy. Six patients achieved an MRD-negative complete response. The study ZUMA-3 continues to enroll more patients.[35] Another phase-II clinical trial of JCAR015 (ROCKET trial; ClinicalTrials.gov identifier: NCT02535364) was initiated in adult patients with relapsed or refractory B-cell ALL. However, the trial was terminated due to safety reasons. This study treated 46 patients using CAR T-cells with cyclophosphamide preconditioning regimen. In this study, complete response was achieved in 91% patients with minimal disease and 75% with morphological disease. No relapse or death occurred in 12 months and 6-month overall survival (OS) rates for all patients were between 57% and 73%.[36]

In all these studies, relapse after CAR T-cell therapy may be observed as over time as CD19 negative B cells may develop that may lead to immune escape of the tumor cells. Hence, CAR T-cell therapy may be combined with stem cell transplantation.[11] CAR T-cells have also been used in B-cell Non-Hodgkin's Lymphomas. CD19 targeted CAR T-cell therapy i.e., axicabtagene ciloleucel, in diffuse large B-cell lymphoma was also approved by the U.S. Food & Drug Administration.

The results of a phase-II multicenter JULIET trial have recently been presented during the Annual Meeting of the American Society of Hematology (ASH) 2017.[37] The study was conducted on patients with relapsed/refractory DLBCL. Primary analysis revealed that 99 patients had progressive disease after at least two lines of chemotherapy or autologous stem cell transplantation. Fludarabine/ cyclophosphamide (73%) or bendamustine (19%) conditioning was performed prior to CAR T-cells therapy. Single CAR T-cell infusion at a median dose of 3.1 × 10^8 cells/kg was administered and the patients were followed-up for 5.6 months. The best overall response rate was 53% and the complete response rate was 40%.[37] The JULIET study also showed that 18 of 28 adult patients with lymphoma had a response (64%). In patients with diffuse large B-cell lymphoma, complete remission occurred in 6 of 14 patients (43%) and in follicular lymphoma complete remission was detected in 10 of the 14 patients (71%). Continued remissions were achieved in the patients during a median follow-up time of 28.6 months. All response rates in patients with diffuse large B-cell lymphoma and those with follicular lymphoma was 86% and 89% respectively. All patients were in complete remission and remained in remission by 6 months after infusion.[38]

In phase I/II ZUMA-1 study, CD19 targeted CAR T-cell therapy (i.e., Axicabtagene ciloleucel) was used in relapsed/refractory diffuse large B-cell lymphoma and showed that 40% of patients achieved complete remission at a median 3 months follow-up and remained in remission for more than a year after infusion.[39] Updated findings of ZUMA-1 study with a median follow-up of 15.4 months enrolled 108 adult (23–76 years) patients with relapse/refractory B-Cell lymphoma. The phase II component included 77 refractory DLBCL and 24 relapse/refractory primary mediastinal B-Cell lymphoma or transformed follicular lymphoma patients. Conditioning regimen was a routine and included fludarabine (30 mg/m^2) and cyclophosphamide (500 mg/m^2) daily for 3 days. The patients received an infusion dose of 2×10^6 CAR T-cells/kg. The complete response rate was 40% while the overall survival and relapse-free survival rates were 46% and 42% respectively.[40] In a multicenter phase-I trial of JCAR017, 13 relapse/refractory diffuse large B-cell lymphoma and 1 mantle cell lymphoma (ClinicalTrials.gov: NCT02631044) cases were treated with CAR T-cells strategy. The mantle cell lymphoma patient had progressive disease. Response rates for the diffuse large B-cell lymphoma cases were 82%.[41]

8.4.2 CAR T-cell Therapy for Acute Myeloid Leukemia (AML)

AML is less immune responsive disease than B-cell malignancies and the relapse/refractory AML has an aggressive prognosis with contemporary methods of treatment and hence necessitates the development of novel therapies. Moreover, finding a specific target antigen in AML is harder than B-cell malignancies due to sharing of the potential targets on the surface of normal cells. For example, CD47 and CD96 are present on AML cells akin to their presence on the surface of normal cells.[19] Due to the potential cytotoxicity, use of CAR T-cell therapy for AML treatment has been limited only in clinical settings. Similarly, CD33 and CD123 are shared between the AML cells and immature normal myeloid cells. Hence, targeting CD133 potentially impairs physiological hematopoiesis. Similarly, CD123 is intensely expressed on AML cells while it is lowly expressed on hematopoietic stem/progenitor cells. Data emanating from AML patient-derived cellular studies and their established grafts in immunodeficient mice have shown that targeting CD33 or CD123 may cause severe myeloablation in AML patients.[42] A phase-I study by Wang *et al.* assessed the feasibility and efficacy of CD133-directed CAR T-cell (CART-33) therapy for a 41-year old AML patient (ClinicalTrials. gov as NCT01864902), the patient had partial response.[43] The patient received 1.12×10^9 autologous CART-33 cell infusion which caused rigorous chills, fever, fluctuating pancytopenia and elevated cytokine levels for IL-6, IL-8, TNF-alpha and INF-gamma. A 2-week follow-up showed decreased blasts in the bone marrow. These data warrant further research to reduce the adverse effects of CAR T-cells targeting CD33 or CD123 for AML patients.

A variant form of CD44 (CD44v6) is expressed on hematologic as well as epithelial tumors and is being considered as a potential target for the treatment of AML patients. CD44 targeting approach using CAR T-cells has already given encouraging results in the experimental animal models; however, it is not without cytotoxic effects on keratinocytes and lethal adverse effects on the epithelium.[44] Hence the data still has to be investigated in the clinical settings. On the same note, Lewis Y (LeY) antigen is present on the surface of AML cells and CAR T-cells targeting LeY antigen may produce anti-tumor activity on AML cells with minimal toxicity. Based on these data, a phase -I study enrolled 5 patients, one of whom died due to complications of sepsis relevant to re-induction of chemotherapy.[45] The remaining four patients received CAR anti-LEY T-cell therapy for AML to assess the safety and post-infusion persistence of the infused T-cells. A total of 1×10^9 T-cells were infused, of which 14–38% cell expressed the LeY specific CAR T-cells. Three of the 4 patients showed a biological response to the treatment which was durable. More studies are ongoing involving LeY antigen targeting on AML cells. Based on experimental animal studies, some of the other potential targets for CAR T-cells to treat AML include CLEC12A, folate receptor β, FLT3 receptor, CD38, CD56, CD117, or Muc-1. Similarly, there are ongoing studies to reduce the toxicity of CAR T-cells targeted to these antigens that involve CAR T-cell termination strategies. These strategies include placing suicide genes in the CAR construct and terminating CAR T-cells with a chemical or using specific antibody that includes cetuximab for truncated epidermal growth factor receptor or rituximab for CD20 or alemtuzumab for CD52.

8.4.3 CAR T-cell Therapy for Hodgkin's Lymphoma

Reed Steinberg cells (RSCs; also called lacunar histiocytes in some types) are malignant cells in Hodgkin's lymphoma. Although the origin of RSCs is B-cells but they lack in B-cell surface antigens expression. Hence, CAR T-cells based treatment targeting B-cell surface antigens may not be effective to treat Hodgkin's lymphoma. On the other hand, RSCs express CD123 and CD30 on their surface that may be exploited for CAR T-cells therapy of Hodgkin Lymphoma.[26,46] An anti-CD30 monoclonal antibody preparation Brentixumab Vedotin is used for treating Hodgkin's lymphoma patients but the approach is not without its challenges in the clinical perspective.[46] The strategy has progressed to clinical studies for safety and efficacy assessment.[47] CD30 targeted CAR T-cells were used in 7 patients with relapse/refractory Hodgkin lymphoma using cell doses of 2×10^7 and 2×10^8 cells/m^2 without conditioning regimen before the infusion of CAR T-cells. Follow-up at 6 weeks after cell infusion showed complete remission in 1 patient, partial remission in 1 patient and stable disease in 4 patients.[47]

8.4.4 CAR T-cell Therapy for Myeloma

Despite advancement in the pharmacological management of multiple myeloma, it remains an incurable disease and necessitates a finding of new treatment options. High dose melphalan followed by infusion of autologous hematopoietic stem cell transplantation is routinely performed in multiple myeloma patients. A combination of autologous hematopoietic stem cell transplantation and CD19 targeted CAR T-cell (CTL019) therapy may be used to reduce tumor load in multiple myeloma. Clinical evaluation of 10 patients with relapsed/refractory multiple myeloma with CTL019 after high dose melphalan and autologous stem cell transplantation showed the safety and feasibility of the approach.[48] Interestingly, concomitant CTL019 improved the duration of response towards the standard multiple myeloma therapies. Multiple myeloma involves plasma cells that express specific antigens on their surface. Well-known amongst these expressed antigens on plasma cells include CD38, CS1 (SLAMF7), CD138 and light chain (kappa or lambda). Anti-CD38 monoclonal antibody daratumumab and anti-CS1 monoclonal antibody elotuzumab are currently used in relapse/refractory multiple myeloma patients. Additionally, plasma cells also express CD56, BCMA and CD44v6. These antigens may be targeted with CAR T-cells for multiple myeloma therapy. However, all these antigens are also expressed by the normal cells. Therefore, targeting these antigens with CAR T-cells to treat multiple myeloma may be associated with adverse effects including toxicity. Data from the pre-clinical studies have shown that CAR T-cells against both CD38 and SLAMF7 are effective for killing plasma cells *in vitro*. However, CD38 and SLAMF7 are also present on other lymphoid and myeloid cells. Therefore, CAR T-cell treatment against CD38 or SLAMF7 may cause myelotoxicity. Moreover, SLAMF7 targeted CAR T-cells have affect on both T-cells and NK cells. To reduce potential toxicity of CD38 targeted CAR T-cells affinity optimization studies are ongoing in preclinical setting.[49]

CD138, also named as syndecan-1, is highly expressed on the surface of plasma cells. However, CD38 is also expressed on normal epithelial cells. Therefore, CD138 targeted CAR T-cells (BT062) leads to adverse effects on epithelial cells, such as inflammation on palms and soles.[50] In phase I/II clinical study involving four confirmed cases of relapse/refractory multiple myeloma and one patient having plasma cell leukemia, the patients were treated with CAR T-cells targeting CD138. No toxicity was observed in any patient while 4 of the multiple myeloma patients had stable disease longer than three months and one patient with plasma cell leukemia had a reduction of the plasma cells in peripheral blood.[51]

Light chains may be an alternative target for CAR T-cells in the treatment of multiple myeloma. In a phase I clinical study including 8 relapsed or refractory multiple myeloma patients (1 patient had Ig A, 4 patients had IgG and 3 patients had free kappa light chains) were treated with autologous CAR T-cells targeting kappa light chain. Five of eight patients received 12.5 mg/kg cyclophosphamide before CAR T-cell infusion and had stable disease lasting 6–17 months. No toxicities were observed.[52] On the same note, BCMA is a protein of tumor necrosis factor receptor family that is expressed on plasma cells but not expressed by any non-lymphoid cells. Therefore, 4 different early phase studies were conducted to treat multiple myeloma with BCMA targeted CAR T-cells. In a phase I study, BCMA targeted CAR T-cells used in 16 multiple myeloma patients out of which 13 (81%) patients achieved hematological response.[53] In another study, 10 patients were treated with BCMA targeted CAR T-cells with cyclophosphamide conditioning regimen. Six of the ten patients (60%) achieved hematological response.[54] *Ex vivo* treatment of the primary myeloma cells lost their colony-forming potential in response to CTL019 and CAR T-cells against plasma cell antigen BCMA.[54] Put together, CTL019 improved the duration of response to standard myeloma therapy. In another study, data of 16 patients who received 9×10^6/kg CAR T-cells expressing a CAR specific for BCMA has been reported. The patients received CAR T-cell therapy after conditioning chemotherapy of cyclophosphamide and fludarabine. The results of the study showed an overall response rate in 81% of patients, with 63% having a very good response or complete response. A median event-free survival was 31 weeks. Although CRS toxicities were severe in some cases, these were reversible in all patients. BCMA targeted CAR T-cells was effective and these patients achieved hematological responses with prolonged duration of response.[55] More recently, a single-arm clinical study has been reported on 19 relapse/refractory multiple myeloma patients to assess the safety and efficacy of CAR-T cell therapy directed against BCMA.[56] In this study median CAR T-cell (LCAR-B38M) dose targeting BCMA was 4.7×10^6 cells/kg and the patients were followed-up until a median duration of 208 days (62–321 days). Without an exception, all 19 (100%) of patients achieved hematological response while 18 out of 19 (95%) patients reached complete remission or near-complete remission without an event of relapse during a median 6 month follow-up time.[56] A prior study has also employed CAR T-cells therapy for multiple myeloma directed against CD44v6 antigen which is expressed on epitheloid cells as well. Studies are ongoing for this target in treating multiple myeloma with CAR T-cells.[57] In the light of the data from the pre-clinical as well as clinical studies, the future of CAR T-cell therapy has some certainties as well as uncertainties. Further studies would be required to enhance its efficacy for the achievement of optimal prognosis with minimal side effects.[58]

8.5 TOXICITY OF CAR T-CELLS

8.5.1 On-Target and Off-Tumor Toxicity

All CAR T-cells must have a target antigen to eliminate cancer cells. When the target antigen is co-expressed in normal and cancer cells, CAR T-cells will fail to differentiate between the two and destroy them indiscriminately. Therefore, it is important that the target antigen should be carefully selected that its expression should be restricted to the cancer cell to avoid harming the normal tissues. This toxicity is also named as "On-target, off-tumor toxicity". On-target, off-tumor toxicity may be predicted before CAR T-cell infusion which is associated with the expression level of the target antigen on normal tissues and may be lowered by using low expressed antigens as targets for CAR T-cells. For example, the success of CD19 targeted CAR T-cells used for B-cell malignancies is associated with the low expression profiles of CD19 in normal cells.[59] Therefore, to find unique targets and develop CAR T-cells against these targets in the treatment of malignancies is imperative and remains an area of logical research in pre-clinical studies before it is used in clinical studies.

8.5.2 On-Target, On-Tumor Toxicities

On-target, on-tumor toxicities can be presented as CRS or tumor lysis syndrome.[59] On-target, on-tumor toxicities are common toxicities of CAR T-cells having a safe target and associated with anti-tumor activity and tumor burden. CRS has been detected in 74–100% of patients having CD19 targeted CAR T-cell therapy.[60] Activated CAR T-cells release effector cytokines, i.e., IL2, IFNγ, TNFα which activate endothelial cells, antigen-presenting cells and macrophages. These cells also produce IL6 which together with the released cytokines cause a systemic inflammatory response (also called CRS). The symptoms of CRS vary between mild symptoms (fever, flu-like findings, and shortness of breath, fatigue, dizziness, and hypotension) and severe symptoms (respiratory failure, coagulopathy, and organ toxicity). Organ toxicity may range from cardiac, respiratory, gastrointestinal, hepatic, renal, dermatological failures to multi-organ failure which is generally observed during the first week after CAR T-cell infusion. The C-reactive protein levels and fever are indicators for the existence and severity of CRS. Early, a prolonged and higher temperature is associated with more severe CRS in patients. IL-6 receptor blocking monoclonal antibody tocilizumab and steroids are used in CRS. Therefore, tocilizumab should be on the ward before the infusion of CAR T-cells.[60,61] Grade-1 CRS is not life-threatening and it is easy to treat. It is characterized by fever, nausea, fatigue, headache, myalgia and malaise. Grade-2 CRS patients are also easy to treat and show symptoms of toxicity, oxygen requirement <40% or mild hypotension (responsive to fluids or low dose of vasopressors). Fluid bolus or low dose noradrenalin, supplementary O2 and tocilizumab administration may be needed to treat these patients. On the other hand, Grade-3 toxicity has severe symptoms including oxygen requirement >40% or severe hypotension (requiring high-dose or multiple vasopressors). Treatment of Grade-3 toxicity CRS patients requires high doses or multiple vasopressors with tocilizumab and/or steroids. Grade-4 toxicity includes life-threatening symptoms and patients require ventilator support. The management of patients requires intensive care unit and mechanical ventilation. Tocilizumab and/or steroids should be used in these patients.[30]

Besides CRS, Tumor Lysis Syndrome (TLS) is another on-target, on-tumor toxicity which is the consequence of tumor cell destruction. It is also observed in patients having high tumor burden. Lysis of tumor cells cause systemic metabolic disturbances and is characterized by elevation of uric acid, phosphate and potassium levels in the patient serum.[59,60] The severity of CRS and TLS is generally associated with tumor burden.

8.5.3 Off-Target Toxicity

Off-target toxicity result from CAR T-cells' activation via irrelevant antigen or substance. The interaction of the Fc part of CAR T-cells with the Fc receptor (FcR) of the innate immune cells (macrophages and NK cells) may cause the antigen-independent activation of innate cells. In a few cases, fatal cardiotoxicity has been reported in CAR T-cells targeted at antigens on the surfaces of solid tumors.[59]

8.5.4 Neurotoxicity

Neurotoxicity or the CAR T-cell related encephalopathy syndrome (CES) is characterized by disorientation, seizures, motor weakness and increased intracranial pressure (papilledema and cerebral edema). Although the exact cause of CES is not well-understood, it is considered to be related to infiltration of cerebrospinal fluid with CAR T-cells. It is usually reversible and self-limiting. A combination of dexamethasone and levetiracetam may be administrated to the patients.[61,62]

8.5.5 Genotoxicity

The use of integrating viral vectors to genetically manipulate the CAR T-cells may potentially cause oncogenic mutagenesis. Pre-clinical studies have shown a risk of carcinogenesis. Integration of the viral DNA into the host cell genome causes such side effects via the vector will remain in the patients' body for a very long time. CRISPR and other non-viral technologies may be used to reduce this adverse effect.[63]

8.6 IMMUNOGENICITY AND IMMUNOSUPPRESSION

As the scFv part of CAR T-cells is mostly derived from mouse monoclonal antibodies, the murine-derived proteins have potential immunogenicity and stimulate the production of anti-mouse antibody to initiate an allergic reaction and/or severe anaphylaxis.[64] In suspected cases, humanizing scFvs should be preferred to mouse-derived ones.[65] Immunosuppressive conditioning regimen has an anti-tumor effect on cancer cells and associated with more effective response to CAR T-cell treatment. Therefore, it is commonly performed before CAR T-cell infusion. However, the immunosuppressive conditioning regimen is also associated with pancytopenia and its complications including anemia, coagulopathy, and neutropenic sepsis.

8.7 THE PERMANENCY OF CAR T-CELLS

Alive CAR T-cells must be maintained in patients' bodies in a dose that would be effective for remission of the tumors. Proliferation of CAR T-cells continues for up to 3 months after infusion.[65] Therefore, complete remission has been reported in patients treated with anti-CD19 CAR T-cells and the observed remission was permanent. In these cases, CAR T-cells survived for years. It has been shown that the first CAR-T-cells have been survived for over 16 years in the body and prevented cancer.[67] However, the effectiveness of CAR T-cells is reduced or even lost over time due to expression of targeted antigens on cancer cells. Cancer cells may lose CD19 expression after response with anti-CD 19 CAR T-cells. It can appear in B-Cell acute lymphoblastic leukemia in approximately 14% of patients.[68]

Reducing the number of circulating T-cells with conditioning regimen before CAR T-cell infusion as part of lymphodepletion, contribute to the *in vivo* proliferation of CAR T-cells. Although the optimal lymphodepletion is not known, it improves the effectivity of CAR T-cells and it is performed before CAR T-cell infusion using bendamustine, pentostatin, fludarabine and/or cyclophosphamide. *In vivo* duration of CAR-T-cells and disease-free survival increases in patients pre-treated with fludarabine and cyclophosphamide. In a meta-analysis, the 6-month progression the free survival in patients with lymphodepletion regimen was 94.6% and 54.5% in patients without lymphodepletion (p < 0.001). Conditioning regimen can also reduce tumor burden and on-target, on-tumor toxicities in patients. Therefore, conditioning regimen is important for both safety and effectiveness of CAR T-cell therapy.[68-70] On the other hand, the permanent CAR T-cells may also cause toxicity including genotoxicity and immunogenicity. Producing CAR T-cells with CRISPR technology BOTH may increase the effectiveness of CAR T-cells and reduce genotoxicity. Rapamycin-induced caspase 9 suicide gene also inactivate CAR T-cells in an event of toxicity. Other techniques including programmable T cell responses with synNotch receptor, the use of zinc finger nuclease gene editing and use of polymeric nanoparticles to modify circulating CAR T-cells are developing.[71,72]

8.8 CONCLUSİON

The divergence between the protocols for the development of CAR T-cells and the study designs makes it difficult to define the outcome of these studies which makes it even more challenging to answer all the questions about CAR T-cell treatment. The best production method, the most well-defined and characterized target antigen, the most suitable costimulatory molecule, the best hinge and the best transmembrane region for the best CAR T-cells are the areas which still stand as further investigation and research areas. Lentiviral and retroviral gene transfer techniques seem less safe for human use but these are far more effective than non-viral options. CAR T-cell expansion protocols vary between studies but still require optimization. The optimal safe dose for an effective CAR T-cell treatment in patients having malignancy is not known either.[19] Nonetheless, we have learned some lessons from CAR T-cell studies:

1. Second and third-generation CAR T-cells are more effective than the first generation.
2. Conditioning regimen before CAR T-cell treatment is advised.
3. Cytokine release and tumor lysis syndromes are associated with the effectivity of CAR T-cells.

Clinical trials are continuing to regenerate CAR T-cells for solid tumors and hematologic cancers. The FDA approved CAR T-cell treatment for relapse/refractory pediatric acute lymphoblastic leukemia and diffuse large B-Cell lymphoma, declaring it as a breakthrough event. Studies continue with patients having multiple myeloma, acute myeloid leukemia, and many solid tumors. CAR T-cell treatment seems to be our new hope to keep patients alive in the near future.

REFERENCES

1. Abbas AK, Lichtman AH, Pillai S. Basic Immunology: Functions and Disorders of the Immune System, 5th edition. Chapter 1. Introduction to the immune system. Missouri: Elsevier, 2016, pp. 1–26.
2. Abbas AK, Lichtman AH, Pillai S. Basic Immunology: Functions and Disorders of the Immune System, 5th edition. Chapter 7. Humoral immune responses. Missouri: Elsevier, 2016, pp. 147–169.
3. Glassman PM, Balthasar JP. Mechanistic considerations for the use of monoclonal antibodies for cancer therapy. *Cancer Biol Med.* 2014; 11(1): 20–33.
4. Giat E, Ehrenfeld M, Shoenfeld Y. Cancer and autoimmune diseases. *Autoimmun Rev.* 2017; 16(10): 1049–1057.
5. Mayor PC, Eng KH, Singel KL, Abrams SI, Odunsi K, Moysich KB, Fuleihan R, Garabedian E, Lugar P, Ochs HD, Bonilla FA, Buckley RH, Sullivan KE, Ballas ZK, Cunningham-Rundles C, Segal BH. Cancer in primary immunodeficiency diseases: Cancer incidence in the United States Immune Deficiency Network Registry. *J Allergy Clin Immunol.* 2018; 141(3): 1028–1035.
6. McCarthy NJ, Bishop MR. Nonmyeloablative allogeneic stem cell transplantation: Early promise and limitations. *Oncologist.* 2000; 5(6): 487–496.
7. Koury J, Lucero M, Cato C, Chang L, Geiger J, Henry D, Hernandez J, Hung F, Kaur P, Teskey G, Tran A. Immunotherapies: Exploiting the immune system for Cancer treatment. *J Immunol Res.* 2018; 2018: 9585614.
8. Grywalska E, Pasiarski M, Góźdź S, Roliński J. Immune-checkpoint inhibitors for combating T-cell dysfunction in cancer. *Onco Targets Ther.* 2018; 11: 6505–6524.

9. Malkovska Mchayleh W, Bedi P, Sehgal R, Solh M. Chimeric antigen receptor T-cells: The future is now. *J Clin Med*. 2019; 8(2) pii: E207.

10. Ataca P, Arslan Ö. Chimeric antigen receptor T Cell therapy in hematology. *Turk J Haematol*. 2015; 32(4): 285–294.

11. Benmebarek M-R, Karches CH, Cadilha BL, Lesch S, Endres S, Kobold S. Killing mechanisms of chimeric antigen receptor (CAR) T Cells. *Int J Mol Sci*. 2019; 20(6): 1283.

12. Ramos CA, Dotti G. Chimeric antigen receptor (CAR)-engineered lymphocytes for cancer therapy. *Expert Opin Biol Ther*. 2011; 11(7): 855–873.

13. Chmielewski M, Hombach AA, Abken H. Antigen-specific T-cell activation independently of the MHC: Chimeric antigen receptor-redirected T-cells. *Front Immunol*. 2013; 4: 371

14. Fesnak AD, June CH, Levine BL. Engineered T-cells: The promise and challenges of cancer immunotherapy. *Nat Rev Cancer*. 2016; 16(9): 566–581.

15. Kochenderfer JN, Feldman SA, Zhao Y, Xu H, Black MA, Morgan RA, Wilson WH, Rosenberg SA. Construction and preclinical evaluation of an anti-CD19 chimeric antigen receptor. *J Immunother*. 2009; 32: 689–702.

16. Tumaini B, Lee DW, Lin T, Castiello L, Stroncek DF, Mackall C, Wayne A, Sabatino M. Simplified process for the production of anti-CD19-CAR engineered T-cells. *Cytotherapy*. 2013; 15(11): 1406–1415.

17. Levine BL, June CH. Perspective: Assembly line immunotherapy. *Nature*. 2013; 498: S17.

18. Jensen MC, Popplewell L, DiGiusto D, Kalos M, Raubitschek A, Forman SJ. A first-in-human clinical trial of adoptive therapy using CD19-specific chimeric antigen receptor re-directed T-Cells for recurrent/refractory follicular lymphoma. *Blood*. 2007; 110: 288.

19. Hartmann J, Schüßler-Lenz M, Bondanza A, Buchholz CJ. Clinical development of CAR T-cells-challenges and opportunities in translating innovative treatment concepts. *EMBO Mol Med*. 2017; 9(9): 1183–1197.

20. Atilla E, Kilic P, Gurman G. Cellular therapies: Day by day, all the way. *Transfus Apher Sci*. 2018; 57(2): 187–196

21. Bozdağ SC, Yüksel MK, Demirer T. Adult stem cells and medicine. *Adv Exp Med Biol*. 2018; 1079: 17–36.

22. Turtle CJ, *et al.* Addition of fludarabine to cyclophosphamide lymphodepletion improves in vivo expansion of CD19 chimeric antigen receptor-modified T-cells and clinical outcome in adults with B cell acute lymphoblastic leukemia. *Blood*. 2015; 3773–3773.

23. Hay KA. CRS and neurotoxicity after CD19 chimeric antigen receptor-modified (CAR-) T cell therapy. *Br J Haematol*. 2018; 183(3): 364–374.

24. Maus MV, Grupp SA, Porter DL, June CH. Antibody-modified T-cells: CARs take the front seat for hematologic malignancies. *Blood*. 2014; 123(17): 2625–2635.

25. Morgan R, Yang JC, Kitano M, Dudley ME, Laurencot CM, Rosenberg SA. Case report of a serious adverse event following the administration of T-cells transduced with a chimeric antigen receptor recognizing ERBB2. *Mol Ther*. 2010; 28 (4): 843–851.

26. Avanzi MP, Brentjens RJ. Emerging role of CAR T-cells in non-Hodgkin's Lymphoma. *J Natl Compr Canc Netw*. 2017; 15(11): 1429–1437.

27. Scheuermann RH, Racila E. CD19 antigen in leukemia and lymphoma diagnosis and immunotherapy. *Leukemia & Lymphoma*. 1995; 18(5–6): 385–397.

28. Haso W, Lee DW, Shah NN, *et al*. Anti-CD22-chimeric antigen receptors targeting B-cell precursor acute lymphoblastic leukemia. *Blood*. 2013; 121(7): 1165–1174.

29. Porter DL, Levine BL, Kalos M, Bagg A, June CH. Chimeric antigen receptor-modified T cells in chronic lymphoid leukemia. *N Engl J Med*. 2011; 365: 725–733.

30. Turtle CJ, Hay KA, Hanafi LA, Li D, Cherian S, Chen X, Wood B, Lozanski A, Byrd JC, Heimfeld S, Riddell SR, Maloney DG. Durable molecular remissions in chronic lymphocytic leukemia treated with CD19-specific chimeric antigen receptor-modified T-cells after failure of ibrutinib. *J Clin Oncol*. 2017; 35: 3010–3020.

31. Maude SL, Frey N, Shaw PA, Aplenc R. Chimeric antigen receptor T-cells for sustained remissions in leukemia. *N Engl J Med*. 2014; 371(16): 1507–1517.

32. Maude SL, Teachey DT, Rheingold SR, Shaw PA, Aplenc R, Barrett DM, *et al*. Sustained remissions with CD19-specific chimeric antigen receptor (CAR)-modified T-cells in children with relapsed/refractory ALL. *J Clin Oncol*. 2016; 34: Suppl 15, Abstract 3011. Retrieved from http://meetinglibrary.asco.org/content/166869-176.

33. Grupp SA, Maude SL, Shaw PA, Aplenc R, Barrett DM, Callahan C, Lacey SF, *et al*. Durable remissions in children with relapsed/refractory ALL treated with T-cells engineered with a CD19-targeted chimeric antigen receptor (CTL019). *Blood*. 2015; 126: 681.

34. Maude SL, Laetsch TW, Buechner J, Rives S, Boyer M, Bittencourt H, Bader P, *et al*. Tisagenlecleucel in children and young adults with B-cell lymphoblastic leukemia. *N Engl J Med*. 2018; 378(5): 439–448.

35. Shah BD, Wierda WG, Schiller GJ, Bishop MR, Castro JE, Sabatino M, Bot A, *et al*. Updated results from ZUMA-3, a phase 1/2 study of KTE-C19 chimeric antigen receptor (CAR) T cell therapy, in adults with high-burden relapsed/refractory acute lymphoblastic leukemia (R/R ALL). *J Clin Oncol*. 2017; 35(15_suppl): 3024–3024.

36. Park JH, Riviere I, Wang X, Purdon T, Sadelain M, Brentjens RJ. Impact of disease burden on long-term outcome of 19-28z CAR modified T-cells in adult patients with relapsed B-ALL. 2016 ASCO Annual Meeting. Abstract 7003.

37. Schuster SJ, Bishop MR, Tam CS, *et al*. Primary analysis of JULIET: A global, pivotal, phase 2 trial of CTL019 in adult patients with relapsed or refractory diffuse large B-cell lymphoma. Presented at the 59th American Society of Hematology (ASH) Annual Meeting. December 9–12, 2017; Atlanta, GA. Abstract 577.

38. Schuster SJ, Svoboda J, Chong EA, Nasta SD, Mato AR, Anak Ö, Brogdon JL, Pruteanu-Malinici I, Bhoj V, Landsburg D, Wasik M, Levine BL, Lacey SF, Melenhorst JJ, Porter DL, June CH. Chimeric antigen receptor T-cells in efractory B-cell lymphomas. *N Engl J Med*. 2017; 377(26): 2545–2554.

39. Neelapu SS, Locke FL, Bartlett NL, *et al*. Long-term follow-up ZUMA-1: A pivotal trial of axicabtagene ciloleucel (Axi-Cel; KTE-C19) in patients with refractory aggressive

non-Hodgkin lymphoma. Presented at the 59th American Society of Hematology (ASH) Annual Meeting. December 9–12, 2017; Atlanta, GA. Abstract 578.

40. Neelapu SS, Locke FL, Bartlett NL, Lekakis LJ, Miklos DB, Jacobson CA, Braunschweig I, Oluwole OO, et al. Axicabtagene ciloleucel CAR T-cell therapy in refractory large B-cell lymphoma. N Engl J Med. 2017; 377(26): 2531–2544.

41. Abramson JS, Palomba L, Gordon LI, Lunning M, Arnason J, Forero-Torres A, Albertson TM, et al. Transcend NHL 001: Immunotherapy with the CD19-directed CAR T-cell product JCAR017 results in high complete response rates in relapsed or refractory b-cell non-Hodgkin lymphoma. Blood. 2016; 128: 4192.

42. Pizzitola I, Anjos-Afonso F, Rouault-Pierre K, Lassailly F, Tettamanti S, Spinelli O, Biondi A, et al. Chimeric antigen receptors against CD33/CD123 antigens efficiently target primary acute myeloid leukemia cells in vivo. Leukemia. 2014; 28: 1596–1605.

43. Wang QS, Wang Y, Lv HY. Treatment of CD33-directed chimeric antigen receptor-modified T-cells in one patient with relapsed and refractory acute myeloid leukemia. Mol Ther. 2015; 23: 184–191.

44. Casucci M, Nicolis di Robilant B, Falcone L, Camisa B, Norelli M, Genovese P, Gentner B, et al. CD44v6-targeted T cells mediate potent antitumor effects against acute myeloid leukemia and multiple myeloma. Blood. 2013; 122(20): 3461–3472.

45. Ritchie DS, Neeson PJ, Khot A. Persistence and efficacy of second generation CAR T-cell against the LeY antigen in acute myeloid leukemia. Mol Ther. 2013; 21: 2122–2129.

46. Grover NS, Savoldo B. Challenges of driving CD30-directed CAR-T cells to the clinic. BMC Cancer. 2019; 19: 203.

47. Ramos CA, Ballard B, Liu E, et al. Chimeric T-cells for therapy of CD30+ Hodgkin and non-Hodgkin lymphomas [abstract]. Blood. 2015; 126: Abstract 185.

48. Garfall AL, Stadtmauer EA, Hwang WT, Lacey SF, Melenhorst JJ, Krevvata M, et al. Anti-CD19 CAR T-cells with high-dose melphalan and autologous stem cell transplantation for refractory multiple myeloma. JCI insight. 2018; 3(8).

49. Cohen AD. CAR T-cells and other cellular therapies for multiple myeloma: Hematologic malignancy-plasma cell dyscrasia. American Society of Clinical Oncology Educational Book 2018 Update; 38 (May 23, 2018) e6–e15.

50. Heffner LT, et al. BT062, an antibody-drug conjugate directed against CD138, given weekly for 3 weeks in each 4 week cycle: Safety and further evidence of clinical activity. Am Soc Hematol Annu Meet Proc. 2012; 120: 653.

51. Guo B, et al. CD138-directed adoptive immunotherapy of chimeric antigen receptor (CAR)-modified T-cells for multiple myeloma. J Cell Immunother. 2016; 2.1(2016): 28–35.

52. Ramos CA, Savoldo B, Torrano V, et al. Clinical responses with T lymphocytes targeting malignancy-associated κ light chains. J Clin Invest. 2016; 126: 2588–2596.

53. Cohen AD, Garfall AL, Stadtmauer EA, Lacey SF, Lancaster E, Vogl DT, et al. Safety and efficacy of b-cell maturation antigen (BCMA)-specific chimeric antigen receptor T-cells (CART-BCMA) with cyclophosphamide conditioning for refractory multiple myeloma (MM). Blood. 2017; 130(Suppl 1): 505.

54. Berdeja JG, Lin Y, Raje N, Munshi N, Siegel D, Liedtke M, et al. Durable clinical responses in heavily pretreated patients with relapsed/refractory multiple myeloma: Updated results from a multicenter study of bb2121 Anti-BCMA CAR T cell therapy. *Blood*. 2017; 130(Suppl 1): 740.

55. Brudno J, Marc I, Hartman SD, Rose JJ, Lam N, Stetler-Stevenson M, Salem D, et al. T-cells genetically modified to express an anti-B-cell maturation antigen chimeric antigen receptor cause remissions of poor-prognosis relapsed multiple myeloma. *Blood*. 2018; 36(22): 2267–2280.

56. Fan F, Zhao W, Liu J, He A, Chen Y, Cao X, Yang N, et al. Durable remissions with BCMA-specific chimeric antigen receptor (CAR)-modified T-cells in patients with refractory/relapsed multiple myeloma. *J Clinical Oncol*. 2017; 35(18_suppl): LBA3001-LBA.

57. Casucci M, Robilant B, Falcone L, Camisa B, Norelli M, Genovese P, Gentner B, et al. CD44v6-targeted T-cells mediate potent antitumor effects against acute myeloid leukemia and multiple myeloma. *Blood*. 2013; 122: 3461–3472.

58. Cornell RF, Costa LJ. The future of chimeric antigen receptor T-cell therapy for the treatment of multiple myeloma. *Biol Blood Marrow Transplant*. 2018, Nov 15; pii: S1083-8791(18)30719-5.

59. Sun S, Hao H, Yang G, Zhang Y, Fu Y. Immunotherapy with CAR-modified T-cells: Toxicities and overcoming strategies. *J Immunol Res*. 2018: 2386187.

60. Titov A, Petukhov A, Staliarova A, Motorin D, Bulatov E, Shuvalov O, Soond SM, et al. The biological basis and clinical symptoms of CAR-T therapy-associated toxicites. *Cell Death Dis*. 2018; 9(9): 897.

61. Neelapu SS, Tummala S, Kebriaei P, Wierda W, Gutierrez C, Locke FL, Komanduri KV, et al. Chimeric antigen receptor T-cell therapy — assessment and management of toxicities. *Nat Rev Clin Oncol*. 2018; 15(1): 47–62.

62. Acharya UH, Dhawale T, Yun S, Jacobson CA, Chavez JC, Ramos JD, Appelbaum J, Maloney DG. Management of cytokine release syndrome and neurotoxicity in chimeric antigen receptor (CAR) T cell therapy. *Expert Rev Hematol*. 2019; 12(3): 195–205.

63. Mollanoori H, Shahraki H, Rahmati Y, Teimourian S. CRISPR/Cas9 and CAR-T cell, collaboration of two revolutionary technologies in cancer immunotherapy, an instruction for successful cancer treatment. *Hum Immunol*. 2018, Sep 24; S0198-8859(18)30201-5.

64. Hege KE, Bergsland EK, Fisher GA, Nemunaitis JJ, Warren RS, McArthur JG, Lin AA, et al. Safety, tumor trafficking and immunogenicity of chimeric antigen receptor (CAR)-T cells specific for TAG-72 in colorectal cancer. *J Immuno Ther Cancer*. 2017; 5:22.

65. Maude SL, Barrett DM, Rheingold SR, Aplenc R, Teachey DT, Callahan C, Baniewicz D, et al. Efficacy of humanized CD19-targeted Chimeric antigen receptor (CAR)-modified T cells in children and young adults with relapsed/refractory acute lymphoblastic leukemia. *Blood*. 2016; 128: 217.

66. Frigault MJ, Lee J, Basil MC, Carpenito C, Motohashi S, Scholler J, Kawalekar OU, et al. Identification of chimeric antigen receptors that mediate constitutive or inducible proliferation of T-cells. *Cancer Immunol Res*. 2015; 3(4): 356–367.

67. Scholler J, Brady TL, Binder-Scholl G, Hwang WT, Plesa G, Hege KM, Vogel AN, *et al*. Decade-long safety and function of retroviral-modified chimeric antigen receptor T-cells. *Sci Transl Med*. 2012; 4(132): 132ra53.

68. Wang Z, Wu Z, Liu Y, Han W. New development in CAR-T cell therapy. *J Hematol Oncol*. 2017; 10: 53.

69. Wei G, Ding L, Wang J, Hu Y, Huang H. Advances of CD19-directed chimeric antigen receptor-modified T-cells in refractory/relapsed acute lymphoblastic leukemia. *Exp Hematol Oncol*. 2017; 6: 10.

70. Zhang T, Cao L, Xie J, Shi N, Zhang Z, Luo Z, Yue D, *et al*. Efficiency of CD19 chimeric antigen receptor-modified T-cells for treatment of B cell malignancies in phase I clinical trials: A meta-analysis. *Oncotarget*. 2015; 6(32): 33961–33971.

71. Wang Z, Wu Z, Liu Y, Han W. New development in CAR-T cell therapy. *J Hematol Oncol*. 2017; 10: 53.

72. Minutolo NG, Hollander EE, Powell Jr. DJ. The emergence of universal immune receptor T Cell therapy for cancer. *Front. Oncol*. 2019, 26 March. https://doi.org/10.3389/fonc.2019.00176

CHAPTER 9

Harnessing stem cell secretome towards cell-free therapeutic strategies

Sharida Fakurazi*,†, Hasfar Amynurliyana A. Ghofar*,
Norshariza Nordin†, Suleiman Alhaji Muhammad*,‡

*Institute of Bioscience
†Faculty of Medicine & Health Sciences
Universiti Putra Malaysia, 43400 UPM, Serdang, Selangor Darul Ehsan, Malaysia
‡Department of Biochemistry, Usmanu Danfodiyo University, P.M.B 2346 Sokoto, Nigeria

ABSTRACT

Recent years have observed the development of stem cell therapy, which appeared as promising treatment strategies for various disease conditions. Cell therapies employ stem cells, or cells grown from stem cells, to replace or rejuvenate damaged tissue. Numerous findings have suggested a significant therapeutic advantage with the utilization of cell therapeutic approaches in various neurological disorders including amyotrophic lateral sclerosis, various heart disorders involving end-stage ischaemic heart diseases, myocardial infarction, or preventing vascular restenosis. Various bone fractures are also reported to benefit from cell therapy including osteogenesis imperfecta. Nonetheless, cell therapy has its drawbacks, which include the risk of tumorigenesis or immune rejection. The therapy also faces ethical and political controversies besides scientific challenges. Consequently, efforts are made towards the development of stem cell secretome which are the secreted factors produced by the stem cells that are responsible for mediating and modulating stem cells effects in the disease condition. The secretome holds various added advantages: it can be manufactured, freeze-dried, packaged, and transported more easily which is some of the numerous advantages of using secretome over the use of stem cells. Besides, as secretome is free from cells, there is no need to match the donors and recipients to avoid the risk of rejection. For that reason, stem cell-derived secretome is a promising possibility to be used as pharmaceuticals for regenerative medicine. Up till now, there have have been limited clinical trials utilizing secretome in certain disease conditions.

* Corresponding author. Email: sharida@upm.edu.my

KEY WORDS

Secretome; Stem cells; Cell-free therapy; Paracrine factors.

LIST OF ABBREVIATIONS

AFSCs	=	Amniotic fluid stem cells
BBB	=	Blood-brain barrier
BDNF	=	Brain-derived neurotrophic factor
bFGF	=	Basic fibroblast growth factor
BMP	=	Bone morphogenetic protein
CM	=	Conditioned medium
ECM	=	Extracellular matrix
EVs	=	Extracellular vesicles
FBS	=	Fetal bovine serum
HDF-D	=	Diabetic human dermal fibroblast
HGF	=	Hepatocytes growth factor
IDO	=	Indoleamine-2,3-dioxygenase
IGF	=	Insulin-like growth factor
IL	=	Interleukin
LIF	=	Leukaemia inhibitory factor
MCP-1	=	Monocyte chemoattractant protein-1
MSCs	=	Mesenchymal stem cells
SHED	=	Stem cells from human exfoliated deciduous teeth
TGF-β	=	Transforming growth factor-β
TNF-α	=	Tumor necrosis factor- α
Treg cells	=	Regulatory T- cells
VEGF	=	Vascular endothelial growth factor

9.1 INTRODUCTION

Revolutionary development in therapeutic strategy has devised some novel approaches in utilizing stem cells to replace or rejuvenate damaged tissues in various disease conditions. Stem cells are cells that can undergo self-renewal and are capable of differentiating into different cell lineages. They can be derived from an embryo or adult tissues. Application of either embryonic, fetal, amniotic, umbilical cord, or adult stem cells has created an innovative transformation in cancer therapies and regenerative medicine by providing the possibility of generating multiple therapeutically useful cell types for the treatment of numerous genetic and degenerative disorders. Due to ethical controversies over how embryonic stem cells are generated, adult stem cells have become an attractive source of stem cells in the field of regenerative medicine. Several studies have shown the therapeutic benefits of stem cells in various pathologies such as Alzheimer's disease, Parkinson's disease, myocardial infarction, osteoarthritis, wounds, and liver fibrosis. Over the years, the therapeutic effect of stem cells is attributed to their differentiation capacity that allows the cells to form different tissues of the body. Contrary to this assertion, recent studies indicate that the tissue regenerative capacity of stem cells could be due to their multiple secretions that modulate the microenvironment to exert the needed therapeutic effect. The secreted factors are soluble bioactive factors, microvesicles and exosomes that are derived from stem cells under different culture conditions. Recent studies utilizing secretome derived from the stem cells showed that these factors exert similar or superior effects in tissue repair compared to stem cells. This chapter highlights the current therapeutic benefits of stem cells and some of their drawbacks and how secretome devoid of cells can be utilized for effective therapies in regenerative medicine. Besides, data on the conditioned medium (CM) derived from full-term amniotic fluid stem cells and its effect on the proliferation and migration of diabetic human dermal fibroblast are discussed.

9.2 STEM CELLS IN REGENERATIVE MEDICINE

Recent evidence suggests that stem cells have evolved as therapeutic strategies for various diseases. Cell-based therapies rely on an understanding of how stem cells develop, differentiate and maintain the formation of new healthy tissue for diseases that require replacement or transplant of the affected tissue. Among them, age-related functional defects and neurodegenerative disorders such as in Parkinson and Alzheimer's diseases have benefited from stem cell transplant therapy. Similarly, the regenerative potential of stem cells has been reported in diseases such as osteoarthritis, pulmonary hypertension, diabetic wound, traumatic brain injury, cerebral ischemia, spinal cord and others. Administration of stem cells obtained from umbilical cord was reported to improve basic motor behavior in Parkinsonian mouse model.[1] Also, stem cells derived from bone marrow have been shown to differentiate into neuron cells and when these cells were transplanted into Parkinsonian rats, they caused a significant improvement on behavior deficit.[2] Transplantation of human neural stem cells improves cognition in a murine model of Alzheimer's disease.[3] It has been demonstrated that administration of stem cells promote cartilage regeneration in animal models of osteoarthritis.[4,5] Clinical trials also supported the therapeutic benefit of stem cells in osteoarthritis.[6,7] Furthermore, stem cells derived from various sources have been demonstrated to mitigate the pathological changes associated with pulmonary hypertension.[8–10] A previous study has also shown that intravenous transplantation of stem cells sourced from bone marrow differentiated into neurons and astrocytes improved neurological function in traumatic brain injury.[11] Several studies have shown the therapeutic benefit of stem cells in accelerating wound contraction in diabetic animals[12,13] and full-skin thickness excisional wound.[14] Bone marrow-derived stem cells administered to spinal cord injury rats improved locomotive and functional recovery.[15] Treatment of cerebral ischaemic animals with stem cells or in combination with mild hypothermia was able to reduce the percentage of infarct area and promote functional recovery through angiogenesis.[16] A result of systematic review and meta-analysis indicates that the treatment of myocardial ischaemic mouse with cardiac stem cells improved left ventricular ejection fraction.[17] The use of stem cells in various studies showed the safety and therapeutic benefit in different disease models.

9.3 DRAWBACKS ASSOCIATED WITH STEM CELL TREATMENT

Regardless of promises and hopes that stem cells are able to offer in the field of regenerative medicine, there are nevertheless many other on-going issues that need to be comprehensively and carefully addressed. The use of embryonic stem cell-based therapy has always raised ethical and safety concerns. Pluripotency nature of the embryonic stem cells permits the cells to undergo unlimited proliferation and they are able to generate any cell type. Due to their ability to produce different types of cells of the body, the cells become difficult to control after *in vivo* implantation.[18] Transplantation of these cells may sometimes result in the formation of teratomas or tumors. Prokhorova *et al.*[19] have shown that implantation of human embryonic stem cells into immunodeficient mice results in the formation of teratoma and the percentage of teratoma appearance was site-dependent. Although reports suggest that postnatal stem cells are safe for regenerative therapies, the use of these cells posed some challenges that undoubtedly need to be addressed before the universal acceptability of cell-based therapy may be attained. Among the challenges are the culture medium used for the expansion of the stem cells. The culture medium is often supplemented with fetal bovine serum (FBS) for the expansion of these cells. FBS is an ill-defined supplement that contains a mixture of an undefined component which possesses the risk of transmission of zoonotic pathogens such as prions and viruses and possible immune rejection.[20,21] This challenge has led to the introduction of xeno-free culture medium for the expansion of clinical-grade stem cells in the field of regenerative medicine. With this development, it is expected that this challenge will be overcome for better production of clinical-grade stem cells. Another challenge is how to avert immunorejection after implantation of stem cells. This challenge is also seen as a major barrier for successful stem cell therapy because the immune system of the recipient may recognize the transplanted cells as a foreign substance that can trigger immune rejection.[22] There is also a challenge of the appropriate cell dose that could yield desired results as various studies used different doses of stem cells. Therefore, it may be important to come up with an appropriate dose for the effective treatment of various diseases. Furthermore, a long-term culture of stem cells is likely to undergo genomic alterations.[23] However, the genomic changes observed in their study suggest the cells did not undergo malignant transformation. It is, therefore, recommended that genomic assessment from time to time is a key for ensuring genomic stability for clinical application of stem cells in regenerative therapy. Also, the use of low passage stem cells may be safer compared to high passage cells. Other challenges such as robust differentiation of stem cells into the specific lineages of interest and understanding the significant role of the host tissue and microenvironment are requisite for clinical translation of stem cell-based therapy.

9.4 PREPARATION AND CHARACTERIZATION OF STEM CELL SECRETOME

The preparation and characterization of secretome involve several stages, starting from isolation and characterization of cells from the tissue biopsies and expansion of the isolated cells in culture medium supplemented with 10% FBS. After thorough washing of the cells with phosphate buffer saline, the culture media is then changed to serum-free medium or the cells are cultivated in hypoxic condition or treated with pharmacological agents for about 24 h or more. During this preconditioned period, the cells secrete soluble bioactive factors (growth factors, cytokines and chemokines) and extracellular vesicles (exosomes and microvesicles) into the culture medium. Subsequently, the cell supernatant is collected and subject to centrifugation and filtration to remove cell debris. The product obtained is the secretome (conditioned medium) and can be stored at −80°C for further analysis. Stem cell secretome consists of a large number of various proteins that may require further separation techniques to reduce the complexity of the conditioned medium.[24] Extracellular vesicles can be isolated from the conditioned medium by differential centrifugation or ultrafiltration to obtained exosomes or microvesicles. The proteomic analysis, which involves identification and quantification of various proteins contained in the secretome can be achieved through mass spectrometry, biochemical and

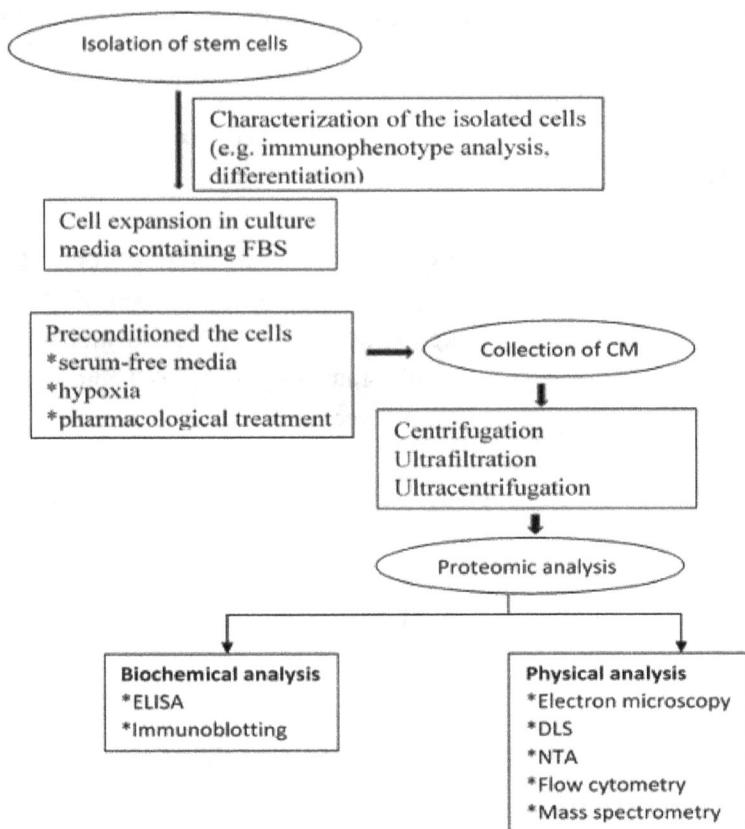

Figure-9.1.   Schematic preparation and characterization of stem cell secretome.

DLS: Dynamic light scattering; ELISA: enzyme-linked immunosorbent assay; NTA: nanoparticle tracking analysis.

immunological analysis (Figure-9.1). Protein functional annotation, secreted protein prediction and computer modelling of pathways and functions of the secretome can be done using bioinformatics.[24]

9.5   GROWTH FACTORS, CYTOKINES AND CHEMOKINES

Stem cells secrete various soluble factors such as growth factors, cytokines and chemokines. The soluble factors secreted by the stem cells include hepatocytes growth factor (HGF), transforming growth factor-β (TGF-β), insulin-like growth factor (IGF), vascular endothelial growth factor VEGF), basic fibroblast growth factor (bFGF), brain-derived neurotrophic factor (BDNF), monocyte chemoattractant protein-1 (MCP-1), interleukin-10 (IL-10), interleukin-6 (IL-6), bone morphogenetic protein (BMP), and indoleamine-2,3-dioxygenase (IDO).[25,26] These bioactive soluble factors are capable of modulating tissue microenvironment

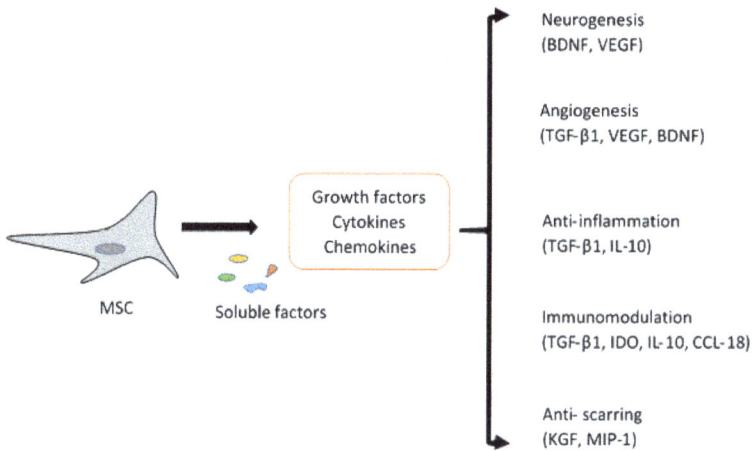

Figure-9.2.   Mechanism of MSC secreted soluble factors in modulating tissue microenvironment for repair and regeneration.

BDNF: brain-derived neurotrophic factor; VEGF: vascular endothelial growth factor; TGF- β1: transforming growth factor- β1; IL-10: interleukin-10; IDO: indoleamine-2,3-dioxygenase; CCL-18: chemokine (C-C motif) ligand 18; KGF: keratinocyte growth factor; MIP: macrophage inflammatory protein 1.

to promote angiogenesis, cell proliferation and differentiation, neurogenesis, immunomodulation and anti-inflammation and anti-apoptotic activity (Figure-9.2). Together, these activities can then initiate tissue repair and regeneration. Several studies have reported these soluble factors in the conditioned medium prepared from mesenchymal stem cells.[27–29] BDNF is an important neurotrophin that promotes neuronal survival and differentiation. Administration of this factor in animal models of stroke reduced infarct volume, promote neurogenesis and functional recovery.[30] Conditioned medium (CM) derived from stem cells from human exfoliated deciduous teeth (SHED) has been shown to inhibit apoptosis and fibrotic scarring in carbon tetrachloride-induced liver fibrosis.[28] However, HGF-silenced SHED-CM did not promote anti-inflammatory and anti-scarring activities, suggesting this soluble protein is a necessary factor in the CM to improve liver fibrosis.[28] Furthermore, Lv et al.[31] revealed that MSCs secrete high amounts of BMP7 in vitro and improved glomerular fibrosis in vivo in diabetic rats. Similarly, MSC-CM suppresses the TGF-β signalling pathway that was activated in a diabetic model in vitro. When the diabetic cells were treated with CM neutralized of BMP7, the inhibitory effect of the CM was reversed. This shows that MSCs contribute to the improvement of renal fibrosis through the paracrine action of secreted proteins. Kano et al. also showed that SHED-CM improved neurological function and nerve regeneration in a rat model of nerve injury.[32] On the other hand, when the CM was depleted of MCP-1 and secreted ectodomain of sialic acid-binding Ig-like lectin-9 (sSiglec-9) lost its ability to restore neurological function in nerve injury. This suggests that these factors are key for SHED-CM-mediated nerve regeneration. In another study by Matsubara et al.,[33] SHED-CM promotes functional recovery and anti-inflammatory activity in a spinal cord contusion rat model. Conversely, the rat treated with CM depleted of MCP-1 and sSiglec-9 showed more tissue damage and less neurological function when compared with rats treated with complete secretome, demonstrating the importance of these two proteins in improving functional recovery in spinal cord injury. Yoo et al.,[34] also demonstrated the role of TGF-β in improving functional recovery in ischaemic brain injury following MSCs transplantation. It was observed that MSC-CM decreased the migration and production of MCP-1 in freshly isolated immune cells in vitro. However, transplantation of MSC-silenced TGF-β did not reduce the infiltration of CD68+ cells into the ischaemic brain and only slight motor function recovery was observed.[34] Immunomodulation of MSCs is due to their ability to secrete soluble bioactive molecules that allow intercellular communication. MSCs are able to regulate both innate and adaptive immune response through inhibition of T cells, reduce B cell activation and proliferation and activate regulatory T (Treg) cells due to the secretion of these soluble bioactive factors.[35,36] TGF-β can activate the Treg cells and macrophages and transmits the immunosuppressive effect to other cells for activation of different mechanisms of immunosuppression.[37] These secreted factors act in a concerted manner for immune regulation. Taken together, the preclinical evidence suggests that soluble growth factors, cytokines, and chemokines play important roles in tissue repair and regeneration.

9.6 EXTRACELLULAR VESICLES

Extracellular vesicles (EVs) are membrane-contained vesicles that are released by MSCs and most cell types.[38] They transfer bioactive molecules such as lipids, proteins and nucleic acids including coding and noncoding ribonucleic acids.[39] EVs are capable of transferring information to other cells and thus, can influence the function of the recipient cells for protection and at the same time deliver various messengers to the distant sites.[40] There are three broad classifications of extracellular vesicles based on their respective origins, size, shape, and composition.[41] The three classes include exosomes, microvesicles, and apoptotic bodies. Exosomes are similar in size ranging from 30–130 nm and are produced by invagination and fusion of the endosomal membrane of multivesicular bodies which are finally released by exocytosis.[41] The size of microvesicles are somewhat heterogeneous, ranging from 100–1000 nm and formed by direct budding or shedding from the plasma membrane.[38] Apoptotic bodies range from 50–4000 nm in size and are also produced by direct shedding from the plasma membrane similar to microvesicles. Even though microvesicles share similar biogenesis with apoptotic bodies, they originate from different cell sources. Apoptotic bodies originate indiscriminately from the apoptotic cells, whereas microvesicles originate from living cells in an orderly manner through the trafficking of cellular molecules to the plasma membrane, resulting in protrusion of the cell's surface, budding and then shedding from the membrane.[42] Apart from the role of EVs in intercellular communication to modulate various pathological processes and maintenance of homeostasis, they also have the potential to remove unwanted molecular materials as a way of maintaining cell functions.[40] Besides, EVs are potential natural drug delivery systems because of their homing capacity that allows them to deliver and release the drug to a specific target.[43] Exosomes can deliver their cargo by fusing with the membrane of the recipient cells and release the cargo for the mediation of physiological, and pathological events.[44] They can recognize the target cells because of their free-floating nature, adhesion and antigen recognition capabilities. Extracellular vesicles from MSCs can promote the production of Treg cells and M2 macrophages, which at the same time are capable of suppressing the maturation of monocytes as well as the proliferation of T and B cells.[45] Furthermore, Cossetti et al.[46] demonstrated that EVs from neural stem cells exposed to inflammatory cytokines induced interferon-gamma pathways through activation of STAT1 signalling in target cells. This suggests that grafted stem cells may utilize this signalling pathway regulated by EVs to communicate with the host immune system. Taken together, EVs possess immunomodulation capacity and can induce immune suppression.

9.7 SECRETOME AS THERAPEUTIC STRATEGIES FOR ALLEVIATING DISEASE CONDITIONS

The use of stem cell-derived secretome is evolving as the mechanism by which stem cells exert their therapeutic benefits. Secretome is a repertoire of secreted soluble factors and extracellular vesicles by the stem cells that are capable of modulating tissue microenvironment for effective repair and regeneration. Harnessing stem cell secretome for tissue repair would provide dynamic advantages over the use of stem cells.[47] The advantages of secretome over stem cell therapy include among others: storage for a long period without the use of cryo-preservatives that may be toxic; able to be freeze-dried and packaged for easy transportation; formation of potentially tumorigenicity and immune rejection could be resolved since it is devoid of cells; production in large scale can be easily achieved within a limited period; and it can be available off-the-shelf that may guarantee self-administration. It was initially suggested that stem cells exert their effect through differentiation when they are transplanted to repair the damaged tissue. However, emerging evidence suggests that the therapeutic strategies by which stem cells act to regenerate damaged tissue are due to paracrine action of their secretions. These secreted factors allow the cells to communicate between themselves and their microenvironment to induce a cascade of signalling events for the promotion of endogenous repair. Reparative processes are achieved through modulation of the inflammatory process, inhibition of apoptosis and immune regulation. Recent studies have strengthened the paracrine hypothesis as the mechanism by which stem cells act to evoke tissue reparative process.[4,48] The soluble factors released by these cells are cytokines, chemokines, growth factors that act in a concerted manner to elicit biological activity without the cells. Extracellular vesicles, which encompass microvesicles and exosomes are other factors released by the stem cells. They are potent vehicles that allow intercellular communication through the transfer of cargo as lipids, proteins and nucleic acids to the recipient cells, thereby influencing several physiological and pathological processes.[40,49] Several preclinical studies have shown that secretome without the stem cells was able to promote repair of damaged tissues such as cartilage,[50,51] liver[52] and heart.[53] Similarly, secretome has also been demonstrated to improve pathological alterations in pulmonary arterial hypertension.[54,55] Treatment of Alzheimer's disease,[56] spinal cord injury,[33] cerebral ischaemia[57,58] and nerve injury[32] with secretome has been demonstrated to improve tissue damage and functional recovery.

Exosomes derived from MSCs reduced infarct size in a mouse model of myocardial ischemia/reperfusion.[59] It was also demonstrated that injection of exosomes derived from human umbilical MSCs mitigates liver fibrosis by reducing the surface fibrous capsules and hepatic inflammation in a carbon tetrachloride-induced model of liver fibrosis.[60] Neuroprotective effects of exosomes derived from MSCs have also been reported. Xin et al.[61] report that systemic injection of exosomes improved functional recovery and promotes neurogenesis and neurite remodelling as well as angiogenesis after stroke in rats. Similarly, functional improvement and reduction in neuro-inflammation have been reported in rats with traumatic brain injury following exosomes intervention.[62] Furthermore, treatment of osteoarthritic chondrocytes with EVs increased the expression of matrix proteins (collagen type 2 and aggrecan) and inhibited the expression of matrix metalloproteinase-13 and a disintegrin and metalloproteinase with thrombospondin motif 5 expression and protect cartilage degradation in mice.[63] These suggest the therapeutic benefits of EVs in various diseases models. However, robust standardized culture conditions and purification of secreted factors are key requirements that if addressed adequately would permit the production of clinical-grade secretome for the treatment of various disease conditions.

9.8 EXTRACELLULAR VESICLES AS DRUG DELIVERY SYSTEMS

Extracellular vesicles hold great potential as drug delivery systems because of their non-immunogenicity that could allow them to be in circulation for longer duration without being destroyed by proteases and immune systems. Also to their non-immunogenicity, EVs are known for their homing capacity that could make these membrane-contained vesicles excellent drug delivery systems. The report indicates that EVs are capable of crossing the blood-brain barrier (BBB) that allow their movement from the peripheral circulation to the central nervous system.[64] The ability of EVs to pass through blood-brain barriers could make them efficient drug delivery vehicles, especially for the treatment of neurodegenerative diseases. Morales-Prieto et al.[65] demonstrated that the injection of exosomes was able to pass BBB and induce the activation of glial cells in mice. It has been shown that intranasal administration of exosomes encapsulated with curcumin or Stat3 inhibitor ameliorates lipopolysaccharide-induced brain inflammation, experimental autoimmune encephalitis and delayed brain tumor growth in mice.[66] Similarly, Tang et al. report that encapsulation of methotrexate and cisplatin in EVs inhibits tumor growth in murine tumor models without any toxic effect.[67] Furthermore, exosomes have also been used as drug delivery vehicles for the treatment of Parkinson's disease. Haney et al.[68] demonstrated that catalase loaded into exosomes improved neuroprotection in both in vitro and in vivo models of Parkinson's disease. EVs may offer specific benefits compared to synthetic drug delivery systems because they are thought to gain from innate mechanisms, which can increase their physicochemical stability and can bypass a number of biological barriers to deliver drugs to challenging target sites.[69] Unlike the liposomes that act through complement activation to trigger degradation and clearance, EVs expressed membrane-bound complement regulators (CD55 and CD59) and this characteristic feature offers benefits for drugs requiring longer circulation times or harsh inflammatory settings.[69] It is also important to acknowledge the fact that despite promising results of EVs as drug delivery vehicles, there are challenges that must be addressed for large-scale production for effective clinical applications. A large number of cells are required to produce adequate EVs for preclinical studies. Most current protocols for purification are not able to remove completely co-isolates or soluble factors[70] and this may present biological side effects.[69]

To overcome these limitations, it is important to standardize and improve the isolation and purification processes for the generation of large quantities of clinical grade EVs for drug delivery.

9.9 SECRETOME ACCELERATES MIGRATION AND PROLIFERATION OF DIABETIC HUMAN DERMAL FIBROBLAST CELLS

The wound healing process in diabetic patients does not follow the mechanism coordinated in normal healing. Besides, bacterial colonization or infection can further disrupt the healing process.[71] It has been previously shown that prolonged wound healing process is associated with extended inflammatory phase, defective angiogenesis and decrease fibroblast proliferation.[72] Wound healing is a multifaceted process that involves an organized cascade of events involving angiogenesis, cell proliferation, migration and deposition of extracellular matrix.[73] This complex process is believed to be regulated by soluble signalling molecules and the local molecular framework of the extracellular milieu, thereby influencing the activity of fibroblasts.[74] A fibroblast is also known to play an important role in maintaining skin integrity through a paracrine mechanism, which is due to the activity of bioactive factors.[75] Here, we presented data on the effect of the secretome derived from full-time amniotic fluid on cutaneous diabetic healing *in vitro*. Full-term rat amniotic fluid stem cells (AFSCs) were used to prepare the conditioned medium. Cells were cultured at a density of 0.25×10^6 in ES medium (GMEM supplemented with sodium bicarbonate, L-glutamine, Sodium Pyruvate, MEM NEAA, 2-mercaptoethanol, 15% fetal bovine serum (Gibco, Life Technologies, USA) and leukaemia inhibitory factor (LIF) (Millipore Singapore Pte Ltd, Singapore) incubated overnight. After incubation, the old medium was discarded and the attached cells were washed 3 times with phosphate-buffered saline (PBS). The media was then changed to serum-free DMEM/F-12 with or without LIF and incubated for 72 h. The cell supernatant was collected, spun at 2500 rpm for 10 min and filtered through 0.22 μm filters and labeled as CM. Subsequently, the cytokine level in the CM was then measured using ELISA. The prepared CM was diluted with fresh serum-free DMEM/F-12 to prepare three different dilutions (25%, 50% and 100%) of CM. The effect of CM on the proliferation and migration of diabetic human dermal fibroblast (HDF-D) was then evaluated. Cytokine level in the prepared CM is depicted in Figure-9.3. The growth factors and cytokines of interest measured are known to be involved in the three phases of wound healing. They include interleukin-6 (IL-6), interleukin-1β (IL-1β), transforming growth factor-β1 (TGF-β1), tumor necrosis factor-α (TNF-α) and vascular endothelial factor-c (VEGF-C.) The addition of LIF to the

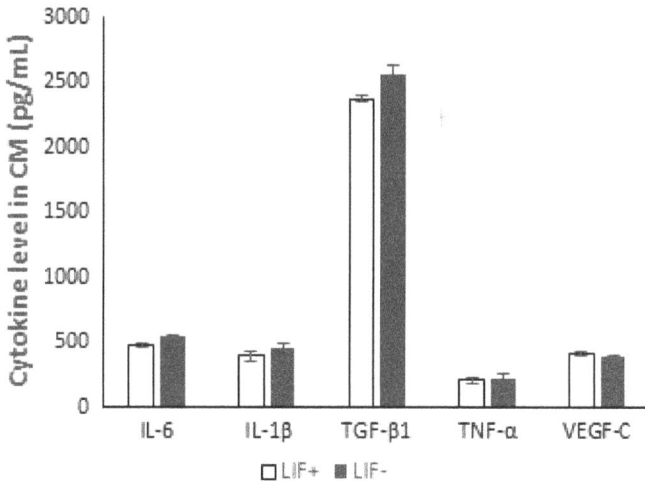

Figure-9.3. The level of cytokine in conditioned medium.

Various growth factors and cytokines were detected in the prepared conditioned medium. IL-6: interleukin-6; IL-1β: interleukin-1β; TGF-β1: transforming growth factor- β1; TNF-α: tumor necrosis factor-α; VEGF-C: vascular endothelial growth factor; LIF: leukaemia inhibitory factor.

culture medium during the production of CM did not contribute in any way to the amount of cytokines released by the cells. TGF-β1 was one of the highest amounts of cytokine detected in the prepared CM.

Conditioned medium supports the proliferation of HDF-D and it was observed to be dose-dependent for the three time points. A similar pattern of results was obtained for both CM prepared with or without LIF. This suggests that LIF is not an important factor for the preparation of CM from AFSCs. It has been previously demonstrated that AFSCs can be cultivated in a media used for embryonic stem cells that are based on a feeder layer of mitotic inactive embryonic fibroblasts to produce LIF.[76] The latter can inhibit the differentiation of embryonic stem cells and maintain their proliferation. The addition of LIF does not seem to contribute to the amounts of growth factors and subsequently to the role of CM in promoting HDF-D proliferation. The CM was able to support cell proliferation better than the serum-free medium (SFM) (Figure-9.4), which can be attributed to an array of secreted factors.

Wound scratch test assay was performed to measure the potential of cell migration of scratched HDF-D incubated with different concentrations of CM over 24 h. Conditioned medium was able to promote cell migration after 24 h of incubation and this seems to be in a dose-dependent manner (Figure-9.5a & b). The wound healing process is facilitated by native wound factors and systemic mediators. An imbalance between these factors could lead to chronic

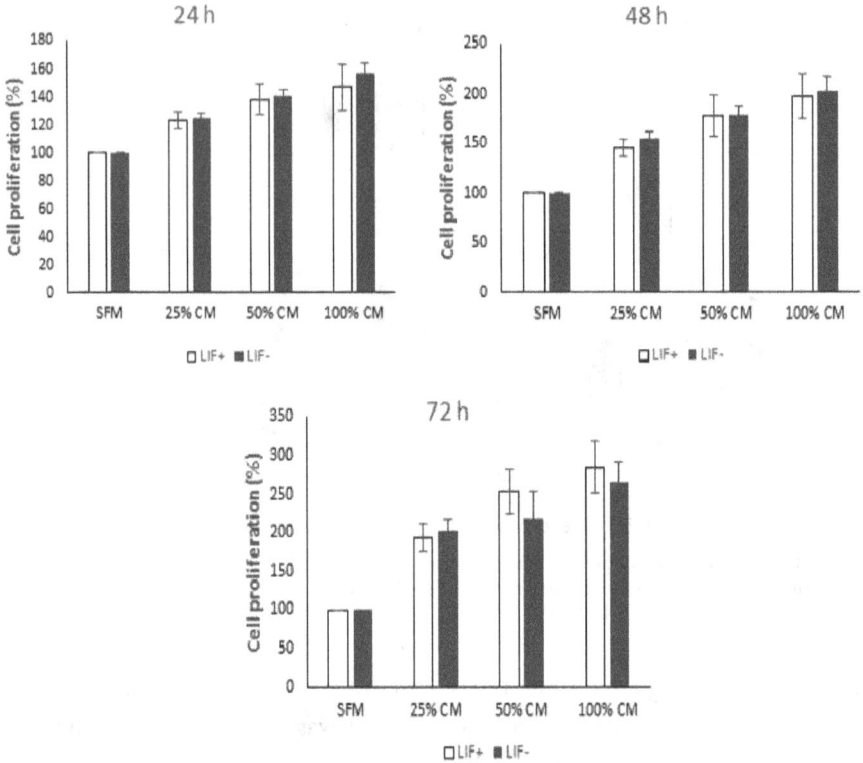

Figure-9.4.   Effect of CM on the proliferation of HDF-D.

SFM: serum-free media; 25% CM: conditioned medium diluted with 75% SFM; 50% CM: conditioned medium diluted with 50% SFM; and 100% CM: complete conditioned medium.

wounds and delay the process of healing.[77] The conditioned medium contains growth factors and cytokine that can trigger the wound healing process. Growth factors and cytokines are important factors in the initiation and final stage of the wound healing process. Notably among the role of these factors are inflammation, angiogenesis, and suppression of inflammation. For example, platelet-derived growth factor and interleukin-1 are important growth factors that attract neutrophils to the wound area for the elimination of possible bacteria contamination.[78,79] Interleukin-1 can activate the expression of hepatocyte growth factor in fibroblasts to promote angiogenesis, re-epithelialization and granulation tissue formation.[77] The conversion of monocytes to macrophages, which augment the inflammatory response is being facilitated by the TGF-β. The latter also plays a vital role in the proliferative stage of wound healing. Furthermore, VEGF and fibroblast growth factor (FGF) promote endothelial cell proliferation and angiogenesis, resulting in the deposition, synthesis as well as the organization

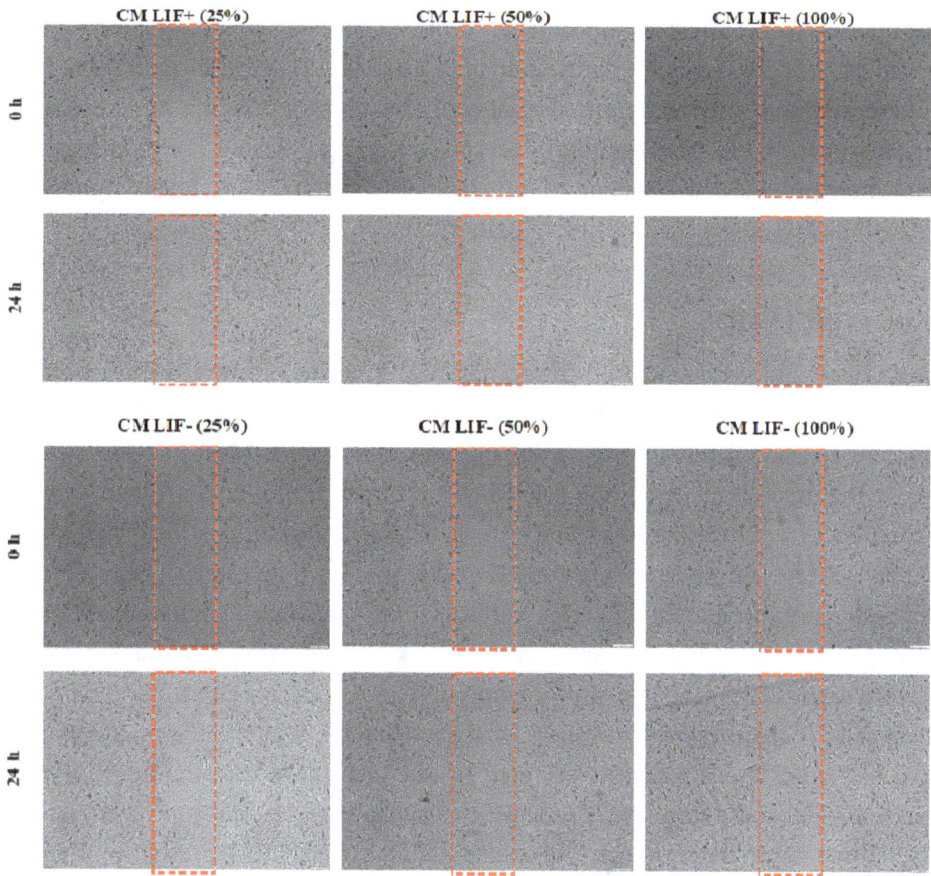

Figure-9.5. In vitro scratch assay showing the migration of HDF-D wound incubated with CM. (a) CM produced with the addition of LIF, (b) CM produced without the addition of LIF. Different concentration of CM was used and the effect seems to be dose-dependent.

of new extracellular matrix.[78] Interleukin-6 can act to promote inflammation, angiogenesis, re-epithelization, deposition of collagen and tissue remodelling. IL-6 also acts indirectly to induce infiltration of neutrophil and macrophage, angiogenesis and epidermal cell proliferation through activation of IL-1, TGF-β1 and VEGF production.[77] Proliferation and migration of cells in the skin are the most important processes in skin regeneration and wound healing. The migration of HDF is essential for wound healing as they play a vital part in cutaneous wound repair, and remodelling.[80] The HDF proliferates and migrates into the wound opening for the synthesis, deposition and remodelling of new extracellular matrix (ECM) as well as the expression of large bundles of actin as myofibroblasts.[80,81] Wound healing is a multifaceted process that involves many cell types such as fibroblasts, keratinocytes, endothelial cells, macrophages and platelets. This complex process is achieved through migration, infiltration, proliferation, and differentiation involving these cells that could result in an inflammatory response, new tissue formation and eventually to wound healing.[78] Besides, the regulation of the whole process is dependent on intercellular communication involving several growth factors, chemokines, and cytokines.

9.10 CONCLUSION

Stem cells hold great potential in regenerative medicine. During the last decade, several studies have demonstrated unprecedented advances in the development of stem cell-based therapy for various disease conditions. Most of these studies are at the preclinical stages and the translation into clinical settings is underway. Selection of appropriate dose and risk of immune rejection and formation of tumorigenesis are among the limitations of stem cell therapy. Earlier studies suggest that transplanted stem cells exert therapeutic benefits by differentiation into multiple cell types. However, an emerging breakthrough has shown that secretome devoid of the cells is capable of producing similar or superior efficacy compared to stem cells. Increasing evidence shows that growth factors, cytokines, microvesicles, and exosomes are the factors secreted by the stem cells that modulate several physiological and pathological processes to evoke tissue repair and regeneration. Secretome-based therapy with CM or extracellular vesicles may present substantial advantages over stem cells in relation to production, storage, handling and shelf life as biotherapeutic agent. The composition of stem cell secretome is complex because cells from different sources and culture condition could yield products with a varied amount of secreted factors. This in itself may not pose an impediment for its wide clinical application as a regenerative milieu if a standardized protocol for the production and characterization is established by the regulatory bodies. Furthermore, it has been demonstrated that CM derived from full-term AFSCs was capable of promoting the proliferation and migration of HDF-D, which could be attributed to multiple secretions. This suggests that CM is a potential regenerative product that can be used for the treatment of the diabetic wound. Most importantly, good manufacturing practice and quality control are essential requirements for the establishment of the safety and efficacy profile of stem cell-derived secretome. If all of these are pursued rigorously in an unbiased manner, it would make secretome, especially extracellular vesicles, not only disease-modulating agents but also a natural drug delivery system.

REFERENCES

1. Kang EJ, Lee YH, Kim MJ, Lee YM, Kumar BM, Jeon BG, Ock SA, Kim HJ, Rho GJ. Transplantation of porcine umbilical cord matrix mesenchymal stem cells in a mouse model of Parkinson's disease. *J Tissue Eng Rege Med*. 2013; 7: 169–182.

2. Levy YS, Bahat-Stroomza M, Barzilay R, Burshtein A, Bulvik S, Barhum Y, Panet H, *et al*. Regenerative effect of neural-induced human mesenchymal stromal cells in rat models of Parkinson's disease. *Cytotherapy*. 2008; 10: 340–352.

3. McGinley LM, Kashlan ON, Bruno ES, Chen KS, Hayes JM, Kashlan SR, Raykin J, *et al*. Human neural stem cell transplantation improves cognition in a murine model of Alzheimer's disease. *Sci Rep*. 2018; 8: 14776.

4. Muhammad SA, Nordin N, Mehat MZ, Fakurazi S. Comparative efficacy of stem cells and secretome in articular cartilage regeneration: A systematic review and meta-analysis, *Cell Tissue Res*. 2019; 375(2): 329–344.

5. Wang W, He N, Feng C, Liu V, Zhang L, Wang F, He J, *et al*. Human adipose-derived mesenchymal progenitor cells engraft into rabbit articular cartilage. *Int J Mol Sci*. 16: 12076–12091.

6. Davatchi F, Abdollahi BS, Mohyeddin M, Nikbin B. Mesenchymal stem cell therapy for knee osteoarthritis: 5 years follow-up of three patients. *Int J Rheum Dis*. 2016; 19: 219–225.

7. Jevotovsky DS, Alfonso AR, Einhorn TA, Chiu ES. Osteoarthritis and stem cell therapy in humans: A systematic review. *Osteoarthr Cartil*. 2018; 26: 711–729.

8. Alencar AKN, Pimentel-Coelho PM, Montes GC, da Silva M, Mendes LVP, Montagnoli TL, Silva AMS, *et al*. Human mesenchymal stem cell therapy reverses Su5416/hypoxia-induced pulmonary arterial hypertension in mice. *Front Pharmacol*. 2018; 9: doi:10.3389/fphar. 2018.01395.

9. Eguchi M, Ikeda S, Kusumoto S, Sato D, Koide Y, Kawano H, Maemura K. Adipose-derived regenerative cell therapy inhibits the progression of monocrotaline-induced pulmonary hypertension in rats. *Life Sci*. 2014; 118: 306–312.

10. Umar S, de Visser YP, Steendijk P, Schutte CI, Laghmani EH, Wagenaar GTM, Bax WH, *et al*. Allogenic stem cell therapy improves right ventricular function by improving lung pathology in rats with pulmonary hypertension. *Am J Physiol Heart Circ Physiol*. 2009; 297: H1606–1616.

11. Anbari F, Khalili MA, Bahrami AR, Khoradmehr A, Sadeghian F, Fesahat F, Nabi A. Intravenous transplantation of bone marrow mesenchymal stem cells promotes neural regeneration after traumatic brain injury. *Neural Regen Res*. 9: 919–923.

12. Falanga V, Iwamoto S, Chartier M, Yufit T, Butmarc J, Kouttab N, Shrayer D, Carson P. Autologous bone marrow-derived cultured mesenchymal stem cells delivered in a fibrin spray accelerate healing in murine and human cutaneous wounds, *Tissue Eng*. 2007; 13: 1299–1312.

13. Wan J, Xia L, Liang W, Liu Y, Cai Q. Transplantation of bone marrow-derived mesenchymal stem cells promotes delayed wound healing in diabetic rats. *J Diabetes Res*. 2013; 2013: 647107.

14. Pratheesh MD, Dubey PK, Gade NE, Nath A, Sivanarayanan TB, Madhu DN, Somal A, et al. Comparative study on characterization and wound healing potential of goat (Capra hircus) mesenchymal stem cells derived from fetal origin amniotic fluid and adult bone marrow. *Res Vet Sci.* 2017; 112: 81–88.

15. Osaka M, Honmou O, Murakami T, Nonaka T, Houkin K, Hamada H, Kocsis JD. Intravenous administration of mesenchymal stem cells derived from bone marrow after contusive spinal cord injury improves functional outcome. *Brain Res.* 2010; 134: 226–235.

16. Bi M, Wang J, Zhang Y, Li L, Wang L, Yao R, Duan S, et al. Bone mesenchymal stem cells transplantation combined with mild hypothermia improves the prognosis of cerebral ischemia in rats. *Plos One.* 2017; 13:e0197405.

17. Lang CI, Wolfien M, Langenbach A, Müller P, Wolkenhauer O, Yavari A, Ince H, et al. Cardiac cell therapies for the treatment of acute myocardial infarction: a meta-analysis from mouse studies. *Cell Physiol Biochem.* 2017; 42: 254–268.

18. Volarevic V, Markovic BS, Gazdic M, Volarevic A, Jovicic N, Arsenijevic N, Armstrong L, et al. Ethical and safety issues of stem cell-based therapy. *Int J Med Sci.* 2018; 15: 36–45.

19. Prokhorova TA, Harkness LM, Frandsen U, Ditzel N, Schrøder HD, Burns JS, Kassem M. Teratoma formation by human embryonic stem cells is site dependent and enhanced by the presence of Matrigel. *Stem Cells Dev.* 2009; 18: 47–54.

20. Hatlapatka T, Moretti P, Lavrentieva A, Hass R, Marquardt N, Jacobs R, Kasper C. Optimization of culture conditions for the expansion of umbilical cord-derived mesenchymal stem or stromal cell-like cells using xeno-free culture conditions. *Tissue Eng Part C Methods.* 2011; 17: 485–493.

21. Trubiani O, Piattelli A, Gatta V, Marchisio M, Diomede F, D'Aurora M, Merciaro I, et al. Assessment of an efficient xeno-free culture system of human periodontal ligament stem cells. *Tissue Eng Part C Methods.* 2015; 21: 52–64.

22. Ikehara S. Grand challenges in stem cell treatments. *Front Cell Dev Biol.* 2013; 1: 2.

23. Wang Y, Zhang Z, Chi Y, Zhang Q, Xu F, Yang Z, Meng L, et al. Long-term cultured mesenchymal stem cells frequently develop genomic mutations but do not undergo malignant transformation. *Cell Death Dis.* 2013; 4:e950.

24. Kupcova SH. Proteomic techniques for characterisation of mesenchymal stem cell secretome. *Biochimie.* 2013; 95: 2196–2211.

25. Cunningham CJ, Redondo-Castro E, Allan SM. The therapeutic potential of the mesenchymal stem cell secretome in ischaemic stroke. *J Cereb Blood Flow Metab.* 2018; 38(8): 1276–1292.

26. Kyurkchiev D, Bochev I, Ivanova-Todorova E, Mourdjeva M, Oreshkova T, Belemezova K, Kyurkchiev S. Secretion of immunoregulatory cytokines by mesenchymal stem cells. *World J Stem Cells.* 2014; 6: 552–570.

27. Ferreire JR, Teixeira GQ, Santos SG, Barbosa MA, Almeida-Porada G, Gonçalves RM. Mesenchymal stromal cell secretome: Influencing therapeutic potential by cellular pre-conditioning. *Front Immunol.* 2018; 9: 2837.

28. Hirata M, Ishigami M, Matsushita Y, Ito T, Hattori H, Hibi H, Goto H, et al. Multifaceted therapeutic benefits of factors derived from dental pulp stem cells for mouse liver fibrosis. *Stem Cells Transl Med.* 2016; 5: 1416–1424.

29. Inukai T, Katagiri W, Yoshimi R, Osugi M, Kawai T, Hibi H, Ueda M. Novel application of stem cell-derived factors for periodontal regeneration. Biochem. *Biophys Res Commun*. 2013; 430: 763–768.

30. Schäbitz WR, Steigleder T, Cooper-Kuhn CM, Schwab S, Sommer C, Schneider A, Kuhn HG. Intravenous brain-derived neurotrophic factor enhances poststroke sensorimotor recovery and stimulates neurogenesis. *Stroke*. 2007; 38: 2165–2172.

31. Lv S, Liu G, Sun A, Wang J, Cheng J, Wang W, Liu X, *et al*. Mesenchymal stem cells ameliorate diabetic glomerular fibrosis in vivo and in vitro by inhibiting TGF-β signalling via secretion of bone morphogenetic protein 7. *Diab Vasc Dis Res*. 2014; 11: 251–261.

32. Kano F, Matsubara K, Ueda M, Hibi H, Yamamoto A. Secreted ectodomain of sialic acid-binding Ig-like lectin-9 and monocyte chemoattractant protein-1 synergistically regenerate transected rat peripheral nerves by altering macrophage polarity. *Stem Cells*. 2017; 35: 641–653.

33. Matsubara K, Matsushita Y, Sakai K, Kano F, Kondo M, Noda M, Hashimoto N, *et al*. Secreted ectodomain of sialic acid-binding Ig-like lectin-9 and monocyte chemoattractant protein-1 promote recovery after rat spinal cord injury by altering macrophage polarity. *J Neurosci*. 2015; 35: 2452–2464.

34. Yoo SW, Chang DY, Lee HS, Kim GH, Park JS, Ryu BY, Joe EH, *et al*. Immune suppression following suppression mesenchymal stem cell transplantation in the ischemic brain is mediated by TGF-β. *Neurobiol Dis*. 2013; 58: 249–257.

35. Wang YH, Wu DB, Chen B, Chen EQ, Tang H. Progress in mesenchymal stem cell–based therapy for acute liver failure. *Stem Cell Res Ther*. 2018; 9: 227.

36. Zhang Y, Cai W, Huang Q, Gu Y, Shi Y, Huang J, Zhao F, *et al*. Mesenchymal stem cells alleviate bacteria-induced liver injury in mice by inducing regulatory dendritic cells. *Hepatol*. 2014; 59: 671–682.

37. Li H, Shen S, Fu H, Wang Z, Li X, Sui X, Yuan M, *et al*. Immunomodulatory functions of mesenchymal stem cells in tissue engineering. *Stem Cells Int*. 2019: 9671206.

38. Qiu G, Zheng G, Ge M, Wang J, Huang R, Shu Q, Xu J. Mesenchymal stem cell-derived extracellular vesicles affect disease outcomes via transfer of microRNAs. *Stem Cell Res Ther*. 2018; 9: 320.

39. Fatima F, Nawaz M. Vesiculated long non-coding RNAs: offshore packages deciphering trans-regulation between cells, cancer progression and resistance to therapies. *Non-Coding RNA*. 2017; 3: 10.

40. Yáñez-Mó M, Siljander PRM, Andreu Z, Zavec AB, Borràs FE, Buzas EI, Buzas K, *et al*. Biological properties of extracellular vesicles and their physiological functions. *J Extracell Vesicles*. 2015; 4: doi:10.3402/jev.v4.27066.

41. Seo Y, Kim HS, Hong IS. Stem cell-derived extracellular vesicles as immunomodulatory therapeutics. *Stem Cells Int*. 2019; 2019: 1–10.

42. Phan J, Kumar P, Hao D, Gao K, Farmer D, Wang A. Engineering mesenchymal stem cells to improve their exosome efficacy and yield for cell-free therapy. *J Extracell Vesicles*. 2018; 7: 1522236.

43. Bari E, Perteghella S, Di Silvestre D, Sorlini M, Catenacci L, Sorrenti M, Marrubini G, *et al*. Pilot production of mesenchymal stem/stromal freeze-dried secretome for

cell-free regenerative nanomedicine: a validated GMP-compliant process. *Cells*. 2018; 7: 190.

44. Bunggulawa EJ, Wang W, Yin T, Wang N, Durkan C, Wang Y, Wang G. Recent advancements in the use of exosomes as drug delivery systems. *J Nanobiotechnology*. 2018; 16: 81.

45. Wang M, Yuan Q, Xie L. Mesenchymal stem cell-based immunomodulation: Properties and clinical application. *Stem Cells Int*. 2018; 2018: 3057624.

46. Cossetti C, Iraci N, Mercer TR, Leonardi T, Alpi E, Drago D, Alfaro-Cervello C, *et al*. Extracellular vesicles from neural stem cells transfer IFN-γ via Ifngr1 to activate Stat1 signaling in target cells. *Mol. Cell* 56:193–204.

47. Vizoso FJ, Eiro N, Cid S, Schneider J, Perez-Fernandez R. Mesenchymal stem cell secretome: Toward cell-free therapeutic strategies in regenerative medicine. *Int J Mol Sci*. 2017; 18: 1852.

48 Muhammad SA, Nordin N, Fakurazi S. Regenerative potential of secretome from dental stem cells: a systematic review of preclinical studies. *Rev Neurosci*. 2018; 29: 321–332.

49. Vishnubhatla I, Corteling R, Stevanato L, Hicks C, Sinden J. The development of stem cell-derived exosomes as a cell-free regenerative medicine. *J Circ Biomark*. 2014; 3: 2.

50. Zhang S, Chu WC, Lai RC, Lim SK, Hui JHP, Toh WS. Exosomes derived from human embryonic mesenchymal stem cells promote osteochondral regeneration. *Osteoarthr Cartil*. 2016; 24: 2135–2140.

51. Zhu Y, Wang Y, Zhao B, Niu X, Hu B, Li Q, Zhang J, *et al*. Comparison of exosomes secreted by induced pluripotent stem cell-derived mesenchymal stem cells and synovial membrane-derived mesenchymal stem cells for the treatment of osteoarthritis. *Stem Cell Res Ther*. 2017; 8: doi:10.1186/s13287-017-0510-9.

52. Matsushita Y, Ishigami M, Matsubara K, Kondo M, Wakayama H, Goto H, Ueda M, Yamamoto A. Multifaceted therapeutic benefits of factors derived from stem cells from human exfoliated deciduous teeth for acute liver failure in rats. *J Tissue Eng Regen Med*. 2017; 11: 1888–1896.

53. Yamaguchi S, Shibata R, Yamamoto N, Nishikawa M, Hibi H, Tanigawa T, Ueda M, *et al*. Dental pulp-derived stem cell conditioned medium reduces cardiac injury following ischemia-reperfusion. *Sci Rep*. 2015; 5: 16295.

54. Rathinasabapathy A, Bruce E, Espejo A, Horowitz A, Sudhan DR, Nair A, Guzzo D, *et al*. Therapeutic potential of adipose stem cell-derived conditioned medium against pulmonary hypertension and lung fibrosis. *Br J Pharmacol*. 2016; 173: 2859–2879.

55. Willis GR, Fernandez-Gonzalez A, Reis M, Mitsialis SA, Kourembanas S. Macrophage immunomodulation: The gatekeeper for mesenchymal stem cell derived-exosomes in pulmonary arterial hypertension? *Int J Mol Sci*. 2018; 19: doi:10.3390/ijms19092534.

56. Mita T, Furukawa-Hibi Y, Takeuchi H, Hattori H, Yamada K, Hibi H, Ueda M, Yamamoto A. Conditioned medium from the stem cells of human dental pulp improves cognitive function in a mouse model of Alzheimer's disease. *Behav Brain Res*. 2015; 293: 189–197.

57. Inoue T, Sugiyama M, Hattori H, Wakita H, Wakabayashi T, Ueda M. Stem cells from human exfoliated deciduous tooth-derived conditioned medium enhance recovery of focal cerebral ischemia in rats. *Tissue Eng Part A*. 2013; 19: 24–29.

58. Yamagata M, Yamamoto A, Kako E, Kaneko N, Matsubara K, Sakai K, Sawamoto K, *et al*. Human dental pulp-derived stem cells protect against hypoxic-ischemic brain injury in neonatal mice. *Stroke*. 2013; 44: 551–554.

59. Lai RC, Arslan F, Lee MM, Sze NSK, Choo A, Chen TS, Salto-Tellez M, *et al*. Exosome secreted by MSC reduces myocardial ischemia/reperfusion injury. *Stem Cell Res*. 2010; 4: 214–222.

60. Li T, Yan Y, Wang B, Qian H, Zhang X, Shen L, Wang M, *et al*. Exosomes derived from human umbilical cord mesenchymal stem cells alleviate liver fibrosis. *Stem Cells Dev*. 2013; 22: 845–854.

61. Xin H, Li Y, Cui Y, Yang JJ, Zhang ZG, Chopp M. Systemic administration of exosomes released from mesenchymal stromal cells promote functional recovery and neurovascular plasticity after stroke in rats. *J Cereb Blood Flow Metab*. 2013; 33: 1711–1715.

62. Zhang Y, Chopp M, Zhang ZG, Katakowski M, Xin H, Qu C, Ali M, Mahmood A, Xiong Y. Systemic administration of cell-free exosomes generated by human bone marrow derived mesenchymal stem cells cultured under 2D and 3D conditions improves functional recovery in rats after traumatic brain injury. *Neurochem Int*. 2017; 111: 69–81.

63. Cosenza S, Ruiz M, Toupet K, Jorgensen C, Noël D. Mesenchymal stem cells derived exosomes and microparticles protect cartilage and bone from degradation in osteoarthritis. *Sci Rep*. 2017; 7: 16214.

64. Matsumoto J, Stewart T, Banks WA, Zhang J. The transport mechanism of extracellular vesicles at the blood-brain barrier. *Curr Pharm Des*. 2017; 23: 6206–6214.

65. Morales-Prieto DM, Stojiljkovic M, Diezel C, Streicher PE, Röstel F, Lindner J, Weis S, *et al*. Peripheral blood exosomes pass blood-brain-barrier and induce glial cell activation. *BioRxiv*. 2018; 2018: 471409.

66. Zhuang X, Xiang X, Grizzle W, Sun D, Zhang S, Axtell RC, Ju S, *et al*. Treatment of brain inflammatory diseases by delivering exosome encapsulated anti-inflammatory drugs from the nasal region to the brain, *Mol Ther*. 2011; 19: 1769–1779.

67. Tang K, Y Zhang, H Zhang, P Xu, J Liu, J Ma, M Lv, *et al*. Delivery of chemotherapeutic drugs in tumour cell-derived microparticles, *Nat Commun*. 2012; 3: 1282.

68. Haney MJ, Klyachko NL, Zhao Y, Gupta R, Plotnikova EG, He Z, Patel T, *et al*. Exosomes as drug delivery vehicles for Parkinson's disease therapy. *J Control Release*. 2015; 207: 18–30.

69. Armstrong JPK, Stevens MM. Strategic design of extracellular vesicle drug delivery systems. *Adv Drug Deliv Rev*. 2018; 130: 12–16.

70. Webber J, Clayton A. How pure are your vesicles. *J Extracell Vesicles*. 2013; 2013: 2.

71. Mendes JJ, Leandro CI, Bonaparte DP, Pinto AL. A rat model of diabetic wound infection for the evaluation of topical antimicrobial therapies. *Comp Med*. 2012; 62: 37–48.

72. Goren I, Müller E, Pfeilschifter J, Frank S. Severely impaired insulin signaling in chronic wounds of diabetic ob/ob mice: a potential role of tumor necrosis factor-alpha. *Am J Pathol.* 2006; 168: 765–777.

73. Jun EK, Zhang Q, Yoon BS, Moon JH, Lee G, Park G, Kang PJ, *et al.* Hypoxic conditioned medium from human amniotic fluid-derived mesenchymal stem cells accelerates skin wound healing through TGF-β/SMAD2 and PI3K/Akt pathways. *Int J Mol Sci.* 2014; 15: 605–628.

74. Tracy LE, Minasian RA, Caterson EJ. Extracellular matrix and dermal fibroblast function in the healing wound. *Adv Wound Care.* 2016; 5: 119–136.

75. Kim WS, Park BS, Sung JH, Yang JM, Park SB, Kwak SJ, Park JS. Wound healing effect of adipose-derived stem cells: a critical role of secretory factors on human dermal fibroblasts. *J Dermatol Sci.* 2007; 48:15–24.

76. Piccoli M, Franzin C, Bertin E, Urbani L, Blaauw B, Repele A, Taschin E, *et al.* Amniotic fluid stem cells restore the muscle cell niche in a HSA-Cre, Smn(F7/F7) mouse model. *Stem Cells.* 2012; 30: 1675–1684.

77. Behm B, Babilas P, Landthaler M, Schreml S. Cytokines, chemokines and growth factors in wound healing. *J Eur Acad Dermatol Venereol.* 2012; 26: 812–820.

78. Barrientos S, Stojadinovic O, Golinko MS, Brem H, Tomic-Canic M. Growth factors and cytokines in wound healing. *Wound Repair Regen.* 2008; 16: 585–601.

79. Hantash BM, Zhao L, Knowles JA, Lorenz HP. Adult and fetal wound healing. *Front Biosci J Virtual Libr.* 2008; 13: 51–61.

80. Li W, Fan J, Chen M, Guan S, Sawcer D, Bokoch GM, Woodley DT. Mechanism of human dermal fibroblast migration driven by type I collagen and platelet-derived growth factor-BB. *Mol Biol Cell.* 2004; 15: 294–309.

81. Walter MNM, Wright KT, Fuller HR, MacNeil S, Johnson WEB. Mesenchymal stem cell-conditioned medium accelerates skin wound healing: an in vitro study of fibroblast and keratinocyte scratch assays. *Exp Cell Res.* 2010; 316: 1271–1281.

Index

small molecule, 8
    inhibitor, 4, 5
solid tumor, 150, 231
somatic cell, 122
Sonic Hegdehog, 153
Sox2, 54, 119, 161
spatiotemporal, 157
specialized, 129
specificity, 216
spectrophotometer, 59
sphingomyelinase 2, 131
STAT3, 154, 164
stem cell, 51, 85, 86, 88, 95, 96, 104, 106,
    107, 109, 117, 118, 237
    therapy, 133
stem-cell like, 162
stemness, 154, 165
stromal cell, 61
stromal cell derived factor-1α (SDF-1α),
    54
surface marker, 148
survival, 159
suspension, 59
systemic lupus erythematosus (SLE), 178,
    180, 181, 193–195, 197, 198

T
targeted, 225
T-cell, 178, 180, 181, 183, 186–190, 192, 211
telomerase, 121
teratoma, 8, 9
therapeutic, 53, 71, 154
    strategy, 17
T-lymphocyte, 213, 214, 217
topography, 158
totipotent, 118
toxicity, 59, 225, 226
toxicology, 119
transcript, 122

transcription factor, 120, 154
transdifferentiation, 7, 8, 11
    hypothesis, 118
transduction, 217, 219
transformation, 238
transforming growth factor-1β (TGF-1β),
    58
transforming growth factor-β (TGF-β),
    132
transgene, 217
transplantation, 221, 222
transplanted, 135
tricarboxylic acid (TCA), 60
triglyceride, 60
tri-lineage, 53
tumor, 146, 149, 156, 216
tumor-associated, 156
tumorigenesis, 153, 163, 237
tumorigenic, 160
tumorigenicity, 124, 155
tumor necrosis factor-α (TNF-α), 65, 227
tumor niche, 163
tumor vaccine, 214
type 1 diabetes mellitus (T1DM), 193, 195,
    197

U
umbilical cord, 70

V
vascular, 157
vascular cell adhesion molecule (VCAM1),
    160
vascular endothelial (VE)-cadherin, 32
vascular endothelial growth factor
    (VEGF), 57, 137, 157
        VEGF-A, 160
vascularization, 161
viral, 127, 217

volatile, 69
volatile organic compound (VOC), 69

W
Warburg effect, 163
Wnt, 4, 6, 7
    Wnt/β-catenin, 153
    Wnt signalling, 153

Y
Yamanaka factor, 122

Z
ZUMA-1 study, 223
ZUMA-3, 222